CAMBRIDGE TRACTS IN MATHEMATICS

General Editors

J. BERTOIN, B. BOLLOBÁS, W. FULTON, B. KRA, I. MOERDIJK,
C. PRAEGER, P. SARNAK, B. SIMON, B. TOTARO

---

233   Category and Measure

CAMBRIDGE TRACTS IN MATHEMATICS

GENERAL EDITORS

J. BERTOIN, B. BOLLOBÁS, W. FULTON, B. KRA, I. MOERDIJK, C. PRAEGER, P. SARNAK, B. SIMON, B. TOTARO

A complete list of books in the series can be found at www.cambridge.org/mathematics. Recent titles include the following:

Category and Measure
Infinite Combinatorics, Topology and Groups

N. H. BINGHAM
Imperial College London

ADAM J. OSTASZEWSKI
London School of Economics and Political Science

CAMBRIDGE
UNIVERSITY PRESS

CAMBRIDGE
UNIVERSITY PRESS

Shaftesbury Road, Cambridge CB2 8EA, United Kingdom

One Liberty Plaza, 20th Floor, New York, NY 10006, USA

477 Williamstown Road, Port Melbourne, VIC 3207, Australia

314–321, 3rd Floor, Plot 3, Splendor Forum, Jasola District Centre,
New Delhi - 110025, India

103 Penang Road, #05–06/07, Visioncrest Commercial, Singapore 238467

Cambridge University Press is part of Cambridge University Press & Assessment,
a department of the University of Cambridge.

We share the University's mission to contribute to society through the pursuit of
education, learning and research at the highest international levels of excellence.

www.cambridge.org
Information on this title: www.cambridge.org/9780521196079

DOI: 10.1017/9781139048057

First published 2025

A catalogue record for this publication is available from the British Library

*A Cataloging-in-Publication data record for this book is available from the Library of
Congress*

ISBN 978-0-521-19607-9 Hardback

To Cecilie

Nick

To Monika, and to Juliana and Konstancja

Adam

Contents

Preface

The origins of this book stem largely from two earlier books, its 'parents', as it were. The first is John Oxtoby's (lovely, succinct) *Measure and Category: A Survey of the Analogies between Topological and Measure Spaces* (1971, 1980) (the term category is used here in the sense of Baire category, not of functors and categories in algebraic topology). Our own title *Category and Measure: Infinite Combinatorics, Topology and Groups* both pays tribute to Oxtoby's book as motivation and inspiration and declares our main theme: for us, it is *category*, rather than measure, that is paramount, while for Oxtoby it is the other way round.

The second is the first author's first book, *Regular Variation* (Bingham, Goldie and Teugels, 1987, 1989) – 'BGT' hereafter, for brevity. Regular variation is described in the preface of BGT as 'essentially a chapter in classical real variable theory, together with its applications . . .'. BGT is now well established, widely used and widely cited; it was intended to be comprehensive in its treatment of the theory and in surveying the applications. It left three 'open ends'. The first two were the *foundational question* (BGT, p. 11: measurability suffices, the Baire property suffices, neither includes the other; what is actually needed?), and the *contextual question* (what is its natural context?). The third was the difficulty of the hardest theorem in the 'theory' part of BGT (Th. 1.4.3/Th. 3.2.5). The challenge there was to dissolve this difficulty. All three matters have now been satisfactorily resolved.

While mathematics progresses by journal *papers*, it is consolidated by *books*. Thus as the number of papers one writes in an area grows, one must confront the fact that one's work will be 'for experts only' unless one synthesizes the corpus into a book, aimed at and addressed to the general mathematical public. While a glance at our extensive References will show that our sources are wide-ranging, our aim here is to bring the unifying essentials to the attention and put them at the disposal of the 'mathematician on the street'.

Doing the above provided the motivation for the first phase of our joint work (the first half, up to 2011) and its unifying connecting thread. Accordingly, we addressed ourselves to writing it up in book form, under the title suggested by the two 'parent texts', *Topological Regular Variation*. But under the momentum generated by this first half of our corpus to date, the second half emerged. We had the sense to realise that we should pause on writing the book, lest we 'write the wrong book', and instead let the papers emerge, let the dust settle, and write the book resulting from 'both halves, rather than just the one'.

The result is the present book. Its title reflects its focus and viewpoint. Its formal prerequisites are those of BGT: 'a good background in analysis at, say, first-year graduate level, including in particular a knowledge of measure theory' (BGT, xviii), set theory – particularly descriptive set theory – and mathematical logic. Any background here will only be helpful, but none is assumed; see for instance our 2019 survey 'Set Theory and the Analyst'.

The most important theorem in the ground we cover is the classic *Baire Category Theorem* of 1899, while measure theory (Lebesgue, 1905) was mentioned above. Each gives a family of *small sets* (meagre for category, null for measure). The meagre sets and the null sets form σ-ideals, $\mathcal{M}$ and $\mathcal{N}$, respectively. These, along with their interplay, similarities and contrasts, pervade the book. So too does the theory of analytic sets, much influenced by C. A. (Ambrose) Rogers, the second author's PhD supervisor.

The other main mathematical ingredient is *infinite combinatorics*, a field which owes much to the work of Paul Erdős (1913–1996). Here we had to fashion some of the tools we needed for ourselves, one reason for the fourteen-year delay since 2011.

Regarding regular variation: this is in no sense a book *about* regular variation. (Despite the large citation count of BGT, the first author has often been struck by how many good, knowledgeable mathematicians have never heard of it – though probabilists like himself use it all the time.) That said, the links between this book and regular variation are too important to be 'written in invisible ink'. We have chosen to use regular variation as a 'framing device' for the book. The book opens with a Prologue, containing a summary of 'regular variation up to BGT'. This can be read as part of the main text, skipped, or referred back to for reference, depending on the taste and background of the reader. It closes with an Epilogue, 'Topological Regular Variation', our summary of our own work here since. (Much else has happened in regular variation since BGT, particularly in probability and in higher dimensions, but this would take us too far afield.) This framing device is unusual in a mathematical monograph but familiar enough in other contexts. It is common to see several minutes of action before the credit titles of a film ('Prologue'). And we recall with pleasure Conrad's use of the narrator Marlow as a framing device in a number of his novels ('Prologue and Epilogue').

We refer for background in several areas to standard works. Our main sources are Engelking (Eng1989) for topology, Kechris (Kec1995) for descriptive set theory and Bogachev (Bog2007a) for measure theory. In addition, we refer repeatedly to Oxtoby's classic above, and to the paper by C. A. Rogers and J. E. Jayne (RogJ1980). This is contained in the proceedings of the London Mathematical Society's Instructional Conference on Analytic Sets, University College London, 1978. This conference led eventually to our collaboration and to this book. Analytic sets and their influence permeate the whole book.

While most of the text is addressed to the mathematical public generally, some parts are more specialized. These are marked with an asterisk.

Long bibliographies when numbered burden the reader with remembering numbers. We use instead 'name plus date', to guide the reader's eye when looking up references. For the name, we always use the first three letters of the first author's name (with more if needed to avoid ambiguity), plus the first letter of the second and third authors' names (thus our 2020 paper with Jabłońska and Jabłoński is BinJJ2020). Dates are given in full. This is consistent with usage in ordinary speech, and needed as there are three centuries involved (going back to, e.g. Baire, 1899).

While chapters (Chapter n), sections (§n.p) and results (Theorem n.p.q) are numbered in full, equations are numbered sparingly, and locally. Thus (*), (**) are the first and second such equations in the current section or chapter.

Unless otherwise stated, e or e_G will denote the group identity.

We cannot resist mentioning Tom Körner's description of Baire's category theorem as 'a profound triviality', and Jean-Pierre Kahane quoting this and describing the use he and Bob Kaufman had made of it as profound. Alas, Rogers (see FalGO2015) and Kahane are no longer with us, but their influence pervades the book, and we salute their memory with gratitude.

To close, we thank all those who have helped us over the years, most particularly our collaborators (some alas no longer with us, including the much-missed Harry I. Miller (1939–2018)), in particular Eliza Jabłońska and Tomasz Natkaniec for a careful reading respectively of §§15.7–15.8 and §§9.5–9.7. We are most grateful to David Tranah and Roger Astley of Cambridge University Press, for their kind forbearance during the book's unconscionably long gestation period. Last and most, we thank our wives and families for their love and support during the writing of this book.

Prologue: Regular Variation

P.1 Introduction

Regular variation is a subject both of theoretical interest and of great use in a variety of applications. These include analytic number theory (asymptotics of arithmetic functions, Prime Number Theorem), complex analysis (entire functions – Levin–Pfluger theory) and, particularly, probability theory (limit theorems). The standard work here, covering theory and applications, is Bin-GoT1987 (BGT below for brevity). As it happens, matters left open in BGT – the foundational question, p. 11 (on measurability and the Baire property), and the contextual question (Appendix 1, on contexts beyond the real analysis to which the bulk of the text is devoted) – motivated our joint work. So this book is motivated by two much earlier and by now well-established texts, Oxtoby (Oxt1980) and BGT. But these serve only as background and motivation here; this book is self-contained and may be read without reference to either.

To make the above more specific, here are some instances of 'what regular variation can do for the mathematician in the street'.

P.2 Probability Theory

The prototypical limit theorem in the subject is Kolmogorov's[1] *strong law of large numbers*: that if (X_n) is a sequence of independent copies of a random variable X drawn from some distribution (or law) F, the averages $S_n/n := \sum_1^n X_k/n$ converge as n increases to some limit c with probability 1 ('almost

---

[1] In 2023, the postal address of Moscow State University became 1 Kolmogorov Street; cf. 2 Stefan Banach Street, Warsaw, the postal address of the mathematics department of Warsaw University.

surely', a.s.) if and only if X has a *mean* (first moment, *expectation*) (meaning $\mathbb{E}[\,|X|\,] < \infty$), and then the limit is the mean $\mu = \mathbb{E}[X]$:

$$S_n/n \to \mu\,{}^{\scriptscriptstyle\bullet} \quad (n \to \infty) \quad a.s., \quad \text{where } \mu = \mathbb{E}[X]. \qquad \text{(LLN)}$$

Second only to this is the *central limit theorem*: if also one has finite variance, σ^2 say (finite second moment), and centres at means (subtracting $\mathbb{E}[S_n] = n\mu$), the right scaling is by $\sqrt{n}\sigma$, and one then has a limit distribution, the standard normal (or Gaussian) law $\Phi = N(0, 1)$:

$$\mathbb{P}((S_n - n\mu)/\sigma\sqrt{n} \le x) \to \Phi(x) := \int_{-\infty}^{x} \frac{e^{-\frac{1}{2}u^2}}{\sqrt{2\pi}}\,du \quad (n \to \infty) \text{ for all } x \in \mathbb{R}.$$
$$\text{(CLT)}$$

Because these results are so important, and because one does not always have a (finite) mean and variance, it was of great interest to find versions of them which held under weaker moment conditions. It was realized by Sakovich in 1956 (Sak1956) that regular variation gave the right language here: what one needs for the first is that the *truncated mean is slowly varying*,

$$\int_{-x}^{x} u\,dF(u) \sim \ell_1(x) \qquad (x \to \infty),$$

and for the second that the *truncated variance is slowly varying*,

$$\int_{-x}^{x} u^2\,dF(u) \sim \ell_2(x) \qquad (x \to \infty),$$

where ℓ_1, ℓ_2 are slowly varying (below).

Note. Oddly, despite their importance, these results were overlooked at the time, and they were re-discovered and given prominence in Feller's book (Fel1966). The first author saw them there then ('love at first sight').

More curiously still, although regular variation dates back to 1930 (below), the classic monograph by Gnedenko and Kolmogorov (GneK1954) (the Russian original is from 1949) made no use of it. So its treatment of these and related matters is unnecessarily complicated, and in particular the analysis and the probability are not properly separated. We note that Sakovich's PhD was supervised by Gnedenko.

For more on Gnedenko's work, and his very interesting life, see Bin2014.[2]

Then one has the third member of the trilogy, LLN–CLT–LIL: the *law of the iterated logarithm*. Here the norming, which gives the result its name, is intermediate between those in (LLN) and (CLT), and the conclusion is of a different type:

$$\limsup\,(S_n - n\mu)/\sigma\sqrt{2n\log\log n} = +1 \quad a.s., \qquad \liminf \cdots = -1 \quad a.s.$$

[2] The text of a talk given by the first author at the Gnedenko Centenary Memorial Meeting, Moscow State University, 2012.

Indeed,

$$(S_n - n\mu)/\sigma\sqrt{2n\log\log n} \to\to [-1, 1] \quad \text{a.s.,} \qquad \text{(LIL)}$$

meaning that all points in $[-1, 1]$, and no others, are limits of subsequences, a.s.

Stable laws (limit laws of centred and normed sums of independent copies) provide another good example. See, e.g., two approaches to the 'domain of attraction' problem here by Pitman and Pitman (PitP2016) and Ostaszewski (Ost2016a).

For more on the early history of regular variation in probability theory, see Bin2007.

Extremes The extreme values in a sample – sample maximum and minimum – have always been of great practical importance (the strength of a chain is that of its weakest link, etc.). The theory here dates back to Fisher and Tippett in 1928, so to before regular variation, though the relevance of regular variation was soon realized. The area is growing in importance nowadays, e.g. because of climate change and global warming. There was pioneering early work by Gnedenko in 1943, but the systematic use of regular variation to study extremes stems from de Haan in 1970 (Haa1970). For background and historical comments, see, e.g., our recent survey BinO2021b and the references there.

While Hardy himself was not interested in probability, the Tauberian theory to which he and Littlewood contributed so much has proved very useful in probability theory; see, e.g., Bin2015b.

P.3 Complex Analysis

Recall (see, e.g., BGT, Ch. 4) that an *Abelian* theorem passes from a stronger mode of convergence to a weaker one (such results are usually easy); a *Tauberian* one gives a converse, under an additional condition (a *Tauberian condition*); *Mercerian* theorems (see, e.g., BGT, Ch. 5) are hybrids, going from a condition on both to a stronger conclusion, under *no* Tauberian condition. A prototypical Abelian result will pass from a function

$$f(x) \sim x^\rho \ell(x) \quad (\ell \in R_0) \quad (x \to \infty)$$

to a Mellin convolution

$$(f * k)(x) := \int_0^\infty k(t) f(x/t) dt/t,$$

where ρ lies in the vertical strip in the complex s-plane where the Mellin transform

$$\hat{k}(s) := \int_0^\infty t^{-s} k(t) dt/t = \int_0^\infty u^s k(1/u) du/u$$

converges absolutely, giving

$$(f * k)(x) \sim \hat{k}(\rho) x^\rho \ell(x) \quad (x \to \infty);$$

the factor $\hat{k}(\rho)$ is to be expected, since if $f(x) = x^\rho$, $(f * k)(x) = \hat{k}(\rho) x^\rho$. The Tauberian converse reverses the implication for kernels satisfying Wiener's condition that $\hat{k}(s)$ be non-vanishing for *Re* $s = \rho$ ((NV) below), under suitable Tauberian conditions on f. The Mercerian (or 'ratio Tauberian', below) statement assumes convergence of the quotient,

$$(f * k)(x)/f(x) \to c \quad (x \to \infty)$$

for some constant c, and deduces regular variation of both as above, with $c = \hat{k}(\rho)$.

For entire functions of finite order, one can look (in discs centre 0 and large radius r) at growth rates of the function, its maximum modulus $M(r)$ and the zero-counting function $n(r)$ (both in discs centre 0 and radius r). Matters split between integer and non-integer order. Functions with real negative zeros are simplest; write $\mathcal{E}_\rho$ for the class of entire f with order $\rho < \infty$ and negative zeros. For $f \in \mathcal{E}_\rho$, one has the Valiron–Titchmarsh theorem (BGT, §7.2, Th. 7.2.2), proved by Tauberian methods involving regular variation (BGT, Ch. 4), based on the linear integral transform (Stieltjes transform)

$$\log f(z) = \int_0^\infty \frac{z n(t)}{t + z} dt/t \quad (|\arg z| < \pi).$$

For non-integral order, regular variation of either of $n(r)$, $\log f(re^{i\theta})$ implies regular variation of the other, and convergence of the quotient to a non-zero limit. For more on the Valiron–Titchmarsh Theorem, see DrasS1970 and the references cited there.

The question of whether convergence of this quotient implies regular variation of both functions has been called of 'ratio Tauberian' type; it is in fact *Mercerian* (BGT, Ch. 5). The first such results are due to Edrei and Fuchs (EdrF1966) and Shea (She1969), for $f \in \mathcal{E}_\rho$ and the Stieltjes transform above. Such results were extended by Drasin (Dras1968) and Drasin and Shea (DrasS1976) to more general kernels, using Wiener Tauberian theory. Drasin and Shea had *non-negative* kernels k, for which the relevant Mellin transforms converge absolutely in their strip of convergence. Matters are more complicated when the kernel can change sign (as with Fourier sine and cosine transforms, and Hankel transforms), as here there can be strips of conditional convergence (or Abel summability) also; see Jor1974. The Fourier and Hankel cases were considered in detail by Bingham and Inoue (BinI1997; BinI1999).

As may be seen from the Wiener Tauberian Theorem, P.7.1: if regular variation of index ρ (membership of R_ρ) is to appear in the hypothesis and conclusion, the key condition on the kernel is the *non-vanishing* condition

$$\hat{k}(s) \neq 0 \qquad (\text{Re } s = \rho). \tag{NV}$$

In the corresponding Mercerian results, the key condition on k is the *no-repeat* condition

$$\hat{k}(s) = \hat{k}(\rho) \qquad \text{on Re } s = \rho \text{ only for } s = \rho \tag{NR}$$

(She1969; Jor1974; cf. PalW1934, IV, (18.09)).

One can also consider the *Nevanlinna characteristic*

$$T(r) = T(r, f) := \frac{1}{2\pi} \int_{-\pi}^{\pi} \log_+ |f(re^{i\theta})| \, d\theta$$

(see, e.g., Haym1964). For $f \in \mathcal{E}_\rho$, this is given by the *non-linear* integral transform

$$T(r) = \sup \left\{ \int_0^\infty P(r/t, \theta) N(t) dt/t : \theta \in (0, \pi) \right\},$$

where

$$N(r) := \int_0^r n(t) dt/t, \qquad P(t, \theta) := \frac{1}{\pi} \frac{\sin \theta}{t + 2\cos\theta + t^{-1}}.$$

Baernstein (Bae1969) obtained a non-linear Tauberian theorem (the passage from N to T is Abelian, and simple, EdrF1966): for f entire of genus 0, if $T(r) \sim r^\lambda \ell(r)$ as $r \to \infty$ for ℓ slowly varying, then

(a) if $\lambda \in [0, \frac{1}{2}]$, then $N(r) \sim r^\lambda \ell(r)$;
(b) if $\lambda \in [\frac{1}{2}, 1]$, and f has only negative zeros, then $N(r) \sim \sin \pi\lambda \, r^\lambda \ell(r)$.

The corresponding Mercerian (or ratio Tauberian) theorem was proved by Edrei and Fuchs (EdrF1966), for $\lambda \in [\frac{1}{2}, 1]$: if the ratio converges, then both $N(r)$ and $T(r)$ are regularly varying.

Baernstein (Bae1969) conjectured that his results extend to $\mathcal{M}_\rho$, the class of meromorphic functions of finite order ρ, negative zeros and positive poles, but this is not the case. For counterexamples and discussion, see Dras2010. However, they do extend to the subclass $\mathcal{J}_\rho$ of $\mathcal{M}_\rho$ whose zeros a_n and poles b_n are symmetrically related, $a_n = -b_n$ (Will1972). Edrei (Edr1969) removes geometric restrictions on the zeros and poles, but at the cost of obtaining only 'locally Tauberian' results, in which $r \to \infty$ only through the union of a well-chosen sequence of large intervals.

One can extend to real zeros. One can use the language of proximate orders, due to Valiron in 1913, which can be shown to be equivalent to that of regular variation (and thus that Valiron may be credited with initiating the subject). The resulting *Levin–Pfluger theory* (A. Pfluger in 1938, B. Ya. Levin in 1964; BGT, Ch. 7) may be regarded as weakening the severe geometric restriction that all zeros lie on one ray, or one line, as far as possible.

The contrasts between key examples throw light on the theory, which they inspired. To quote BGT (end of §7.6): 'It is instructive to compare the two examples $\sin \pi z$ and $1/\Gamma(z)$. Their rates of growth differ, as above; their zeros differ not so much in their density as in their geometry. An extensive study of the integer-order case has been given by Pfluger (1946), motivated by the contrasts between these examples.'

As well as the maximum modulus, the minimum modulus of an entire function is of interest:

$$M(r) := \sup\{|f(z)| : |z| \le r\}, \qquad m(r) := \inf\{|f(z)| : |z| \le r\}.$$

One has the $\cos \pi \rho$ *theorem* (see BGT, §7.7; Bae1974; Ess1975 and the references there for details): if f is entire of order $\rho \in [0, 1)$,

$$\limsup \frac{\log m(r)}{\log M(r)} \ge \cos \pi \rho.$$

Functions extremal here are particularly interesting; see DrasS1969. Here one encounters *exceptional sets*, of logarithmic density 0, which cannot be avoided (Haym1970).

Pólya Peaks. Pólya peaks (of the 'first and second kinds') are a device in real analysis, introduced by Pólya (Poly1923) for the study of entire functions. They were named by Edrei in the 1960s. Their use was extended to meromorphic functions by Hayman (Haym1964, §4.4); for details, see DrasS1972. It turns out that they are intimately linked to the *Matuszewska indices* (BGT, §2.1) $\alpha(f)$, $\beta(f)$ of regular variation: both kinds of peak exist in the interval $[\beta(f), \alpha(f)]$ and nowhere else (the Pólya Peak Theorem; BGT1987, Th. 2.5.2).

In fact, the use of Pólya peaks in the results above (Edrei, Drasin and Shea, Jordan) may be avoided; see Bingham and Inoue (BinI2000a). This may be preferred on thematic grounds in complex analysis, as well as to simplify the proofs.

Recently, essential use of O-regular variation has been made in the theory of *ultraholomorphic functions*; see JimSS2019 for background and details.

P.4 Analytic Number Theory

For background on Abelian, Tauberian and Mercerian theorems as mentioned above, see, e.g., BGT, §4.5; Kor2004.

Tauberian theorems such as the Hardy–Littlewood–Karamata theorem are extensively used in analytic number theory (see, e.g., Ten2015; BGT, Ch. 4). So too is e.g. Kohlbecker's Tauberian Theorem on asymptotics of partitions (BGT, Th. 4.12.1). Tauberian (and Mercerian) theorems can be used to give short proofs of the Prime Number Theorem (BGT, §6.2).

The *prime divisor functions*,

$$\omega(n) := \# \text{ distinct prime divisors of } n,$$

$$\Omega(n) := \# \text{ prime divisors of } n \qquad \text{(counted with multiplicity)}$$

(Ten2015, I.2.2), illustrate our approach well. The classical estimates are (Hardy and Ramanujan in 1917; Ten2015, I.3.6,7)

$$\frac{1}{x} \sum_{n \leq x} \omega(n) = \log\log x + c_1 + O(1/\log x) \qquad (x \to \infty),$$

with c_1 a known constant, and similarly for $\Omega(n)$ with a different known constant $c_2 > c_1$. One also has the classical Erdős–Kac central limit theorem of 1939,

$$\frac{1}{x} |\{n \leq x : \omega(n) \leq \log\log x + \lambda\sqrt{\log\log x}\}| \to \Phi(\lambda) \quad (x \to \infty \text{ for all } \lambda \in \mathbb{R}),$$

the beginning of probabilistic number theory, and its refinement of Berry–Esseen type, due to Rényi and Turán in 1958, with error term $O(1/\sqrt{\log\log x})$ uniform in λ (Ten2015, III.4.15). Our methods give (BinI2000b)

$$\frac{1}{\lambda x} \sum_{n \leq \lambda x} \omega(n) - \frac{1}{x} \sum_{n \leq x} \omega(n) \sim \frac{\log \lambda}{\log x} \quad (x \to \infty \text{ for all } \lambda \in \mathbb{R}).$$

This is a statement of regular-variation type, so it has a representation theorem, namely

$$\frac{1}{x} \sum_{n \leq x} \omega(n) = C + \int_2^x (1 + o(1)) \frac{dt}{t \log t} + o(1/\log x)$$

(note the *two* error terms, one inside the integral, one outside). This is not comparable to the classical results. There, it is the size of the error terms that counts, but there is no information on behaviour under differencing; here, matters are reversed. Similarly for results of Mertens (Ten2015, I.1.4; HarW2008, Th. 425, 427) on sums over primes p (BinI2000b),

$$\sum_{p \leq x} \frac{\log p}{p}, \qquad \sum_{p \leq x} 1/p.$$

P.5 Regular Variation: Preliminaries

The subject of regular variation originates with the Yugoslav mathematician Jovan Karamata (1902–1967) in 1930 (Kar2009). It concerns limiting relations of the form

$$f(\lambda x)/f(x) \to g(\lambda) \qquad (x \to \infty) \quad \forall \, \lambda > 0, \tag{K}$$

for positive functions f on $\mathbb{R}_+$. Relevant here is the multiplicative group of positive reals, $(\mathbb{R}_+, \times)$, with Haar measure dx/x. While this formulation is the one useful for applications, for theory it is more convenient to pass to the additive group of reals, $(\mathbb{R}, +)$, Haar measure Lebesgue measure dx, where we write this as

$$h(u + x) - h(x) \to k(u) \qquad (x \to \infty) \quad \forall \, u \in \mathbb{R}. \tag{K$_+$}$$

We can pass at will between these two formulations via the exp/log isomorphism. The core of the resulting theory is treated in full in Chapter 1 of BGT, with further results (e.g. with lim replaced by lim sup – where one may lose measurability, by 'character degradation') in Chapter 2 of BGT.

The limit function g in (K) satisfies the *Cauchy functional equation*

$$g(\lambda\mu) = g(\lambda)g(\mu) \qquad (\lambda, \, \mu > 0). \tag{CFE}$$

For background on functional equations, see AczD1989; the classic context is Ban1920.

P.6 Topological Regular Variation

Solutions to (CFE), as is typical with functional equations, exhibit a sharp dichotomy: they are either *very nice* (continuous, here) or *very nasty* (pathological – unbounded above and below on every interval, or even on any non-meagre Baire set or non-null measurable set). Since (as we shall see below) such 'bad' solutions can be manufactured at will from a Hamel basis (of the reals, as a vector space over the rationals), we will call this the *Hamel pathology*.

Under mild regularity conditions (such as measurability, the Baire property, No Trumps **NT**, etc.), this gives

$$g(\lambda) \equiv \lambda^\rho$$

for some $\rho \in \mathbb{R}$. Then f is called *regularly varying* with *index* ρ, $f \in R_\rho$. Functions in class R_0 are called *slowly varying*, written ℓ (for *lente*, or *langsam*).

By taking logarithms, (K) may be written in terms of the limit of the difference of a function at arguments λx and x. It turns out that this may be fruitfully generalized by introducing a denominator $\ell \in R_0$:

$$[f(\lambda x) - f(x)]/\ell(x) \to k(\lambda) \qquad (x \to \infty) \qquad \forall \lambda > 0 \qquad \text{(BK/DH)}$$

(using a denominator in R_ρ for $\rho \neq 0$ gives nothing new; see, e.g., BGT, §3.2). This study goes back to Bojanic and Karamata (BojK1963), and independently to de Haan (Haa1970), whence the name (BK/DH); see BGT, Chapter 3 for a full account.

There are three key theorems that underlie any form of regular variation (there are two forms above; more will follow). These are (under mild conditions):

the *Uniform Convergence Theorem*, **UCT**: the convergence in (K), (BK/DH) takes place *uniformly* on compact λ-sets in $\mathbb{R}_+$;

the *Representation Theorem*: giving that $\ell \in R_0$ if and only if it may be written in the form

$$\ell(x) = c(x) \exp\{ \int_1^x \epsilon(u)\,du/u \} \qquad (x \geq 1) \qquad \text{(RT)}$$

where

$$c(x) \to c \in \mathbb{R}_+, \qquad \epsilon(x) \to 0 \qquad (x \to \infty)$$

(here $\epsilon(.)$ may be taken as smooth as desired, while $c(.)$ may be as rough as the mild regularity conditions allow);

the *Characterization Theorem*: giving the form of $g(\lambda)$ as λ^ρ as above and that of k in (BK/DH) as

$$k(\lambda) = ch_\rho(\lambda), c \in \mathbb{R}_+; \quad h_\rho(\lambda) := \int_1^\lambda u^{\rho-1}\,du = (\lambda^\rho - 1)/\rho \ (\lambda > 0),$$

with the usual 'l'Hospital convention' that the right-hand side above is taken as $\log \lambda$ when $\rho = 0$.

Even with the simplest functional equation that arises here (the Cauchy), some mild regularity condition is required. There is a dichotomy: as above solutions are either very nice (powers λ^ρ or the $h_\rho(\lambda)$ as above) or very nasty – pathological (e.g. unbounded above and below on any non-negligible set).

Exceptional Sets. There are situations in which the passage to the limit in slow and regular variation needs to avoid some *exceptional set*; see BGT, §2.9. Examples arise in complex analysis: BGT, Th. 7.2.4 (a result of Titchmarsh in 1927), and in work of Drasin and Shea (DrasS1976) on functions extremal for the $\cos \pi\rho$ theorem of Wiman and Valiron mentioned above. We will need such exceptional sets below, in dealing with sequential aspects of regular variation.

Thinning (Quantifier Weakening). Another key question, going back to a conjecture of Karamata, concerns *weakening the quantifier*, ∀, in (K), (BK/DH): requiring the convergence to take place for *some but not all* $\lambda > 0$. Rather than having a continuum of conditions to check, one may be able to reduce this to finitely many, or even to just *two* (of course, one could not expect just one to suffice!). Results of this kind – which one might call *thinning* results, as they involve thinning of the λ-set on which convergence is required (cf. BinO2010a) – go back to Heiberg (Hei1971) and Seneta (Sen1973). Interestingly, given the side-condition of Heiberg–Seneta type, one no longer needs to impose the regularity condition needed above to eliminate pathology.

Matters were taken further in Bingham and Goldie (BinGo1982a) and Bingham and Ostaszewski (BinO2018a; BinO2020a): with

$$g^*(\lambda) := \limsup_{x\to\infty} f(\lambda x)/f(x),$$

assume

$$\limsup_{\lambda\downarrow 1} g^*(\lambda) \le 1.$$

Then the following are equivalent (for positive f):

(i) there exists $\rho \in \mathbb{R}$ such that

$$f(\lambda x)/f(x) \to \lambda^\rho \qquad (x \to \infty) \qquad \forall \lambda > 0;$$

(ii) $g(\lambda) := \lim_{x\to\infty} f(\lambda x)/f(x)$ exists, finite, for a λ-set of positive measure [a non-meagre Baire set];

(iii) $g(\lambda)$ exists, finite, in a λ-set dense in $\mathbb{R}_+$;

(iv) $g(\lambda)$ exists, finite, for $\lambda = \lambda_1$, λ_2 with $(\log \lambda_1)/\log \lambda_2$ irrational.

The reader may recognize that *Kronecker's Theorem* (HarW2008, Ch. XXIII) lies behind (iv) here.

There are corresponding thinning results for (BK/DH); see BGT, Th. 3.2.5, Th. 1.4.3. As remarked there, the result for (K) is no easier, despite its simpler context. This is because the thinning takes place in the quantifier over λ, which affects only the numerator in (BK/DH). The effect is that the general ℓ in the denominator is no harder to handle than $\ell \equiv 1$, when (BK/DH) reduces to the logarithmic form of (K).

Remark The proofs of these two thinning results from BGT are the hardest in that book on the theory of regular variation as such (that of the Drasin–Shea–Jordan theorem, BGT, §5.2, is harder, but belongs rather to Mercerian theory). The search for simpler proofs was the motivation behind several of our papers, on thinning (BinO2010a) and quantifier weakening (BinO2018a; BinO2020a).

Frullani Integrals. The *Frullani integral* (G. Frullani, 1828; see Ostr1976 for the history) of a locally integrable function ψ on $\mathbb{R}_+$ is the improper integral

$$I = I(\psi; a, b) := \int_{0+}^{\infty-} \{\psi(\lambda t) - \psi(t)\}dt/t = \lim_{\epsilon \downarrow 0, X \uparrow \infty} \int_{\epsilon}^{X} \{\psi(\lambda t) - \psi(t)\}dt/t.$$

Writing $bt = u$, we see that $I(\psi; a, b) = I(\psi; a/b, 1) = I(\psi, a/b)$, say. So we may restrict attention to

$$I = I(\psi; \lambda) := \int_{0+}^{\infty-} \{\psi(\lambda t) - \psi(t)\}dt/t = \lim_{\epsilon \downarrow 0, X \uparrow \infty} \int_{\epsilon}^{X} \{\psi(\lambda t) - \psi(t)\}dt/t.$$

Now

$$\int_{\epsilon}^{X} \{\psi(\lambda t) - \psi(t)\}dt/t = \int_{X}^{\lambda X} \psi(t)dt/t - \int_{\epsilon}^{\lambda \epsilon} \psi(t)dt/t.$$

So the two limits, concerning behaviour at ∞ and at 0, may be handled separately. One obtains (BinGo1982b, §6; BGT Th. 1.6.6):

Theorem *For ψ and its Frullani integral $I(\psi; \lambda)$ as above, the following are equivalent:*

(i) *$I(\psi; \lambda)$ exists for all $\lambda \in \mathbb{R}_+$;*
(ii) *$I(\psi; \lambda)$ exists for λ in a set of positive measure [a non-meagre Baire set];*
(iii) *$I(\psi; \lambda)$ exists for λ in a dense set in $\mathbb{R}_+$ (or for $\lambda = \lambda_1, \lambda_2$ with $(\log \lambda_1)/\log \lambda_2$ irrational), and*

$$\liminf_{\lambda \downarrow 1} \liminf_{x \to \infty} \int_{x}^{\lambda x} \psi(t)dt/t \geq 0,$$

$$\liminf_{\lambda \downarrow 1} \liminf_{x \to \infty} \int_{x}^{\lambda x} \psi(1/t)dt/t \geq 0.$$

Each of (i)–(iii) holds if and only if both of the following finite limits exist for some (all) $\sigma > 0$:

$$M = M(\psi) = \lim_{x \to \infty} \sigma x^{-\sigma} \int_1^x u^\sigma \psi(u) du/u,$$

$$m = m(\psi) = \lim_{x \to \infty} \sigma x^{-\sigma} \int_1^x u^\sigma \psi(1/u) du/u.$$

Then the Frullani integral is given by

$$I(\psi, \lambda) = (M - m) \log \lambda \qquad (\lambda \in \mathbb{R}_+).$$

This result extends those of Hardy and Littlewood (HarL1924), where $\sigma = 1$.

Frullani integrals occur in probability theory, e.g. in the fluctuation theory of Lévy processes; see, e.g., Bert1996, III.1, p. 73.

Convergence and Cesàro Convergence. The mathematics of the Frullani integral above yields as a by-product the results below (BinGo1982b, §6), showing exactly what is needed for a Cesàro convergent function or sequence to converge. For functions: for ϕ locally integrable on $[0, \infty)$,

$$\frac{1}{x} \int_0^x \phi(t) dt \to c \qquad (x \to \infty)$$

if and only if

$$\phi(x) = a(x) + b(x), \qquad \text{where } a(x) \to c, \quad \int_1^\infty b(t) dt/t \text{ is convergent.}$$

For sequences:

$$\frac{1}{n} \sum_1^n s_k \to c \qquad (n \to \infty)$$

if and only if

$$s_n = a_n + b_n, \qquad \text{where } a_n \to c, \quad \sum_1^\infty b_n/n \text{ is convergent.}$$

We include the proof (due to G. E. H. Reuter) as it is so short.

Proof If $\sum b_n/n$ converges, $(b_1 + \cdots + b_n)/n \to 0$ by Kronecker's Lemma. So if $s_n = a_n + b_n$ as above, $s_n \to c$ (C_1).

Conversely, if $s_n \to c$ (C_1), set $a_{n+1} := (s_1 + \cdots + s_n)/n$. Then $s_n = a_n + n(a_{n+1} - a_n)$, and this is the required decomposition, since if $b_n := n(a_{n+1} - a_n)$, $\sum b_n/n$ converges. $\qquad \square$

Beurling Slow Variation. For $\phi \colon \mathbb{R} \to \mathbb{R}_+$ Baire or measurable, ϕ is called *Beurling slowly varying* if

$$\phi(x) = o(x) \quad \text{and} \quad \phi(x + t\phi(x))/\phi(x) \to 1 \text{ for all } t \in \mathbb{R}, \ (x \to \infty). \quad \text{(BSV)}$$

This originated in unpublished lecture notes of Beurling in 1957 on his Tauberian theorem (below); see Kor2004, IV.11.

If also the convergence in (BSV) is locally uniform in t, ϕ is called *self-neglecting*, written $\phi \in \mathrm{SN}$. The term and the concept arose in probability theory; see, e.g., BinO2021b and the references there.

It was shown by Bloom (Blo1976) that for ϕ *continuous*, the convergence in (BSV) is indeed locally uniform. In fact Bloom's proof needs only the *Darboux property*, or *intermediate-value property*, that ϕ takes every value between any two values it attains. For this and other results, see BinO2014; Ost2015a. Here it is enough to require that ϕ takes a dense set of values between any two values attained, but the question of whether a Darboux-like property can be dropped altogether remains open.

The Representation Theorem gives the Beurling slowly varying ϕ as those *positive* functions of the form

$$\phi(x) = c(x) \int_0^x \epsilon(u)\,du \qquad (x \in \mathbb{R}),$$

where ϵ is C^∞ with $\epsilon(x) \to 0$ as $x \to \infty$, and c is Baire/measurable with $c(x) \to c \in (0, \infty)$ as $x \to \infty$ (BGT Th. 2.11.3; BinO2014, §9; Ost2015a, p. 731).

P.7 Tauberian Theorems

We first recall *Wiener's Tauberian Theorem* (see, e.g., Har1949, XII; Kor2004, II):

Theorem P.7.1 (Wiener's Tauberian Theorem) *Suppose $K \in L_1(\mathbb{R})$ with Fourier transform $\hat{K}$ non-vanishing on $\mathbb{R}$, and $H \in L_\infty(\mathbb{R})$. If*

$$\int K(x - y)H(y)\,dy \ \to \ c \int K(y)\,dy \qquad (x \to \infty),$$

then, for all $G \in L_1(\mathbb{R})$,

$$\int G(x - y)H(y)\,dy \ \to \ c \int G(y)\,dy \qquad (x \to \infty).$$

Here integrals are over $\mathbb{R}$ and we use additive notation; one may work multiplicatively with $\int_0^\infty K(x/y)H(y)dy/y$, etc. The *Tauberian condition* here is of O-type: $H \in L_\infty(\mathbb{R})$, or $H = O(1)$.

Beurling's Tauberian Theorem generalizes Wiener's Tauberian Theorem, to which it reduces in the special case $\phi \equiv 1$:

Theorem P.7.2 (Beurling's Tauberian Theorem)   *Suppose $K \in L_1(\mathbb{R})$ with Fourier transform $\hat{K}$ non-vanishing on $\mathbb{R}$, with ϕ Beurling slowly varying and $H \in L_\infty(\mathbb{R})$. If*

$$\int K\left(\frac{x-y}{\phi(x)}\right) H(y)\,dy/\phi(x) \;\to\; c\int K(y)\,dy \qquad (x \to \infty),$$

then, for all $G \in L_1(\mathbb{R})$,

$$\int G\left(\frac{x-y}{\phi(x)}\right) H(y)\,dy/\phi(x) \;\to\; c\int G(y)\,dy \qquad (x \to \infty).$$

Note that the arguments $x - y$ in Theorem P.7.1 involve the additive *group* $(\mathbb{R}, +)$, and the x/y of its multiplicative version that of the multiplicative *group* $(\mathbb{R}_+, \times)$, while the $(x - y)/\phi(x)$ of Theorem P.7.2 involve the *ring* structure of $\mathbb{R}$. The two results are thus structurally distinct.

For a short and elegant reduction of Beurling's Tauberian Theorem to Wiener's, see Kor2004, IV, Th. 11.1. For an early use of Beurling's Tauberian Theorem in probability theory, see Bin1981.

The Borel–Tauber Theorem

The two most basic families of summability methods are the *Cesàro C_α* ($\alpha > 0$) and *Abel* methods A; see, e.g., Har1949, V–VII; Kor2004, I. Perhaps next in importance, though harder, are the *Euler E_p* ($p \in (0, 1)$) and *Borel* methods; see, e.g., Har1949, VIII, IX; Kor2004, VI. The Euler and Borel methods (plus those of Taylor and Meyer–König) belong to the family of *circle methods* (German: Kreisverfahrung; see MeyK1949). The name derives from the circle of convergence of a power series; such methods were used for analytic continuation by power series.

The key Tauberian theorem for the Borel method – 'Borel–Tauber Theorem' – is:

Theorem P.7.3 (Borel–Tauber Theorem)   *For $s_n := \sum_0^n a_k$: if*

$$e^{-x} \sum_0^\infty s_n x_n/n! \;\to\; s \qquad (x \to \infty)$$

and $a_n = O(1/\sqrt{n})$, then

$$s_n \;\to\; c \qquad (n \to \infty).$$

The setting of Theorem P.7.3 is discrete, involving a sum, while that of Theorems P.7.1 and P.7.2 is continuous, involving an integral. To pass between them, one may either use 'Wiener's second theorem': see, e.g., Har1949, 12.7, which demands less of the integrand (so $H(y)dy$ becomes a Stieltjes integral, $dU(y)$, say) but more of the kernel K; or use an auxiliary approximation argument. There is much more to be said here, but we must refer for further detail to, e.g., Kor2004, VI.

The Tauberian Condition. Here the Tauberian condition is $a_n = O(1/\sqrt{n})$, and this is best-possible in that no weaker O-condition would suffice here. But because the weights $e^{-x}x^n/n!$ (of course those of the *Poisson* distribution $P(x)$ with parameter x) are non-negative, a *one-sided* Tauberian condition suffices: $a_n = O_L(1/\sqrt{n})$, meaning that $\sqrt{n}a_n$ is bounded below (or with O_R and bounded above). In fact such 'pointwise' conditions on the individual a_n are not needed, but rather 'averaged' forms of them involving differences of the s_n. The classical one is of 'slow-decrease' type, due to R. Schmidt in 1925 (see Kor2004, VI.12):

$$\liminf (s_m - s_n) \geq 0 \qquad (m, n \to \infty, \ 0 \leq \sqrt{m} - \sqrt{n} \to 0).$$

Such one-sided Tauberian conditions are studied at length in Bingham and Goldie (BinGo1983).

Valiron Methods. For $\beta \in (0, 1)$, write V_β for the *Valiron* summability method (Bin1984b), given by writing

$$\frac{1}{x^\beta \sqrt{2\pi}} \sum_0^\infty s_k \exp\left\{ -\frac{1}{2}(x - k)^2/x^{2\beta} \right\} \to s \qquad (x \to \infty)$$

as

$$s_n \to s. \tag{V_β}$$

Our principal concern is with the case $\beta = \frac{1}{2}$ (see BinT1986). One sees that the sum above is a discrete form of the condition in Theorem P.7.2 with $K(x) = e^{-x^2/2}/\sqrt{2\pi}$. This is the standard normal probability density Φ or $N(0, 1)$, with Fourier transform (characteristic function) $\exp\{-\frac{1}{2}t^2\}$, which is non-vanishing as in Theorems P.7.1 and P.7.2. This K may thus serve as a Wiener kernel. In the notation of Theorems P.7.1 and P.7.2, one can obtain boundedness of $H(.)$ from the other conditions; see Har1949, p. 220 for the pointwise Tauberian condition $a_n = O(1/\sqrt{n})$ and Har1949, p. 225 for a reference to Vijayaraghan's method of monotone minorants for the slow-decrease Tauberian condition. This allows an easy proof of Theorem P.7.3 from Theorem P.7.2.

That the methods $V_{\frac{1}{2}}$ and B are intimately linked has been known since Hardy and Littlewood in 1916 (HarL1916). In probabilistic language, this link reflects the *central limit theorem*: the Poisson law $P(x)$ with large parameter x is an n-fold convolution of $P(x/n)$ with itself, and so approaches normality. Much more is true: the rapid tail-decay of the Poisson laws allows the use of large-deviation methods. See Kor2004, VI, Th. 6.1, where the range $|n - x| < x^\gamma$ occurs, where $1/2 < \gamma < 2/3$. As Korevaar remarks, this parameter range is the 'signature' of large deviations.

Jakimovski and Karamata–Stirling Methods. There are other summability methods whose weights exhibit central-limit behaviour. We consider independent random variables X_n, integer-valued (so that the weights will form a matrix, below), with partial sums $S_n = \sum_1^n X_k$; write

$$a_{nk} := \mathbb{P}(S_n = k),$$

and write $A = (a_{nk})$ for the summability matrix. The classical case is of *Jakimovski methods* (Jak1959; ZelB1970, §70); here the X_n are Bernoulli (0–1 valued), with

$$\mathbb{P}(X_n = 1) = p_n, \qquad P(X_n = 0) = q_n := 1 - p_n.$$

Writing $p_n = 1/(1 + d_n)$ $(d_n \geq 0)$, this gives

$$\prod_{j=1}^n \left(\frac{x + d_j}{1 + d_j} \right) = \sum_{k=0}^n a_{nk} x^k,$$

and the *Jakimovski method* $[F, d_n]$. The motivating examples are:

(i) the Euler methods, with $d_n = 1/\lambda$, say written $E(\lambda)$;
(ii) the *Karamata–Stirling methods* KS(λ), with $d_n = (n - 1)/\lambda$. Here

$$a_{nk} = \lambda^k S_{nk}/(\lambda)_n,$$

with (S_{nk}) the Stirling numbers of the second kind and

$$(\lambda)_n := \lambda(\lambda + 1) \cdots (\lambda + n - 1).$$

See Bin1988 for their Tauberian theory and BinS1990 for LLN and LIL results.

Turning from the non-identically distributed Bernoulli case to the identically distributed general integer-valued case gives the *random-walk* methods (Bin1984a).

All the summability methods considered here are closely enough linked to be *equivalent for bounded sequences* (as are Euler and Borel, and indeed as are Cesàro and Abel).

Riesz Means and Moving Averages. With K as above, taking $H(x) = H_a(x) :=$ $a^{-1} I_{[0,a]}(x)$ gives conclusions of the form

$$\frac{1}{a\sqrt{n}} \sum_{n \le k < n + a\sqrt{n}} s_k \to s \qquad (n \to \infty),$$

passing back from integrals to sums as above. These are *Riesz means* (HarR1915, IV; Har1949, §4.16, §5.16); 'typical means' there and in ChanM1952, or *moving averages* in the language of probability and statistics. For more on Riesz means and Beurling moving averages, see Bin1981; Bin2019. For related moving averages, see BinG2015; BinO2016a; BinG2017.

The Fourier transform of H_a here is $\hat{H}_a(t) = (\exp(iat) - 1)/(iat)$, which has real zeros, so H cannot be used as a Wiener kernel. But two such H_a with a_1/a_2 irrational may be used, as their Fourier transforms have no common zeros (see, e.g., Wie1933, §10 Th. 6; BinI2000b).

In addition to Riesz means and moving averages, there is a third mode of convergence relevant here, 'perturbed Cesàro convergence with rate'. For $\beta \in (0, 1)$, one has (BinT1986, Th. 3) the equivalence as $n \to \infty$ of

$$s_n \to \sigma \qquad R(\exp\left(n^{1-\beta}\right), 1),$$

$$\frac{1}{un^\beta} \sum_{n \le k < n + un^\beta} s_k \to s, \quad \text{for some (all) } u > 0,$$

$$\frac{1}{n+1} \sum_0^n (s_k + \epsilon_k) = s + o\left(1/n^{1-\beta}\right) \quad \text{for some } \epsilon_n \to 0.$$

The most important case, $\beta = \frac{1}{2}$, is in Bin1981, Th. 2 and BinGo1983, Th. 3. It has distinguished antecedents. That the third statement is *sufficient* for Borel (and so Euler) convergence without the ϵ_n terms (so is clearly sufficient with them) is due to Hardy (Har1904, p. 55; Har1949, Th. 149: the first predates Karamata's work, the second does not). An approach via regular variation gives sufficiency of the general result: the relevant Representation Theorem gives the ϵ_n, which plays the role of the error term within the sum or integral there.

P.8 General Regular Variation

One can usefully combine and generalize all three forms of regular variation (Karamata, Bojanic–Karamata/de Haan, Beurling) encountered above. In BinO2020a we study *general regular variation*, in which one has

$$[f(x + t\phi(x)) - f(x)]/h(x) \to K(t) \quad \text{locally uniformly in } t. \qquad \text{(GRV)}$$

Here f is the function under study, $\phi \in BSV$ and h are auxiliary, and the limit K is called the kernel. By using the algebraic machinery of *Popa groups*, one can substantially reduce the theory to those of the earlier three. In addition, one encounters a number of *functional equations*: Cauchy, Gołąb–Schinzel, Chudziak–Jabłońska, Beurling–Goldie, Goldie. See BinO2020a for further detail (and the planned sequel to this book).

Sequential Results: Kendall's Theorem. As above, regular variation is a *continuous-variable* theory, while our preferred tool, the Baire Category Theorem, is a *discrete-variable* theorem about sequences. But it has long been recognized that sequential results are possible and useful; see, e.g., BGT, §1.9. One finds there reference to early work by Croft (Cro1957), Kingman (Kin1964) and Kendall (in particular Ken1968, Th. 16):

Theorem (Kendall's Theorem)   *If*

$$\limsup_{x \to \infty} x_n = \infty, \qquad \limsup_{x \to \infty} x_{n+1}/x_n = 1$$

and, for some continuous positive functions f and g, interval $I = (a, b)$, $0 < a < b < \infty$ and sequence (a_n),

$$a_n f(\lambda x_n) \to g(\lambda) \in \mathbb{R}_+ \quad (n \to \infty) \quad \text{for all} \quad \lambda \in (a, b),$$

then f varies regularly.

If then $f(x) \sim x^\rho \ell(x)$, one has (BinO2020b)

$$a_n \sim c x_n^{-\rho} \ell(x_n).$$

Because of the importance of Kendall's Theorem in applications, we should thus generalize this result as far as possible, in the light of what is now known. It turns out that one can generalize all three of f, g, I above, but at the cost of introducing an *exceptional set* (BGT, §2.9; DrasS1976). For a function f, say that $f(x)$ has *essential limit* $L = L(f)$, finite, as $x \to \infty$,

$$\text{ess-lim} f(\,.\,) = L,$$

if for all $\epsilon > 0$ there exist $X = X(\epsilon, f) \in \mathbb{R}$ and meagre $M = M(\epsilon, f)$ such that

$$|f(x) - L| < \epsilon \quad \text{for all } x > X, \ x \notin M.$$

Then (BinO2020b, Th. 2.3) one can weaken continuity of f to being Baire, continuity of g to being positive, and I an interval to being a non-meagre Baire set. The weakened conclusion is that

$$K(s) := \text{ess-lim}_{x \to \infty} f(s\lambda)/f(\lambda)$$

exists, finite and multiplicative. One calls such f *weakly quasi-regularly varying*. If further g is Baire, then (BinO2020b, Th. 2.5)

$$K(s) \equiv s^\kappa \quad \text{for some } \kappa \in \mathbb{R};$$

one calls such f *strongly quasi-regularly varying*.

There is also a character-degradation theorem (BinO2020b, Th. 8.1): if $k = k(.,.)$ is Borel, then K, where

$$K(s) := \text{ess-lim}_{x \to \infty} k(s, x)$$

is of ambiguous analytic class Δ_2^1 (see Chapter 7).

Functional Equations: Hamel Bases. The definition

$$f(\lambda x)/f(x) \to g(\lambda) \quad (x \to \infty) \quad \text{for all } \lambda \in (0, \infty)$$

leads immediately to

$$g(\lambda \mu) = g(\lambda) g(\mu) \quad \text{for all } \lambda, \mu \in (0, \infty)$$

(BGT, 1.4.1). This is the *Cauchy functional equation*, in multiplicative form. While this is the form preferred for applications, for theory it is better to change from this multiplicative setting in $(\mathbb{R}_+, \times)$ to the corresponding additive setting in $(\mathbb{R}, +)$ by writing $h(x) := \log f(e^x)$, $k(x) := \log g(e^x)$, giving

$$k(u + v) = k(u) + k(v) \quad \text{for all } u, v \in \mathbb{R}, \tag{CFE}$$

the Cauchy functional equation on the line. Such functions k are called *additive*. From (CFE), one obtains

$$k(mu) = m k(u), \quad k(u/n) = k(u)/n \quad \text{for all } u \in \mathbb{R}, \ m \in \mathbb{N}, \ n \in \mathbb{N} \setminus 0,$$

so

$$k(qu) = q \, k(u) \quad \text{for all } u \in \mathbb{R}, \ q \in \mathbb{Q}.$$

Thus, writing $c := k(1)$,

$$k(x) = c \, x \quad \text{for all } x \in \mathbb{R}$$

if k is continuous, by approximation. So, continuous additive functions are linear.

One can easily extend this result vastly beyond continuity (BGT1987, 1.1.3). One obtains (Ostr1929, for measurable k; Meh1964, for the Baire case) that if an additive function k is bounded above or below on a non-null measurable set [a non-meagre Baire set], k is linear. Thus, an additive function k is linear or (highly) pathological.

To proceed, we need to invoke the Axiom of Choice, AC, in some form (e.g. Zorn's Lemma); that is, to extend the axiom system we work with from ZF (Zermelo–Fraenkel) to ZFC (i.e. ZF + AC). One can now prove easily that every vector space has a basis (see, e.g., Jec1973, 2.2.2). Conversely, it was shown by Blass in 1984 that existence of bases implies AC (Bla1984).

Regarding the real line $\mathbb{R}$ as a vector space over the rationals $\mathbb{Q}$ as ground field, $\mathbb{R}(\mathbb{Q})$ say, if we work in ZFC this shows (G. Hamel in 1905, Ham1905) that we have a *basis*, H say ('H for Hamel', below) for $\mathbb{R}$ over Q. Of course, H is uncountable; indeed, it has the power $\mathfrak{c}$ of the continuum (Kucz1985, Th. IV.2.3, p. 82).

We may now define, at will, *any* function $g : H \to \mathbb{R}$. This may be extended uniquely to a homomorphism $f : \mathbb{R} \mapsto \mathbb{R}$: each $x \in \mathbb{R}$ may be written uniquely as a finite linear combination

$$x = \sum \alpha_i b_i \qquad (c_i \in \mathbb{Q}, \ b_i \in H).$$

Then

$$f(x) := \sum \alpha_i g(b_i).$$

If $x \in H$, the above representation of x reduces to $x = x$, so

$$f \mid H = g.$$

Also, if $y \in \mathbb{R}$ has the representation

$$y = \sum \beta_i b_i,$$

$$f(y) = \sum \beta_i g(b_i).$$

Then $x + y$ has the representation $\sum_i (\alpha_i + \beta_i) b_i$ (the range of summation here being the union of those in the two finite sums for x and y), so

$$f(x + y) = \sum (\alpha_i + \beta_i) g(b_i) = \sum \alpha_i g(b_i) + \sum \beta_i g(b_i) = f(x) + f(y) :$$

f is additive. Were f continuous, we could make it discontinuous by changing its value at one point. But then by the Ostrowski and Mehdi results, f would be unbounded above and below on every interval, say. As no such change can be induced in a continuous function by changing its value at one point, we conclude that f is already discontinuous. Thus a Hamel basis gives us a way of manufacturing pathological (discontinuous) additive functions at will. We call this behaviour the *Hamel pathology*. Such pathological functions – or, identifying a function with its graph, functions with graph a Hamel basis in the plane – are called *Hamel functions*.

The argument above can be presented for additive functions $k \colon \mathbb{R}^d \to \mathbb{R}$ (Kucz1985, V.2); we take $d = 1$ here for simplicity.

Additive functions thus have the property that even a little regularity forces great regularity (the form $c\ x$). Ostaszewski (Ost2015a, p. 729) lists ways in which this can happen: additive functions are continuous if they are:

- Baire (Ban1932, I §3 Th.4);
- measurable (Fre1913; Fre1914);
- bounded on a non-null measurable set (Ostr1929);
- bounded on a non-meagre Baire set (Meh1964).

See BinO2011a for details and references.

P.9 Hamel Bases

Despite the pathological behaviour of the Hamel *functions* above, Hamel *bases* as sets may not themselves be pathological. In 1920 Sierpiński (Sie1920) showed that:

- a Hamel basis H can be (Lebegue-)measurable;
- (Th. I) any measurable Hamel basis has measure 0;
- any Hamel basis has inner measure 0;
- a Hamel basis can be non-measurable.

Thus the classes $\mathcal{H}_1$, $\mathcal{H}_2$ of measurable and non-measurable Hamel bases are both non-empty. Sierpiński also showed (Th. II) that no Hamel base can be an analytic set – indeed, it cannot even be a Borel set. He ends with a corollary of his proofs: There exist two measurable sets $X, Y \subseteq \mathbb{R}$ such that the set of sums $\{x + y \colon x \in X, y \in Y\}$ is non-measurable.

Being a Hamel basis is a purely algebraic concept, while we can switch between the measure and category cases by switching between the density topology (Chapter 7) and the Euclidean topology. We conclude that the classes of Hamel bases with and without the Baire property are both non-empty:

- a Hamel basis may or may not have the Baire property, both cases being possible.
- Sierpiński (Sie1935) also showed this, assuming the Continuum Hypothesis, CH, for part of it.

F. Burton Jones showed in 1942 that an additive function continuous on a set T which is analytic and contains a Hamel basis is continuous (Jon1942b); see also Jon1942b. Kominek proved in 1981 the analogous result with 'continuous' replaced by 'bounded' (Kom1981). Motivated by the analogy between these two results, the present authors (BinO2010a) gave a result with both the Jones and Kominek theorems as corollaries, using Choquet's Capacitability Theorem. They also deduced Jones' theorem from Kominek's and gave another proof of the Uniform Convergence Theorem for slowly varying functions.

Płotka (Plo2003) showed that *every* function $f: \mathbb{R} \to \mathbb{Q}$ can be represented as the pointwise sum of two Hamel functions.

Recall that a *perfect* set is a non-empty closed set with no isolated points. A subset A of a Polish space X is called *Marczewski measurable* if for every perfect set $P \subseteq X$ either $P \cap A$ or $P \setminus A$ contains a perfect set. If every perfect set P contains a perfect subset which misses A, then A is called *Marczewski null*. Marczewski (Mar1935) (writing as E. Szpilrajn) showed that the Marczewski measurable sets form a σ-field, and the Marczewski null sets form a σ-ideal.

Miller and Popvassilev (MillP2000) show:

- (Th. 10) There exists a Hamel basis H for $\mathbb{R}$ which is Marczewski null.
- (Th. 8) There exists a Hamel basis H for $\mathbb{R}^2$ which is Marczewski null.
- (Th. 14) There exists a Hamel basis H for $\mathbb{R}$ which is Marczewski measurable and perfectly dense.

Dorais, Filipów and Natkaniec (DorFN2013) show (Th. 4.2) 'deep differences between Lebesgue or Baire measurability and Marczewski measurability by constructing a discontinuous additive function that is Marczewski measurable'. They also show (Ex. 4.1) that there exist additive (discontinuous) functions that are not Marczewski measurable. For further background, see Kha2004.

P.10 Scaling and Fechner's Law

Fechner's Law (Gustav Fechner (1801–1887) in 1860) may be viewed as stating that, when two related physically meaningful functions f and g have no natural scale in which to measure their units, and are reasonably smooth, then their relationship is given by a power law:

$$f = c g^{\alpha}. \tag{F}$$

For background, see, e.g., Bin2015a; Han2004, §5.6.

Fechner's Law emerges naturally from regular variation, as follows (we restrict attention to the basic case, with f, g positive, increasing and unbounded). They satisfy some unknown functional relationship, say,

$$f(x) = \phi(g(x)): \qquad f = \phi \circ g.$$

As there is no natural scale, then at least asymptotically this relationship should be scale-independent regarding x. So changing scale by λ,

$$f(\lambda x) \sim \psi(\lambda) f(x) \quad \text{for all } \lambda > 0 \tag{RV}$$

for some function $\psi(.) > 0$. Under a minimal smoothness assumption (the Baire property or measurability suffice), f is regularly varying with index $\alpha > 0$, and ψ is a power:

$$f \in R_\alpha \subseteq R := \bigcup_{\alpha > 0} R_\alpha.$$

Similarly from $g = \phi^\leftarrow \circ f$, $g \in R$, and from $\psi = f \circ g^\leftarrow$ with $f, g \in R$, $\phi \in R$ also:

$$\phi(\lambda) = \lambda^\alpha \ell(\lambda) \in R_\alpha,$$

with $\ell \in R_0$.

The classically important special case is the simplest one, ℓ constant, $\ell \equiv c$:

$$\phi(x) = cx^\alpha; \qquad f(x) = cg(x)^\alpha: \qquad f = cg^\alpha,$$

giving Fechner's Law.

Illustrative Example: Athletics Times. For aerobic running below ultra distances (800 m to the marathon, say), time t and distance d show Fechner dependence:

$$t = cd^\alpha.$$

Here c (time per unit distance) reflects the quality of the athlete, while α is approximately constant between athletes. This is illustrated on a real data set (the first author's half-marathon and marathon times) in Bingham and Fry (BinF2010, §8.2.3).

The statistics needed (regression) extends to the study of ageing also. The Rule of Thumb for ageing athletes (over 40, say) is: expect to lose a minute a year on your marathon time through ageing alone. It is well borne out by this data set (BinF2010, Ex. 1.3, Ex. 9.6).

1

Preliminaries

1.1 Littlewood's Three Principles

We shall be dealing extensively with measurable sets and functions, and begin by recalling Littlewood's three principles (J.E. Littlewood (1885–1977) from 1944; see Lit1944, §4), according to which a general situation is 'nearly' an easy situation:

(i) any measurable set is nearly a finite union of intervals;

(ii) any measurable function is nearly continuous;

(iii) any convergent sequence of measurable functions is nearly uniformly convergent.

These statements can be made precise, as follows.

Littlewood's first principle is essentially the regularity of Lebesgue measure. That is, with $|.|$ denoting Lebesgue measure, for A (Lebesgue) measurable $|A|$ is the infimum of $|U|$ over open sets $U \supseteq A$ (open supersets of A) and the supremum of $|K|$ over compact subsets K of A. So one can approximate to within any $\epsilon > 0$ from without by open sets and from within by compact sets; taking $\epsilon = 1/n$ (with $n = 1, 2, \ldots$), one can find a $\mathcal{G}_\delta$ set $G \supseteq A$ and a $\mathcal{F}_\sigma$ set $F \subseteq A$ with $|G \setminus A| = 0$ and $|A \setminus F| = 0$. (For the place of $\mathcal{G}_\delta$ and $\mathcal{F}_\sigma$ sets in the Borel hierarchy, see pages 32 and 51.) As each open set on the line is a countable disjoint union of open intervals, for each $\epsilon > 0$ one can find a finite (disjoint) union U of open intervals whose symmetric difference with A has measure $|U \Delta A| < \epsilon$. See, e.g., Bog2007a, §1.5 or Roy1988, §3.3.

Littlewood's second principle is essentially Lusin's Theorem (N.N. Lusin, or Luzin (1883–1950), in 1912; see, e.g., Bog2007a, Th. 2.2.10 or Hal1950, §55).

Theorem 1.1.1 (Lusin's Continuity Theorem, or Almost-Continuity Theorem) *Given a regular measure and a finite measure space (e.g. Lebesgue measure on* *a compact interval), for f measurable and a.e. finite, f is* almost continuous: *for $\epsilon > 0$, there exists a closed set F on which f is continuous and whose complement has measure $|F^c| < \epsilon$.*

This property of almost continuity in fact characterizes measurability.

Littlewood's third principle is essentially Egorov's Theorem (D.F. Egorov (1869–1931) in 1911; see, e.g., Bog2007a, Th. 2.2.1; Hal1950, Th. 21A).

Theorem 1.1.2 (Egorov's Theorem) *For A measurable of finite measure, and f_n a sequence of measurable functions convergent a.e. to f on A, f_n converges almost uniformly: for each $\epsilon > 0$ there exists $B \subseteq A$ with $|A \setminus B| < \epsilon$ and $f_n \to f$ uniformly on B.*

Thus almost everywhere convergence (in particular, pointwise convergence) implies almost uniform convergence.

For textbook accounts of Littlewood's three principles, see SteS2005, §4.3; Roy1988, §3.6.

Our standard reference texts for measure theory will be Bogachev (Bog2007a; Bog2007b) and Fremlin (Fre2000b; Fre2001; Fre2002; Fre2003; Fre2008).

1.2 Topology: Preliminaries and Notation

We gather here a variety of results which we will need later. These are all known to the experts; others may prefer to return to this for reference as may be needed.

Our standard references for general topology, as already mentioned in the Preface, will be Engelking (Eng1989) and also the *Handbook of Set-Theoretic Topology* (KunV1984).

In the earlier parts of the book we will, for the most part, work in metric spaces. But we may need to use alternative metrics, for instance to take advantage of completeness. So it will be convenient to work from the start with topological spaces (Hausdorff, by assumption), and these may often turn out to be *metrizable*. Under such circumstances there may also often be a second stronger, or as we shall say *finer*, topology in play (i.e. one with more open sets) – which we call *submetrizable*. A helpful analogy is the interplay of weak and strong topologies in function spaces.

In general we develop topological machinery when needed. But here we briefly review basic concepts to establish conventions and notation – for details and proofs of results mentioned below we refer to Eng1989. So below we are concerned with:

(i) separation properties (regular, completely regular, normal, etc.);
(ii) covering and refinement properties (compactness, countable compactness, local finiteness and paracompactness, etc.);
(iii) base properties (second countability and σ-local finiteness, etc.);
(iv) neighbourhood base properties (first countability, regular bases).

Notation. Inclusion will be denoted by $A \subseteq B$, proper inclusion by $A \subset B$, complement by A^c, symmetric difference by $A \Delta B := (A \setminus B) \cup (B \setminus A)$. By $\bar{A}$ we will denote the closure of A when there is only one topology in play. Otherwise the closure will be written $\mathrm{cl}_{\mathcal{T}}(A)$ or just $\mathrm{cl}_{\mathcal{T}} A$ to imply the relevant topology $\mathcal{T}$.

By $\omega := \mathbb{N} \cup \{0\}$ we denote the set of *finite* ordinals.

A subset A of natural numbers ω will often be identified with a real number in $(0,1)$. Indeed, A is identified uniquely by its *indicator function* (on the natural numbers) with $1_A(n) = 1$ or 0, according as $n \in A$ or not. In turn, 1_A, as a binary sequence, determines a real number in $(0,1)$ with binary expansion of that sequence.

Separation Properties. A topological space X is *regular* if for each neighbourhood U of any point x there is a neighbourhood V with $x \in V \subseteq \bar{V} \subseteq U$. Further X is *completely regular* (or Tychonoff) if for each neighbourhood U of any point x there is a continuous function f with $f(x) = 1$ and $f = 0$ outside U and it is *normal* (or Urysohn) if for any disjoint pair of closed sets A, B there is a continuous function f with $f = 1$ on A and $f = 0$ on B.

A space is *pseudonormal* if matters are as in the last definition but one of the two closed sets is countable (e.g. a convergent sequence). Thus a pseudonormal space is completely regular.

Covering and Refinement Properties. A space X is *compact* if every *open* covering, i.e. open family $\mathcal{U}$ covering X (family of open sets with union X), contains a finite subcovering (finite subfamily $\mathcal{U}'$ covering X). The space is *countably compact* if any countable open covering has a finite subcovering. This is to be contrasted with *sequential compactness*, which demands that every sequence $\langle x_n \rangle$ must have a convergent subsequence; such a space is countably compact (see Eng1989, Th. 3.10.30). A space X is *pseudocompact* if every real-valued continuous function is bounded. A countably compact completely regular space is pseudocompact. Every normal pseudocompact space is countably compact. In a metrizable space these three concepts are equivalent.

A space is *Lindelöf* if any open covering contains (has) a countable subcovering.

A family $\mathcal{F}$ is *locally finite* if each point x has a neighbourhood U meeting (intersecting) only finitely many members of $\mathcal{F}$. If each x has a neighbourhood U meeting at most one member of $\mathcal{F}$, then $\mathcal{F}$ is *discrete*.

The family is σ-locally finite (or σ-discrete) if $\mathcal{F} = \bigcup_{i \in \omega} \mathcal{F}_i$ and each $\mathcal{F}_i$ is locally finite (resp. discrete). Any open σ-locally finite covering has a locally finite refinement (by sets not necessarily open) – see, e.g., Eng1989, Lemma 5.1.10.

A family $\mathcal{V}$ *refines* the family $\mathcal{U}$ if each member of $\mathcal{V}$ is included in a member of $\mathcal{U}$.

A space is *paracompact* if every open covering $\mathcal{U}$ has a locally finite open refinement. Examples are Lindelöf (and in particular compact) spaces and metrizable spaces. By Stone's Theorem (Eng1989, Th. 4.4.1) every open cover of a metrizable space has an open refinement that is both locally finite and σ-discrete.

Every paracompact space is normal. A regular space is paracompact if and only if every open cover has an open σ-locally finite refinement (Eng1989, Th. 5.1.11).

By analogy with countable compactness, X is *countably paracompact* if every countable open covering $\mathcal{U}$ has a locally finite open refinement.

Dowker's Theorem (Eng1989, Th. 5.2.8) asserts that X is normal and countably paracompact if and only if $X \times [0, 1]$ is normal. This links normal spaces to Borsuk's Homotopy Extension Theorem (see, e.g., Eng1989, §5.5.21 and note in particular Starbird's Theorem).

Base Properties. The simplest base property is its countability: a topology is called second countable (i.e. satisfies the second axiom of countability) if it has a countable base. The topology of a metric space is second countable if and only if the space is separable (has a countable dense subset). Generalizations of this simplest of all topological countability properties include σ-discrete bases and σ-locally finite bases.

For instance, the Nagata–Smirnov Theorem (Eng1989, Th. 4.4.7) asserts that a space is metrizable if and only if it is regular and has a σ-locally finite base. In the same spirit is Bing's Theorem (Eng1989, Th. 4.4.8) that a space is metrizable if and only if it is regular and has a σ-discrete base.

A further generalization is provided through regular neighbourhood bases below.

Neighbourhood Base Properties. A *neighbourhood base* is an assignment to each point x in space of a family $\mathcal{B}(x)$ of neighbourhoods of x such that for every neighbourhood U of any point x there is a smaller neighbourhood

B in $\mathcal{B}(x)$. The simplest neighbourhood base property is again countability (for each x): a topology is called *first countable* (i.e. satisfies the first axiom of countability) if it has a neighbourhood base assigning to each point x a countable family $\mathcal{B}(x)$ as above.

One obtains a neighbourhood base from a base $\mathcal{B}$ by setting $\mathcal{B}(x) := \{B \in \mathcal{B} : x \in B\}$. This leads to the notion of a base $\mathcal{B}$ that is *point-regular* by requiring that every point x has a neighbourhood U such that all but finitely many members of $\mathcal{B}(x)$ lie in U (i.e. all but finitely many of those members B of $\mathcal{B}$ that contain x lie in U). This notion motivates a 'localized' version.

A base $\mathcal{B}$ is *regular* if for every neighbourhood U of any point x there is a smaller neighbourhood V such that all but finitely many members B of $\mathcal{B}$ meeting V lie in U. Arhangelskii's Theorem (Eng1989, Th. 5.4.6) asserts in particular that a Hausdorff space is metrizable if and only if it has a regular base.

1.3 Convergence Properties

We shall need various modes of convergence, which we now discuss briefly. Let $\langle \Omega, \mathcal{S}, m \rangle$ be a measure space. For Φ a property of subsets of Ω we write $m\{\Phi\}$ as an abbreviation for $m(\{\omega \in \Omega : \Phi(\omega)\})$. In particular, if m is finite we can divide by $m(\Omega)$ to make m a probability, so without loss of generality m is a probability if finite. For $\langle X, \mathcal{T} \rangle$ a topological space, recall that in this latter context a random variable with values in X is an $\mathcal{S}$-measurable map $Y \colon \Omega \to X$; we write $L^0(\Omega, X)$ for the set of random variables.

Modes of Convergence.   For $\langle X, d \rangle$ a metric space and $\langle Y_n : n \in \omega \rangle$, a sequence of random variables with values in X the sequence converges to Y_0:

 (i) m-a.e. or almost surely (a.s.) if $Y_n(\omega) \to Y_0(\omega)$ almost everywhere;
 (ii) in measure/in probability if, for every $\epsilon > 0$, $m\{d(Y_n, Y_0) > \epsilon\} \to 0$ as $n \to \infty$.

Convergence a.e. implies convergence in measure/in probability (Bog2007a, 2.2.3), but not conversely. The standard example here is constructed from the subintervals $I_j := [j/2^{n-1} - 1, (j+1)/2^{n-1} - 1]$ for $2^{n-1} \le j < 2^n$ of $[0, 1]$. As these have lengths shrinking to zero, their indicator functions $f_j(\omega)$ regarded as random variables on $\Omega = [0, 1]$, equipped with Lebesgue measure, converge to zero in probability. They do not converge almost surely: for any irrational ω there is an infinite sequence $n_j(\omega)$ where the n_jth function is 1 and an infinite sequence $m_j(\omega)$ where the n_jth function is 0. So we do not have a.e. convergence, but do have a.e. convergence along a subsequence. This example is canonical, as a result below (the 'subsequence theorem', due to F. Riesz in 1909) shows.

Thus convergence a.e. (or a.s. in the probability case) is a strong mode of convergence and convergence in measure (or in probability) a weaker one. This is reflected in the terminology of the basic limit theorems of probability theory, the strong law of large numbers (SLLN) and the weak law of large numbers (WLLN) (see, e.g., Dud1989, Ch. 8). Other strong modes of convergence are convergence in L_p (or in pth mean; we shall only need $p = 1$ – convergence in mean – and $p = 2$ – convergence in mean square). These are not comparable to convergence a.e. At the other extreme is *convergence in distribution*, or *in law*, as in the central limit theorem (CLT) of probability theory; this is implied by convergence in measure/probability, but not conversely unless the limit is constant. See, e.g., Dud1989, Ch. 9.

Convergence in pth mean is metric and generated by the norm of L_p. Convergence in measure/probability is metric and generated by the *Ky Fan metric*, α say. Convergence in distribution is metric and generated by the *Prohorov metric*, ρ say. That convergence in probability implies convergence in distribution follows from $\rho \le \alpha$; see, e.g., Dud1989, Th. 11.3.5. These results also show that convergence a.e. is metrizable only when it coincides with convergence in measure. This does not happen in general, as the example below shows – indeed, in general convergence a.e. is not even topological (see, e.g., Dud1989, Problem 9.2.2). But it does happen if the measure space is purely atomic, as then there are no non-trivial null sets; examples such as the one above, which take place on $[0, 1]$ under the Lebesgue-measurable sets and Lebesgue measure, do not then apply.

Say that the sequence $\langle Y_n : n \in \omega \rangle$ has the *sub-subsequence property* relative to the null sets if for every subsequence $\langle Y_{n(k)} \rangle$ there is a sub-subsequence $\langle Y_{n(k(j))} \rangle$ converging a.e. to Y_0.

Theorem 1.3.1 (Subsequence Theorem; Dud1989, Th. 9.2.1; Bog2007a, 2.2.5(i)) *For $\langle Y_n : n \in \omega \rangle$ a sequence of random variables with values in a separable metric space $\langle X, d \rangle$, the sequence $\langle Y_n : n \in \omega \rangle$ converges to Y_0 in probability if and only if $\langle Y_n : n \in \omega \rangle$ has the sub-subsequence property relative to the null sets.*

Proof Suppose that $\langle Y_n : n \in \omega \rangle$ does not converge to Y_0 in probability. Then for some $\varepsilon > 0$ there is a subsequence $n(k)$ such that

$$P\{d(Y_{n(k)}, Y_0) > \varepsilon\} > \varepsilon, \text{ for all } k.$$

Thus for any sub-subsequence $\langle Y_{n(k(j))} \rangle$ we have also

$$P\{d(Y_{n(k(j))}, Y_0) > \varepsilon\} > \varepsilon, \text{ for all } j,$$

and so $Y_{n(k(j))}(\omega)$ does not converge to $Y_0(\omega)$ for a non-null set of ω, i.e. $\langle Y_n : n \in \omega \rangle$ does not have the sub-subsequence property.

Now suppose that $\langle Y_n : n \in \omega \rangle$ converges in probability to Y_0. Then so does $\langle Y_{n(k)} : k \in \omega \rangle$ for any given subsequence $n(k)$. Choose for each j an integer $k(j) > j$ such that

$$P\{d(Y_{n(k(j))}, Y_0) > 1/j\} < 1/j^2.$$

Hence, by summability of $1/j^2$, the set

$$\bigcap_k \bigcup_{j>k} \{\omega : d(Y_{n(k(j))}(\omega), Y_0(\omega)) > 1/j\}$$

has P-measure 0, i.e.

$$\bigcup_k \bigcap_{j>k} \{\omega : d(Y_{n(k(j))}(\omega), Y_0(\omega)) \le 1/j\}$$

has P-measure 1. So for ω in this latter set and any $\varepsilon > 0$ there is some $k = k(\omega)$ such that for $j > \max\{k(\omega), 1/\varepsilon\}$

$$d(Y_{n(k(j))}(\omega), Y_0(\omega)) \le 1/j < \varepsilon.$$

That is, $\langle Y_{n(k(j))} \rangle$ converges almost surely to Y_0.   □

Definition   The sequence $\langle Y_n \rangle$ is *Cauchy* or *fundamental* in measure if

$$P\{\sup_{k \ge n} d(Y_n, Y_k) \ge \varepsilon\} \to 0 \text{ as } n \to \infty.$$

Lemma 1.3.2 (Almost Sure Convergence Criterion; Dud1989, 9.2.4; cf. Bog2007a, 2.2.5 (ii))   *If $\langle Y_n \rangle$ is a Cauchy/fundamental sequence, then Y_n a.s. converges to some Y_0 a.s. (and so also in measure).*

Remark   Wagner and Wilczyński (WagW2000) proved that the category version also holds.

1.4 Miscellaneous

We will use $\mathcal{F}$, $\mathcal{G}$ for the families of closed and open sets ('f for *fermé*, g for *geöffnet*'), $\mathcal{H}$ as our usual letter for a family of sets in general, $\mathcal{K}$ for the family of compact sets ('k for *kompakt*'). A family $\mathcal{H}$ is *multiplicative* if it is closed under finite intersection, and so analogously a set-valued map $F : \mathcal{H} \to \mathcal{H}$ is *multiplicative* if $F(A \cap B) = F(A) \cap F(B)$. Dually, say that the family $\mathcal{H}$ is *additive* if it is closed under finite unions, and so also a set-valued map F is *additive* if $F(A \cup B) = F(A) \cup F(B)$. A family $\mathcal{H}$ is *maximally additive* if whenever $A \cup B$ is in $\mathcal{H}$, then at least one of A or B is in $\mathcal{H}$; this is motivated by a context in which a family $\mathcal{H}$ with some property P may be extended to a *maximal* such family, and the extension process permits the inclusion for each $A \cup B$ one of A or B, without violating P. We write $\sigma(\mathcal{H})$ for the σ-algebra generated by (the smallest σ-algebra containing) $\mathcal{H}$. Given a σ-algebra, we will often need to consider a sub-σ-algebra of 'small sets' (prototypical examples: null sets or meagre sets). Such a sub-σ-algebra will be closed under intersection with any set in the σ-algebra, and so will have the structure of an *ideal* (using the viewpoint of Boolean algebra, with intersection as product). We will use $\mathcal{I}$ to denote an ideal generically, qualifying this to distinguish one ideal from another; we write $\mathcal{M}$ for the ideal of meagre sets (see Chapter 2), and $\mathcal{N}$ for the ideal of null sets. Here the measure μ will be Lebesgue measure, n-dimensional Lebesgue or Haar measure. Given a sequence $\sigma := \langle \sigma_1, \sigma_2, \ldots \rangle$, write $\sigma \mid n$ for the first n terms. For $\mathcal{H}$ a family of sets, write $\mathbf{S}(\mathcal{H})$ for the class of sets of the form

$$\bigcup_{\sigma} H(\sigma), \quad \text{where } H(\sigma) := \bigcap_{n=1}^{\infty} H(\sigma \mid n),$$

with each $H(\sigma \mid n) \in \mathcal{H}$. Here $\mathbf{S}$ is the *Souslin operation* (M. Ya. Souslin, or Suslin (1894–1919), in 1917), and the sets in $\mathbf{S}(\mathcal{H})$ are the Souslin-$\mathcal{H}$ sets (the term analytic and notation A are also used; see, e.g., Bog2007a, §1.10). The Souslin operation is important in descriptive set theory, and we shall return to it later in §3.1. Meanwhile we note the following:

(i) The Souslin operation is idempotent: $\mathbf{S}(\mathbf{S}(\mathcal{H})) = \mathbf{S}(\mathcal{H})$.

(ii) The class of measurable sets is closed under the Souslin operation (Lusin and Sierpiński in 1918, Nikodym in 1925, Marczewski (as Szpilrajn) in 1929 and 1933 – see, e.g., RogJ1980, Cor. 2.9.3). We will meet the sets with the Baire property in Chapter 2; for them we have the dual result, also due to Marczewski (see, e.g., RogJ1980, Cor. 2.9.4):

(iii) the class of sets with the Baire property is closed under the Souslin operation.

Evidently $\mathbf{S}(\mathcal{H})$ includes the family of all countable intersections of members of $\mathcal{H}$, i.e. $\mathbf{S}(\mathcal{H}) \supseteq \mathcal{H}_\delta$; with Hausdorff's δ for Durchschnitt notation, writing $\mathbf{i} \mid n = (i_1, \ldots, i_n)$, and noting that

$$\bigcup_{\mathbf{i} \in \mathbb{N}^{\mathbb{N}}} \bigcap_n H(i_1, \ldots, i_n) = \bigcup_{i_1=1}^{\infty} \left(\bigcup_{\mathbf{i} \in \mathbb{N}^{\mathbb{N}}} \bigcap_{n=2}^{\infty} H(i_2, \ldots, i_n) \right),$$

we see also the inclusion of countable unions, i.e. $\mathbf{S}(\mathcal{H}) \supseteq \mathcal{H}_\sigma$, so that

$$\mathcal{H}_\sigma \subseteq \mathcal{H}_{\sigma\delta} \subseteq \mathcal{H}_{\sigma\delta\sigma} \subseteq \cdots \subseteq \mathbf{S}(\mathcal{H}).$$

Beyond the initial levels displayed, the Borelian-$\mathcal{H}$ hierarchy is thus obtained, by transfinite induction through the countable ordinals, by alternating the σ and δ operations at successor ordinals and at limit ordinals 'coalescing' (taking unions of) the preceding families. The hierarchy is thus included in $\mathbf{S}(\mathcal{H})$. We shall refer to the case $\mathcal{H} = \mathcal{K}$ where $\mathcal{K} = \mathcal{K}(X)$ denotes the family of compact sets in X.

Taking $\mathcal{H} = \mathcal{G}$, the open sets of a space (recall 'G for geöffnet'), and $\mathcal{H} = \mathcal{F}$, the closed sets ('F for fermé'), we see that the Borelian-$\mathcal{H}$ and Borelian-$\mathcal{F}$ hierarchies are also included in the corresponding $\mathbf{S}(\mathcal{H})$:

$$\mathcal{G}_\sigma = \mathcal{G} \subseteq \mathcal{G}_{\delta\sigma} \subseteq \mathcal{G}_{\delta\sigma\delta} \subseteq \cdots \subseteq \mathbf{S}(\mathcal{G}),$$
$$\mathcal{F}_\sigma \subseteq \mathcal{F}_{\sigma\delta} \subseteq \mathcal{F}_{\sigma\delta\sigma} \subseteq \cdots \subseteq \mathbf{S}(\mathcal{F}).$$

Evidently, in a metric space $\mathcal{G} \subseteq \mathcal{F}_\sigma$, and similarly $\mathcal{F} \subseteq \mathcal{G}_\delta$. Continuing in this way, $\mathcal{F}_\sigma \subseteq \mathcal{G}_{\delta\sigma}$, $\mathcal{G}_\delta \subseteq \mathcal{F}_{\sigma\delta}$ and so on. So taking the union the two hierarchies amalgamate, creating the *Borel hierarchy*, which is closed under complementarity. We note, for future comparison, the capitalized Greek notation, a compromise between the two languages, emanating from mathematical logic using bold-face Σ for sum and Π for product:

$$\Pi^0_0 \subseteq \Pi^0_1 \subseteq \Pi^0_2 \subseteq \cdots$$
$$\Sigma^0_0 \subseteq \Sigma^0_1 \subseteq \Sigma^0_2 \subseteq \cdots$$

with Π^0_0 for $\mathcal{F}$ and Σ^0_0 for $\mathcal{G}$ and later levels beyond these initial ones indexed by ordinals. The notation stresses the implied reference to the existential quantifier $\exists n$ use of the integers (0-level objects) in the countable union operation over a sequence of sets and the complementary universal quantifier $\forall n$ in the countable intersection. The bold-face signals the implied coding of the sequences involved which are in general unconstructively enumerated. When constructive (more accurately 'effectively enumerated') one drops down to light-face notation.

The sets in $\mathbf{S}(\mathcal{F})$ are the Souslin-$\mathcal{F}$ sets; in the context of the closed subsets of a complete metric space, these are called *absolutely analytic* (subsets) since such subsets when embedded in any 'enveloping' metric space are Souslin-$\mathcal{F}$ in any enveloping metric space. We address their properties in later chapters.

We use 'measure' to mean 'countably additive measure'. If a measure is defined on the power set $\wp(X)$ of all subsets of a countable set X, and vanishes on singletons, it vanishes identically by countable additivity (note that a finitely additive measure may very well vanish on singletons but not identically).

The same statement holds for X of cardinality that of the first uncountable ordinal (Ulam's Theorem of 1930 – see Bog2007a, Th. 1.12.40).

A cardinal κ is called *non-measurable* if whenever a measure is defined on all subsets of a set of cardinality κ, and vanishes on singletons, it vanishes identically. Other cardinals are called *measurable*. Whereas one thinks of measurable sets as being 'nice', here it is non-measurable cardinals that are 'nice'. For background, see, e.g., Bog2007a; Bog2007b; Fre2008. We will meet non-measurable cardinality in connection with results of Pol, 2.1.6 and §12.5*, and in §16.3.

We will study category–measure duality in Chapter 9, and a certain non-metric topology, the *density topology* $\mathcal{D}$, in Chapter 7. This is convenient for our purposes because using it one can bring the category and measure aspects together. We note here that this can only happen because the topology is non-metric. For decompositions of metric measure spaces into two parts, one meagre (small in category) and one null (small in measure), see MarS1949. See also Oxt1980, Ch. 16.

2

Baire Category and Related Results

2.1 Baire's Theorem

In measure theory, as in Chapter 1, one has a natural notion of smallness of a set A, namely A having measure zero (A null, $|A| = 0$), and an approximation to it in having $|A| < \epsilon$ for $\epsilon > 0$ small. We now turn to a topological analogue, which will turn out to be even more important.

As we will work throughout with topological spaces, we abbreviate 'topological space' to 'space' below. A *neighbourhood* of a point x is a set containing an open set containing x – that is, a set containing x in its interior. In particular, neighbourhoods are non-empty. We say that a set A in a space is *dense* if its closure $\overline{A}$ is the whole space, *nowhere dense* if $\overline{A}$ has empty interior. In particular, if A is nowhere dense, then so is $\overline{A}$. Following René Baire (1874–1932) in 1899 (Bai1899; Bai1909), we say that A is *meagre* (or of the first category) if it is contained in a countable union of nowhere dense sets, *non-meagre* (or of the second category) otherwise. Call a set *co-meagre* (or *residual*) if its complement is meagre. Note that *a countable union of meagre sets is meagre*. This is the topological analogue of the measure-theoretic statement that *a countable union of null sets is null*.

We consider in tandem various notions of largeness defined by intersection with target sets (having members in prescribed target sets). Thus a dense set A has members in any neighbourhood (i.e. meets any non-empty open set). A set of positive measure meets (has members in) any co-null set. Likewise a non-meagre set A meets any co-meagre set.

A first test that a set is large (in whatever sense) is to give a construction identifying at least one member in some target set. Two related classic constructions are provided by compactness and completeness. When X is (countably) compact and $F_{n+1} \subseteq F_n$ is a nested sequence of non-empty closed sets, their intersection is non-empty. Likewise, when X is a complete metric space and

$F_{n+1} \subseteq F_n$ is a nested sequence of non-empty closed sets with diameters tending to 0, their intersection is non-empty, in fact a singleton. (The latter is Cantor's Nested Sets Theorem, see below; cf. Mun1975, Lemma 7.3′.)

The two results are closely connected and motivate the notion of topological completeness.

One way to state Cantor's Theorem (omitting the uniqueness aspect), or equivalently to assert completeness (see, e.g., Eng1989, 4.3.9), is to identify a sequence of open families $\mathcal{G}_n$ (i.e. families of open sets), e.g. the family of balls of radius $1/n$ under a complete metric, with the property that if $U_n \in \mathcal{G}_n$ and $\overline{U_{n+1}} \subseteq U_n$, then

$$ \emptyset \neq \bigcap_n \overline{U_{n+1}} \subseteq \bigcap_n G_n. $$

It is clear from here that a $\mathcal{G}_\delta$ subspace of a complete metric space satisfies Cantor's Theorem. More is in fact true: a $\mathcal{G}_\delta$ subspace Y of a complete metric space X may be re-metrized so as to be a complete metric space in its own right. This is *Alexandroff(-Hausdorff) Remetrization Theorem* (see, e.g., Oxt1980, Ch. 12).

Suppose that $\mathcal{G}_n$ is a sequence of open families in a compact space X each covering a subset Y, with

$$ Y = \bigcap_n G_n, \text{ where } G_n := \bigcup \mathcal{G}_n. $$

Then for $U_n \in \mathcal{G}_n$ and $\overline{U_{n+1}} \subseteq U_n$, one has by compactness $\emptyset \neq \bigcap_n \overline{U_{n+1}} \subseteq \bigcap_n G_n = Y$. Motivated by this similarity, one says that a space is *topologically complete* (or *Čech-complete*) if it is a $\mathcal{G}_\delta$ subset of some compact space (e.g. in its Stone–Čech compactification). Thus a $\mathcal{G}_\delta$ subset of a topologically complete space is itself topologically complete. Likewise a closed subset of a topologically complete space is topologically complete (cf. Eng1989, Th. 3.9.6).

The connection between topological completeness and complete metrizability is that a metric space is completely metrizable (i.e. has an equivalent complete metric) if and only if regarded as a metrizable space it is topologically complete. (See Eng1989, 4.3.26; cf. Mazurkiewicz's Theorem, Dug1966.)

Either of these two contexts gives rise directly to a proof of Baire's Theorem when stated in the form that the intersection of a sequence of dense open sets is dense. This may be paraphrased as saying that an intersection of large open sets is large. The proof in either context proceeds by exhibiting a member of the intersection within a target set – in this case a neighbourhood, as follows. It is typical that it is just as easy to find a point in the intersection as finding it in any neighbourhood. The following pair of statements are together known as the *Baire Category Theorem*, or more briefly as *Baire's Theorem*, due to Baire (in 1899, Bai1899). It is amazing to note how conceptually similar his proof is to Cantor's first proof (in 1874) that the reals are uncountable (see Kan2009).

Theorem 2.1.1 (Baire's Theorem; Kec1995, 8B; Eng1989, 3.9.3; Kel1955, 6.34) *In a topologically complete space, in particular in a locally compact Hausdorff space or a completely metrizable space, the intersection of a countable sequence of dense open sets is dense.*

Proof Let V_n be dense and open and U any neighbourhood. As V_1 is dense and open it meets U, so by regularity there is a neighbourhood U_1 with $\overline{U_1} \subseteq U \cap V_1$. Continuing inductively, choose non-empty open sets $\overline{U_n} \subseteq U$ such that $\overline{U_{n+1}} \subseteq U_n \cap V_n$ (the latter is possible because, as before, V_n is open and has points in common with U_n by density). Then $\emptyset \neq \bigcap_n \overline{U_{n+1}} \subseteq U \cap \bigcap_n V_n$. □

An important corollary is its equivalent re-statement, dualized to the closed sets.

Theorem 2.1.2 (Baire's Theorem – Variant) *If a topologically complete space, in particular a locally compact Hausdorff space or a completely metrizable space, is a union of closed sets F_n, then one of the sets F_n has non-empty interior.*

Proof Suppose otherwise. Then each F_n has as complement a dense open set G_n, and any point common to these open sets is not covered by the F_n. □

One may check that the following three properties of a topological space X, each of which may be thought of as giving a sense in which X resembles familiar spaces such as Euclidean space, are equivalent:

(i) each non-empty open set in X is non-meagre;
(ii) each co-meagre set in X is dense;
(iii) the intersection of countably many dense open sets in X is dense.

One calls such a topological space a *Baire space*. We note in passing that a $\mathcal{G}_\delta$ (in particular, an open) subset of a Baire space is Baire. So:

(a) A completely metrizable space is Baire.
(b) A locally compact Hausdorff space is Baire.

In (b) above, the point of the (Hausdorff) space being locally compact is that it has a one-point (Alexandroff) compactification.

(a) Any topologically complete space is Baire.

For background on Baire spaces, see, e.g., AarL1974; HawoM1977.

We think of meagre sets as small in the category (or topological) sense, just as null sets are small in the measure sense. The two concepts of smallness are, however, quite distinct. The real line, a prototypical big set, may be written as the disjoint union of a null set and a meagre set, each small in one way (but big in the other); see, e.g., Oxt1980, Th. 1.6.

Baire's Theorem is a powerful tool in producing non-constructive existence proofs. One shows that things exist not by exhibiting examples but by showing that they exist *generically*. This is a topological counterpart to the *probabilistic method*, for which see AloS2008; Spe1994.

A set A is said to have the *property of Baire* if it is the symmetric difference of an open set with a meagre set: $A = G\Delta M$ with G open and M meagre. We shall abbreviate this by saying simply that A *is Baire*. It can easily be shown that the Baire sets are also the sets $A = F\Delta M$ with F closed and M meagre. The class of Baire sets is closed under complements, and under disjoint unions. It is a σ-algebra – the smallest σ-algebra containing the open sets and the meagre sets (and so, is the σ-algebra generated by the Borel σ-algebra and the meagre sets). The Baire sets are those of the form $U \cup M_1$ with $U \in \mathcal{F}_\sigma$ and M_1 meagre, and those of the form $V \setminus M_2$ with $V \in \mathcal{G}_\delta$ and M_2 meagre; see, e.g., Oxt1980, Ch. 4. In particular, open, closed, $\mathcal{F}_\sigma$ and $\mathcal{G}_\delta$ sets are Baire. One sees that the Baire sets are analogous to the measurable sets, in view of Littlewood's first principle (§1.1).

We recall from §1.4 that Souslin-$\mathcal{F}$ sets, and in particular Borel sets, are Baire (by Nikodym's Theorem; see §1.4). A set S in a space X may have the Baire property in X, but its restriction to a subspace Y, namely $S \cap Y$, does not necessarily have that property in Y. Indeed, although open-ness is retained under restriction, the same is not the case with 'nowhere denseness': if N is nowhere dense in X, then, for any neighbourhood U meeting N, there is a non-empty sub-neighbourhood V missing N, and, passing to the subspace $Y = N$ itself, we no longer have $V \cap Y$ non-empty. Nevertheless, some sets S do *preserve the Baire property under restriction*, and in so doing exhibit a stronger form of the Baire property. Such sets have come to be called *restricted Baire*, an unfortunate term as the more appropriate 'hereditarily Baire' has taken on the meaning of '$\mathcal{F}$-hereditarily Baire'. Following Sto1963, one might also speak of the *completely hereditarily Baire* sets. Fortunately, if S is a Borel set in X, then $S \cap Y$ is a Borel set in Y; but Borel sets are generated from open sets (which are Baire) by operations which preserve the Baire property, and so Borel sets preserve the Baire property under restriction. More generally, this is the case with the larger family of Souslin-$\mathcal{F}$ sets (for definition see §1.4).

Theorem 2.1.3 (Kur1966, §11.VI and VII)   *Souslin-$\mathcal{F}$ sets, and in particular Borel sets, preserve the Baire property under restriction, i.e. inherit the Baire property on arbitrary subspaces.*

A function f is said to have the property of Baire, or more briefly (as above) to be *Baire*, if the inverse image under f of any open set is Baire. One sees that the Baire functions are the category (or topological) analogues of the measurable functions of §1.1.

The alternative usage 'Baire measurable' refers to the smallest class of functions including the continuous functions and closed under pointwise limits. We will study the relation between this concept and that considered here in Chapter 12 on Continuity and Coincidence Theorems.

One reason why Baire functions are of interest is the following Continuity Theorem, a category analogue of Lusin's Continuity Theorem. As usual, we write $f \mid C$ for the function f *restricted* to a subset C of its domain.

Theorem 2.1.4 (Baire's Continuity Theorem; Kec1995, Th. 8.38)   *If $f \colon X \to Y$ is Baire-measurable with Y second countable, then f is continuous off a meagre set, i.e. for some co-meagre subset C the restriction $f \mid C$ is continuous.*

Proof   Let $\mathcal{B}_Y$ be a countable base for the topology of Y. For $B \in \mathcal{B}_Y$, as $f^{-1}(B)$ is Baire, we may choose $U = U_B$ open in X such that $N_B := U_B \triangle f^{-1}(B)$ is meagre; so $U_B \backslash N_B \subset f^{-1}(B)$. Then $C := X \backslash \bigcup_B N_B$ is co-meagre and $f \mid C$ is continuous on C. Indeed, for $c \in C$ and $f(c) \in B \in \mathcal{B}_Y$ we have $c \in U_B$ (for otherwise $c \in f^{-1}(B) \backslash U_B$, and so $c \in U_B \backslash N_B \subset f^{-1}(B)$). $\square$

As an illustration of how countability conditions in topology yield generalizations of classical theorems beyond their original separable metric context, we now give a generalization of Baire's Continuity Theorem. We need the following definition, which extends the usage of §1.2 (cf. Bing's Theorem cited there).

Definition   A map $f \colon X \to Y$ is *σ-discrete* if there is a σ-discrete collection $\mathcal{B}_f \subseteq \wp(X)$, called a *$\sigma$-discrete base* for f, such that, for each open V in Y, there is $\mathcal{B}_f(V) \subseteq \mathcal{B}_f$ with $f^{-1}(V) = \bigcup \mathcal{B}_f(V)$.

Remarks   1. For X or Y metrizable a *continuous* map $f \colon X \to Y$ is σ-discrete. Indeed, for X metrizable it is enough to take $\mathcal{B}_f$ to be any σ-discrete base for the open sets of X. On the other hand, for Y metrizable, and $\mathcal{B}^Y$ any σ-discrete base for the open sets of Y, the set $\mathcal{B}_f := \{f^{-1}(V) : V \in \mathcal{B}^Y\}$ is also σ-discrete.

2. If Y is second countable with *countable* base $\mathcal{B}^Y$, then $\mathcal{B}_f := \{f^{-1}(B) : B \in \mathcal{B}^Y\}$ is countable, so σ-discrete, and of course $\mathcal{B}_f(V) := \{f^{-1}(B) : B \subseteq V\}$.

To motivate Hansell's Theorem below, we note that if $f \mid (A \backslash M)$ is continuous and V is open in Y, then $f^{-1}(V)$ is an open set modulo some subset of M.

Theorem 2.1.5 (Hanse1971, §3.7, Th. 9) *For $f: X \to Y$ a σ-discrete map on a metrizable space, f has the Baire property if and only if there is a meagre set M in X such that $f \mid (X \backslash M)$ is continuous.*

A function may have the stronger property that inverse images of open sets have the (stronger) restricted Baire property. When this is so the function itself is said to have the restricted Baire property. The following is a characterization of such functions in a fairly broad category of spaces. See §12.5.

Theorem 2.1.6 (Pol's Theorem; Pol1976) *A function f from a compact space X to a metric space Y of non-measurable cardinality has the restricted Baire property if and only if for every subspace A of X there is a subset $M \subseteq A$ meagre in A such that $f \mid (A \backslash M)$ is continuous.*

This yields a far-reaching consequence for functions that are known to have the restricted Baire property. Below, X, Y are as in Pol's Theorem, 2.1.6.

Corollary 2.1.7 *For X a Borel subspace in some compactification, Y a metric space of non-measurable cardinality, and f a function from X to Y with a Souslin-$\mathcal{F}(X \times Y)$ graph, there is a meagre subset M in X such that $f \mid (X \backslash M)$ is continuous.*

Proof Let f have Souslin-$\mathcal{F}(X \times Y)$ graph G. As X is Souslin-$\mathcal{F}(K)$, it is analytic. Hence the image set $f(X)$, being the projection of $G \cap (X \times Y)$, is analytic. Likewise $f^{-1}(U)$ is the projection of $X \times (U \cap f(X)) \cap G$ and so Souslin-$\mathcal{F}(X)$. So $f^{-1}(U)$ inherits the Baire property on arbitrary subspaces. That is, f has the restricted Baire property. The conclusion now follows from Pol's Theorem, 2.1.6. □

Remark It was previously shown by Hansell (Hanse1971, §3.7, Cor. 8) that in the case of X *metrizable* and Borel in some compactification (i.e. absolutely Borel), this conclusion holds without any cardinality restriction on Y.

Recall that $\mathcal{B}$ is a *pseudo-base* if every non-empty open set contains a member of $\mathcal{B}$; such a collection $\mathcal{B}$ is *locally countable* if each member of $\mathcal{B}$ contains only countably many members of $\mathcal{B}$. The product of two Baire spaces, one of which has a locally countable pseudo-base, is Baire. An arbitrary product of Baire spaces, each of which has a locally countable pseudo-base, is Baire (Oxt1960, Th. 3). Without the countability assumption, the product need not be Baire (Oxt1960, §4). See also §9.7.

Remark We say that a topological space satisfies *Blumberg's Continuity Theorem* (or has *Blumberg's property*, or is *Blumberg*) if for any real-valued function on X there is a dense set D such that $f \mid D$ is continuous. Note that Blumberg implies Baire (see Bradford and Goffman 1961, BraG1960). The converse is true in a metric space, or more generally a space with a σ-disjoint pseudo-base (as defined above). See Whi1974.

We have noted that the property of being a Baire space is hereditary for the *open* subspaces of a Baire space X; the following is a generalization to a less restricted class of subspaces (as it applies to all dense subspaces).

Theorem 2.1.8 (ComN1965, L. 1.2) *If X is Baire and $A \subseteq X$ meets each non-empty $\mathcal{G}_\delta$ of X, then A is a Baire subspace, so in particular A is non-meagre.*

Proof Let G be open in A. For a sequence of sets U_n that are dense and open in A, choose W and V_n open in X such that $U_n = V_n \cap A$ and $G = W \cap A$. Then V_n is dense in A and so, as X is Baire, $W \cap \bigcap_n V_n$ is a non-empty $\mathcal{G}_\delta$. By assumption, the latter set meets A and so $G \cap \bigcap_n U_n$ is non-empty. □

Theorem 2.1.9 (ComN1965, L. 1.3) *For an arbitrary index set I, if each X_i is Baire and each $A_i \subseteq X_i$ meets each non-empty $\mathcal{G}_\delta$ of X_i, then the (Tychonoff) product space $\prod_i A_i$ meets every non-empty $\mathcal{G}_\delta$ subset of the product $\prod_i X_i$. So in particular the product is Baire.*

Proof For V_n open in $\prod_i X_i$, write

$$V_n = \prod_i V_{i,n,m}, \quad \text{and} \quad V := \bigcap_n V_n,$$

with each $V_{i,n,m}$ open in X_i and with $V_{i,n,m} = X_i$ for cofinitely many m. Put

$$V_i := \bigcap_{n,m} V_{i,n,m}.$$

Then V_i is a $\mathcal{G}_\delta$ in X_i, and so meets A_i, in a_i, say. Then $a = \langle a_i \rangle \in \prod_i A_i \cap V$. □

We close with an extension of Baire's Theorem obtained by weakening its assumption of completeness and so its reliance on open sets in a useful way. We recall from §1.2 that a topological space X is *regular* if for each neighbourhood U of any point there is a neighbourhood V of that point with $\bar{V} \subseteq U$; further, X is *completely regular* if for each neighbourhood U of any point x there is a continuous function f with $f(x) = 1$ and $f = 0$ outside U.

In the definition below, we have a sequence $\langle \mathcal{B}_n \rangle$ of pseudo-bases, which can be thought of as playing the role of 'shrinking diameters'.

Definition (Oxt1960, Section 5) A space is *pseudocomplete* if

(i) there is a sequence of pseudo-bases $\mathcal{B}_n$ such that every nested sequence of sets $B_n \in \mathcal{B}_n$ with $\bar{B}_{n+1} \subseteq B_n$ has non-empty intersection;

(ii) the space is quasi-regular, i.e. every neighbourhood contains the closure of some neighbourhood.

Thus a pseudocomplete space is Baire. Furthermore, an arbitrary product of pseudocomplete spaces is pseudocomplete.

In this connection recall from §1.2 that a completely regular space is *pseudo-compact* if continuous functions are bounded. Significant here is the following equivalent definition in terms of nested sequences of open sets.

Theorem 2.1.10 (GilJ1960, Lemma 9.13)   *A completely regular space is pseudocompact if and only if every nested sequence of open sets V_n has $\bigcap_n \bar{V}_n$ non-empty.*

Proof   Suppose X is pseudocompact, but $\bigcap_n \bar{V}_n$ is empty for some nested sequence of open sets V_n. Pick distinct points x_n and open sets W_n with $x_n \in W_n \subseteq \bar{W}_n \subseteq V_n$; then $\{x_n : n \in \omega\}$ is discrete and also closed, since any limit point would be in $\bigcap_n \bar{V}_n$. In fact more is true: every point x of X has a neighbourhood that meets at most finitely many sets W_n; otherwise, x would be in $\bigcap_n \bar{V}_n$. Now let f_n be continuous with $f_n(x_n) = n$ and $f_n = 0$ on $X \backslash W_n$. Then $f(x) = \sum f_n(x)$ is a series with at most one non-zero term and for any x the function f coincides with some g_n on a neighbourhood of x. So f is continuous and unbounded, contradicting pseudocompactness.

For the converse, note that any unbounded continuous function f yields the nested sequence of open level sets $V_n := \{x : |f(x)| > n\}$ for which $\bigcap_n \bar{V}_n$ is empty. □

We note that completely regular countably compact spaces are pseudocompact; conversely, normal pseudocompact spaces are countably compact (see, e.g., Eng1989, 3.10).

One thus has

Theorem 2.1.11 (ComN1965, Cor. 2.8)   *Any product of pseudocompact spaces, and so in particular of countably compact spaces, is Baire.*

Proof   A pseudocompact space is pseudocomplete – for $\mathcal{B}_n$ take the non-empty open sets and refer to the preceding theorem. □

We obtain better stability behaviour in §2.3 by introducing somewhat stronger properties which lie between Baire and complete.

2.2 Banach's Category Theorem

Banach's Category Theorem (Oxt1980, Ch. 16) or Banach's Union Theorem (Kur1966, §10III) is also known as Banach's Localization Principle. See Kel1955, Th. 6.35; RogJ1980. This is also a result we constantly rely on throughout the book. The proof below illustrates how an elementary proof, applicable in the second-countable case (outlined in the Remark below), may easily be lifted to the more general metric setting by the use of σ-discrete open coverings (for which see §1.2). That notion was not yet available when Banach based his ingenious proof on an open disjoint *almost covering* (i.e. omitting only a nowhere dense set). We need some definitions.

Definitions   Say that T is *locally meagre* at a point s (not necessarily in T) if there is a neighbourhood U_s of s with $T \cap U_s$ meagre, and write $L_{\mathcal{M}}(T)$ for the set of points at which T is locally meagre.

Say that T is *everywhere non-meagre (in T)* if there are no points of T where T is locally meagre; then for each neighbourhood U_t of $t \in T$ the set $U_t \cap T$ is non-meagre.

Recall that s is an *essential limit point* of T if for each neighbourhood U_s of s the set $T \cap U_s$ is non-meagre.

Theorem 2.2.1 (Banach's Category Theorem, or Localization Principle)   *If a topological space S is locally meagre, then S is meagre. That is, a union of any family of meagre open sets is meagre.*

Proof   For $\mathcal{U}$ any maximal disjoint family of open sets U with $U \cap S$ meagre, put $U_S = \bigcup \mathcal{U}$. By maximality, $S \setminus \mathrm{cl}\, U_S$ is *empty*. (Otherwise, $U \cap S$ is meagre for some $U \subseteq X \setminus \mathrm{cl}\, U_S$.) It suffices to show $S \cap U_S$ is meagre, as $(\mathrm{cl}\, U_S) \setminus U_S$ is nowhere dense.

For $U \in \mathcal{U}$ choose $N_n(U)$ nowhere dense so that

$$U \cap S = \bigcup\nolimits_{n \in \omega} N_n(U).$$

Put

$$N_n := \bigcup \{N_n(U) : U \in \mathcal{U}\},$$

which is nowhere dense, because $\mathcal{U}$ disjoint implies for $U \in \mathcal{U}$ that

$$N_n \cap U = N_n(U), \quad \text{and so} \quad S \cap U_S = \bigcup\nolimits_{n \in \omega} N_n.$$

So, if an open V meets N_n, then V meets $N_n(U)$ for some unique $U \in \mathcal{U}$, and so $V \cap U$ contains a non-empty subset W disjoint from $N_n(U)$. Then W is disjoint from N_n, as W meeting N_n gives that V meets N_n, and so meets $N_n(U)$, implying that W meets $N_n(U)$. This is a contradiction, since W is disjoint from $N_n(U)$. □

Theorem 2.2.2 (Banach's Category Theorem – Variant) *In a metrizable space, for T a Baire set, the set $L_\mathcal{M}(T)$ of points at which T is locally meagre is itself meagre. So quasi all points of T are essential limit points of T, and if T is non-meagre, then $T \backslash L_\mathcal{M}(T)$ is everywhere non-meagre.*

Proof By Bing's Theorem (§1.2), select a σ-discrete base for the open sets, say $\mathcal{B} = \bigcup_{i \in \omega} \mathcal{B}_i$ with each $\mathcal{B}_i$ discrete. For $L = L_\mathcal{M}(T)$ and each point $s \in L$ choose an open set $U_s \in \mathcal{B}$ with $T \cap U_s$ meagre. Then $L_i := \{T \cap U_s : U_s \in \mathcal{B}_i\}$ is covered by the discrete family $\{U_s \in \mathcal{B}_i\}$. For fixed i, put $S(i) := \{s : U_s \in \mathcal{B}_i\}$ and, for $s \in S(i)$, write $T \cap U_s := \bigcup_m N_m^{s,i}$ with $N_m^{s,i}$ nowhere dense.

We now show that each set $N_m^i := \bigcup_{s \in S(i)} N_m^{s,i}$ is nowhere dense, so that $L = \bigcup_{i \in \omega} \bigcup_{m \in \omega} N_m^i$ is meagre.

Fixing m, by discreteness, each x has a neighbourhood W_x meeting at most one set U_s with $s \in S(i)$. If W meets no such set, then W avoids L_i, so assume now that it meets such a set, say U_p. Since $N_m^{p,i} \subseteq U_p$ and N_m^{pi} is nowhere dense, there is a neighbourhood $W' \subseteq W$ with W' avoiding N_m^{pi}. Then W' avoids all the sets N_m^{si} for $s \neq p$ in $S(i)$, because W does so, and it avoids N_m^{pi} by construction. So W' avoids N_m^i, so N_m^i is nowhere dense.

For the rest: if now T is non-meagre, then $S := T \backslash L_\mathcal{M}(T)$ is non-meagre and each point of S is an essential limit point of T. So quasi all points of T are essential limit points of T, as asserted. Furthermore, $L_\mathcal{M}(S) = \emptyset$: for if $t \in L_\mathcal{M}(S)$, then for some U_t the set $U_t \cap S$ is meagre, and so also is $U_t \cap (T \cup L(T)) \supseteq U_t \cap T$, i.e. $t \in L_\mathcal{M}(T)$, contradicting that S is disjoint from $L_\mathcal{M}(T)$. □

Remark Were the space second countable with countable base $\mathcal{B}$, one would argue that, since for each s there is $U_s \in \mathcal{B}$ with $T \cap U_s$ meagre, the countable family $\{T \cap U : T \cap U \text{ is meagre and } U \in \mathcal{B}\}$ covers L, so has meagre union, and so L is meagre. The proof above embellishes this 'by localization'.

We next pass to other notions of countability that yield results of the Baire Category Theorem type.

2.3 Countability Conditions – Topological

Countability is present rather obviously in measure theory via σ-additivity, and this is the basis of the measure–category duality in Oxt1980. We extended the scope of category methods in the last section, replacing metric completeness with compactness (to obtain topological, or Čech, completeness). Here we improve on the generalization by weakening the role of compactness. By contrast, the next section brings to bear the mechanics of game theory. Our aim is to extend the scope of category–measure duality.

The over-riding concern is to relax all countability aspects of metrizability as far as possible. There are two modes of thought:

(i) distill local properties shared by all metric spaces into bases, or into types of covering refinements, such as point-finiteness or local finiteness (see, e.g., Eng1989, §4.4) – this we delay;

(ii) impose global structures less rigid than a metric – this we now do.

Recall from the last section that we were guided by the sequence of open covers $\langle \mathcal{G}_n \rangle$ comprising balls of radius shrinking to zero with n, say 2^{-n}. They form the core of metrizability theory; see Gru1984 for a wide overview of different structures motivated by covers $\langle \mathcal{G}_n \rangle$ as a vehicle of countability in a topological space X.

We start with a particularly simple condition: the requirement that the diagonal set in X^2 be a $\mathcal{G}_\delta$-set, i.e. that $\{(x, x) : x \in X\} = \bigcap_n V_n$ with V_n open in X^2. This is reminiscent of 'uniformities qua neighbourhoods of the diagonal' (cf. Eng1989), but the pertinent features which this yields are:

(i) the sequence of open covers $\langle \mathcal{G}_n \rangle$ defined by $\mathcal{G}_n := \{G \text{ open} : G \times G \subseteq V_n\}$, known as the $\mathcal{G}_\delta$-*sequence*; and

(ii) their 'stars' defined by

$$\operatorname{st}(x, \mathcal{G}_n) := \bigcup \{G : x \in G \in \mathcal{G}_n\}.$$

For an illustration of the power of this structure, we mention Chaber's Theorem (see Gru1984); a countably compact space with a $\mathcal{G}_\delta$-diagonal is metrizable. See also Ost1980.

Our next idea introduces a very weak form of compactness: this is the notion of an M-space (M for Morita; again see Gru1984). One demands the existence of open covers $\langle \mathcal{G}_n \rangle$ with:

(i) *clustering*: if a sequence of points $\langle x_n \rangle$ satisfies $x_n \in \operatorname{st}(x, \mathcal{G}_n)$ for each n, then the set $\{x_n : n \in \omega\}$ has a cluster point; and

(ii) *star-refinement*: $\{\operatorname{st}(x, \mathcal{G}_{n+1}) : x \in X\}$ refines $\mathcal{G}_n$ for each n.

As an illustration, we cite Morita's result that a space is metrizable if and only if it is an M-space and has a $\mathcal{G}_\delta$-diagonal. For an application of star-refinement, we recall first that a topology τ on X is called *submetrizable* if there is a coarser metrizable topology τ' on X (i.e. $\tau' \subset \tau$): in brief, τ refines some metrizable topology. (Thus the density topology on $\mathbb{R}$, to be defined in Chapter 7, is submetrizable since it refines the usual Euclidean metric topology.) A connection between the new ideas established so far is given by the following theorem.

Theorem 2.3.1 (See, e.g., Gru1984, Th. 2.5) *A space X is submetrizable if and only if X has a $\mathcal{G}_\delta$-sequence such that $\{\text{st}(x, \mathcal{G}_{n+1}) : x \in X\}$ refines $\mathcal{G}_n$ for each n.*

2.4 Countability Conditions – Games of Banach–Mazur Type

Under what circumstances will a nested sequence have non-empty intersection? This is an existence question (on the existence of a common element). So far we have two answers: for closed sets completeness, or compactness; for open sets Baire's Theorem identifies density. We now introduce generalizations which yield non-empty intersections, but need to be formulated in the language of game theory. These existence assertions concern the existence of a winning strategy, or more accurately the absence of a winning strategy of the counterparty (see the comments by Jean Saint-Raymond in Sai1983). The idea of defining classes of topological spaces via games was initiated by Choquet (see Cho1969, p. 116), but the approach goes back to Banach and Mazur, who worked in a simpler context: the real line. They formulated their game as generating a decimal expansion of a real number. This idea proved to be potent in two ways. As stated, it eventually revolutionized our understanding of 'definable sets of reals' (see §16.1). Restated as generating a nested sequence of intervals, it gave topologists novel ways to introduce useful countability conditions into topology.

The games considered below are related and so have a similar structure. All have *stages* labelled by the natural numbers and all have two players I and II (also referred to, in opposite order, as α and β, and sometimes He and She, a practice we occasionally follow below). Their actions/choices, taken at the various stages, generate in particular a nested sequence of open sets. Each game has a rule specifying the winning player as a function of the actions and referring to the intersection of the nested sequence. A strategy for the player i is a function which assigns at each stage n a choice for player i given the actions taken so far. (So this is a game of perfect information, like chess.)

Interest focusses on characterizing when a particular player, say i, has a winning strategy, i.e. a strategy under which i wins no matter what the choices of the other party are.

For the *Banach–Mazur* game in a space X with target A Players I and II alternately choose nested non-empty open sets U_n and V_n respectively at stage $n = 0, 1, 2, \ldots$, with Player I choosing first. Thus

$$U_0 \supseteq V_1 \supseteq U_1 \supseteq \cdots V_n \supseteq U_{n+1} \supseteq V_{n+1} \supseteq \cdots.$$

Player II wins the game if

$$\bigcap\nolimits_n V_n \subseteq A.$$

Otherwise I wins (i.e. $\bigcap_n V_n$ meets the complement of A). In particular, II wins if the intersection is empty. The fundamental theorem in this context, due to Banach and Mazur, is that a set A is co-meagre in X if and only if II has a winning strategy (see Oxt1980, Ch. 6). Mazur proposed the game (this is Problem 43 in the *Scottish Book*; Mau1981), and Banach responded in 1935 by demonstrating its determinacy for target sets A having the Baire property. Further development is needed to characterize a winning strategy for I.

The special case with $A = \emptyset$ is known as the *Choquet game* $J(X)$ ('J for jeu'). So here the choices are exactly as in the Banach–Mazur game, and now Player II wins if

$$\bigcap\nolimits_n U_n = \emptyset, \quad \text{i.e.} \quad \bigcap\nolimits_n V_n = \emptyset,$$

as in the special case above. Here I is called β.

Theorem 2.4.1 (Oxt1980)   *The space X is a Baire space if and only if X is β-unfavourable (i.e. I has no winning strategy) in the Choquet game $J(X)$.*

The theorem is mute about whether Player II has a winning strategy. The space is said to be a *Choquet space* (or an α-favourable space) if in fact Player II (Player α) has a winning strategy.

Theorem 2.4.2 (Kec1995, 8.11)   *The space X is Choquet if and only if X is co-meagre in its completion.*

The choices in the *strong Choquet game* are as in the Banach–Mazur game with the additional stipulation that at each stage n, Player I chooses also a point $x_n \in U_n$ and Player II's choice is constrained so that

$$x_n \in V_n.$$

Here Player II wins if

$$\bigcap\nolimits_n V_n \neq \emptyset, \quad \text{i.e.} \quad \bigcap\nolimits_n U_n \neq \emptyset.$$

The space is called *strongly Choquet* if Player II has a winning strategy in the strong Choquet game. (Note that if the game is played without recall of the previous moves and Player II has a winning strategy, then instead one speaks of strongly α-favourable spaces.)

Theorem 2.4.3 (Kec1995, 8.17(i)) *The space X is strongly Choquet if and only if X is a $\mathcal{G}_\delta$ in its completion, i.e. X is completely metrizable.*

Theorem 2.4.4 (Kec1995, 8.33(ii)) *For the Banach–Mazur game on a Choquet space which is metrizable, Player I has a winning strategy if and only if A is meagre in the neighbourhood of some point.*

The Christensen games, for which see Chr1974, used to great effect by J. Saint-Raymond in Sai1983, generalize the Choquet game, bringing the latter closer to the original Banach–Mazur game by allowing Player II (at least in the limit) to select a target set. In the points version, $J_{\mathrm{pt}}(X)$, Player II at stage n selects a point x_n. Now Player I wins if

$$\bigcap_n U_n \cap \overline{\{x_n : n \in \omega\}} \neq \emptyset.$$

Two further generalizations arise from replacing point selection by set selection. In one version, J_{cpt}, Player II creates an increasing sequence of compact sets by selecting at stage n a compact set K_n with $K_n \supseteq K_{n-1}$. Now Player I wins if

$$\bigcap_n U_n \cap \overline{\bigcup_n \{K_n : n \in \omega\}} \neq \emptyset.$$

In an alternative version, denoted J_{analytic}, an increasing sequence of K-analytic sets A_n (see §3.1.1 for a definition) replaces the compact sets with II selecting at stage n an analytic set A_n with $A_n \supseteq A_{n-1}$. This time Player I wins if

$$\bigcap_n U_n \cap \overline{\bigcup_n \{A_n : n \in \omega\}} \neq \emptyset.$$

One may compare various games on X. Following Deb1986, we say that J_2 is a *stronger* game than J_1, or is *less favourable* to α than J_1, and write $J_1 \prec J_2$, provided that for each space X

(i) if $J_2(X)$ is α-favourable, then so too is $J_1(X)$; and
(ii) if $J_2(X)$ is β-unfavourable, then so too is $J_1(X)$.

Thus

$$J \prec J_{\mathrm{analytic}} \prec J_{\mathrm{cpt}} \prec J_{\mathrm{pt}}.$$

Note that for a given X the game $J_i(X)$ need not be determined; however, if X is regular and K-analytic, then the games $J_P(X)$ for P = pt, cpt, analytic are determined. One area of application is concerned with the issue of when separate continuity implies joint continuity (see Chapter 12). For the time being we content ourselves with indicating that the game-theoretic approach properly extends the category of 'Baire-like' spaces.

We need a notion of point separation as follows. When, as with $\mathcal{B}a(X)$, the family of Baire sets of the space X, the family $\mathcal{H}$ is closed under complementation, then the separation is symmetric.

Definition   For $A, B \subseteq X$, say that A is *countably distinguished* from B by a family of sets $\mathcal{H}$ if there is a countable subfamily $\mathcal{H}'$ such that for every pair of points $a \in A$, $b \in B$ there is $H \in \mathcal{H}'$ with $a \in H$ and $b \notin H$.

Theorem 2.4.5 (Deb1987, Th. 3)   *For a compact space X and subspaces $A \subseteq A'$ with A dense in X and Baire, if A is countably distinguished from $X \setminus A'$ by $\mathcal{B}a(X)$, then A' is β-unfavourable for the game J_{pt}.*

We refer to the preceding subsection for the notion of complete regularity.

Theorem 2.4.6 (Deb1987, Cor. 7)   *For X completely regular, if X is K-analytic, then $J_{\mathrm{pt}}(X)$ is α-favourable.*

In fact a weaker hypothesis, that instead X be countably determined (i.e. is the image of a compact-valued upper semi-continuous image of a metrizable space), renders $J_{\mathrm{pt}}(X)$ a β-unfavourable game, for which see again Deb1987.

We close this section by noting that products of Choquet spaces are Choquet and open subspaces of Choquet spaces are Choquet. Strongly Choquet spaces are Choquet. Products of strongly Choquet spaces are strongly Choquet, and their non-empty $\mathcal{G}_\delta$ subspaces are strongly Choquet. (For these, see Kec1995, §8.)

2.5 Notes

1. Plumed Spaces.

Intermediate between the two notions of §2.3 in the use of compactness is Arhangelskii's notion of a *p*-space ('plumed' space), which is a simultaneous generalization of both topological completeness and metrizability. We give two related versions under the simplifying assumption that X is completely regular (for which notion see §1.2), an assumption that ensures that X has a compactification (namely the Stone–Čech compactification).

Say that X is a *p-space* if there is a compactification Y in which there are covers $\langle \mathcal{U}_n \rangle$ of X by sets open in Y such that the star-refinement satisfies

$$\bigcap_n \mathrm{st}(x, \mathcal{U}_n) \subseteq X, \ \forall x \in X.$$

Here, for x a point in the space X and $\mathcal{U}$ a family of its subsets, $\mathrm{st}(x, \mathcal{U}) := \bigcup \{U : x \in U \in \mathcal{U}\}$ (see p. 44).

A *p-space* space is called a *strict p-space* if, in addition,

$$\bigcap_n \text{st}(x, \mathcal{U}_n) = \bigcap_n \overline{\text{st}(x, \mathcal{U}_n)}.$$

All metrizable spaces and all topologically complete spaces are *p*-spaces (see, e.g., Arh1995). We refer to Baire *p*-spaces in Bouziad's Theorem in Theorem 2.5.1.

Internal characterization is given by reference to a type of finite-intersection property, as follows (again see, e.g., Gru1984).

(a) X is a *strict p*-space if and only if there exist open covers $\langle \mathcal{G}_n \rangle$ such that $K_x := \bigcap_n \text{st}(x, \mathcal{G}_n)$ is compact for each x, and for each open G with $K_x \subseteq G$ there is n with $\text{st}(x, \mathcal{G}_n) \subseteq G$ (Arh1963).

(b) X is a *p*-space if and only if there exist open covers $\langle \mathcal{G}_n \rangle$ such that if $x \in G_n \in \mathcal{G}_n$, then $K_x := \bigcap_m \bar{G}_m$ is compact for each x, and for each x and open G with $K_x \subseteq G$ there is n such that $\bigcap_{m \le n} \bar{G}_m \subseteq G$ (Arh1963).

As examples of their usefulness, observe that a countable product of *p*-spaces, or of *M*-spaces, is again a *p*-space, or respectively an *M*-space.

See Arh1995, §7 for a wide-ranging discussion of the significance of *p*-spaces.

2. An Application of Games to *p*-Spaces.

The following result, noticed by Bouziad, illustrates how a *p*-space structure on a Baire space makes it almost as good as a topologically complete space.

Theorem 2.5.1 (Bouz1993, Prop. 3.6) *A Baire p-space is β-unfavourable for the Christensen game J_{pt}, i.e. Player I has no winning strategy.*

Proof Let $\langle \mathcal{G}_n \rangle$ be a plumage for X with the properties given in the definition above. Suppose that at stage n II has played $V_1, \ldots, V_n$. Pick $z_n \in G_n \cap V_n$ with $G_n \in \mathcal{G}_n$ and U_n with $z_n \in \bar{U}_n \subseteq G_n \cap V_n$. Recall from Theorem 2.4.1 that, since X is Baire, the strong Choquet game is unfavourable for I. Let I play U_n and let Player II apply her winning strategy against U_n. That determines her play of say V_n. Then

$$\bigcap_n U_n = \bigcap_n V_n \neq \emptyset.$$

As $\bar{U}_{n+1} \subseteq U_n$ and $\bar{U}_n \subseteq G_n$, this implies that $K := \bigcap_n \bar{G}_n$ is non-empty; by the definition of *p*-space K is compact. Now some point of K is a cluster point of $\langle x_n \rangle$; otherwise, each point $k \in K$ has a neighbourhood U_k avoiding $Z := \{z_n : n \in \omega\}$ and so, by compactness, some finite union of these yields an open set U with $U \supseteq K$ and avoiding Z. But $\bar{G}_m \subseteq U$ for some m and this implies that U_{m+1} avoids Z, a contradiction since $z_m \in \bar{U}_m \subseteq \bar{G}_m \subseteq U$ which avoids Z. So Z has a cluster point $z \in K$. But $z_n \in \bar{U}_n$ for all z, so $z \in \bar{U}_n$ for all n; so

$$z \in \bigcap_n U_n \cap \overline{\{z_n : n \in \omega\}} \neq \emptyset,$$

guaranteeing a win in the Christensen game. $\square$

For fixed-point theorems in the context of game theory, see Bor1989.

3

Borel Sets, Analytic Sets and Beyond: $\mathbf{\Delta^1_2}$

3.1 Borel Sets and Analytic Sets

The theory of analytic sets dates from work of Souslin in 1916, Luzin in 1917 (Lus1917) and Luzin and Sierpiński in 1918 (LusS1923). For mono-graph treatments, see Lus1930; RogJ1980. The historical origins, in an error of Lebesgue in 1905, are given there – in Lebesgue's preface to Lus1930 and in RogJ1980, §1.3: projections of Borel sets need not be Borel, whence the degradation studied above. Modern interest in analytic sets dates from its use in measure theory in work in the 1950s by, e.g., Roy Davies and Choquet – see Del1980.

In the Euclidean, and also in a metrizable topological, context the open sets and the closed sets both generate the same σ-algebra – the Borel sets. Indeed, one passes from the open sets $\mathcal{G}$ – the topology – to the closed sets $\mathcal{F}$ via complements; but one may avoid complementation altogether by viewing the closed sets as a subfamily of the $\mathcal{G}_\delta$ sets (or the open sets as the subfamily of the $\mathcal{F}_\sigma$). At this point we signal that on occasion it is worth drawing a distinction between σ-algebras generated from a family $\mathcal{H}$ by reference only to the 'positive' operations of countable intersections and unions – that is precluding the 'negative' operation corresponding to complementation. Making this distinction, one obtains for $\mathcal{H} = \mathcal{G}$ or $\mathcal{F}$ the Borelian-$\mathcal{G}$ and Borelian-$\mathcal{F}$, which coincide in a metric setting, and more importantly, corresponding to the compact sets $\mathcal{K}$ there are also the Borelian-$\mathcal{K}$ sets, which coincide with the earlier σ-algebras in a locally compact (and so in the Euclidean) setting.

The *Borelian hierarchy* has a parallel development as sets of points satis-fying some property naturally described by a formula written out in a logical symbolism using (in addition to the single real number playing the role of the argument) only the usual connectives (& and 'or') and universal and existential quantifiers over the natural numbers – typically identified as a function $\sigma \in \omega^\omega$

(e.g. via the continued-fraction expansion). The function σ acts as a *coding device* in two natural contexts: coded reference to a countable ordinal α (needed to construct a Borel set at the αth level in the hierarchy) and reference to a given open set G through a listing of the indices n of basic open sets B_n (taken from a fixed basis $\mathcal{B}$) with $B_n \subseteq G$ (so that G is the union of the coded basic open sets). One singles out the use of an effectively computable σ from among all possible σ.

Some Euclidean sets rely in their natural (rigorous, but informal) definition on the use of a universal or existential quantifier ranging over the reals. The use of an existential quantifier may be interpreted as *projection*; thus a set on the line described as

$$E := \{x \in \mathbb{R} : (\exists y)\Phi(x, y)\}$$

referring to the plane Borel set

$$F := \{(x, y) \in \mathbb{R}^2 : \Phi(x, y)\}$$

described using logical symbolism may also be written

$$E = \mathrm{proj}_1 F.$$

The question arises of *when a Borel set F has a Borel projection*. The standard answers are: when

(i) each vertical $\{x\} \times \mathbb{R}$ meets F in at most one point: here projection is a one-to-one continuous mapping; more generally

(ii) each vertical $\{x\} \times \mathbb{R}$ meets F in a σ-compact set (e.g. a countable set). This is the *(Rogers-)Arsenin–Kunugui Theorem* (see, e.g., RogJ1980, Thms. 5.9.1, 5.9.2).

Beyond this, in general E is not Borel. In the Euclidean context sets of the general form E above are Souslin-$\mathcal{K}$ sets (see §1.4), i.e. of the form

$$\bigcup_{\sigma \in \omega^\omega} K(\sigma), \quad \text{where} \quad K(\sigma) := \bigcap_{n \in \omega} K(\sigma \mid n),$$

where this representation has $\mathrm{diam}\, K(\sigma|n) \leq 2^{-n}$ and $K(\sigma \mid n + 1) \subseteq K(\sigma \mid n)$. In view of this, sets of the form E are simply known as *Souslin* sets or *analytic* sets. See RogW1966 for their projections.

Remark While in $\mathbb{R}$ the concepts of being analytic and Souslin coincide, this is not so in general. They coincide here because of upper semi-continuity (see §1.4) of the mapping $K \colon \omega^\omega \to \mathbb{R}$; indeed, for open G, if

$$K(\sigma) \subseteq G, \quad \text{i.e.} \quad \bigcap_{n \in \omega} G^c \cap K(\sigma \mid n) = \emptyset,$$

then, by (countable) compactness, $G^c \cap K(\sigma \mid N) = \emptyset$ for some N, and so, for τ with $\tau \mid N = \sigma \mid N$,

$$K(\tau) \subseteq G.$$

Regarding ω as a discrete space metrized so that distinct points are unit distance apart, the countable product topology on ω^ω is metrized by discounted first differences:

$$d(\sigma, \tau) = 2^{-n}, \quad \text{where} \quad n = \inf\{m : \sigma_m \neq \tau_m\}.$$

Under this structure on ω^ω, the mapping $\sigma \mapsto K(\sigma) := \bigcap_{n \in \omega} K(\sigma \mid n)$ is continuous on the closed set $\{\sigma : K(\sigma) \neq \emptyset\}$.

Finally we mention a result of particular significance to the hierarchy of projective sets presented in §3.2.

Theorem 3.1.1 (Souslin's Theorem) *In a Polish space, a set is Borel if and only if it and its complement are both analytic.*

The projection of a Baire set may be, but need not be, Baire: under Gödel's Axiom of Constructibility,

$$V = L,$$

asserting that all sets are constructible (see Dev1973 and Chapter 1), there are non-Baire sets which are projections of co-analytic sets (which are themselves Baire), as the example (BinO2010b, Th. 4) demonstrates. However, the weaker concept of *shift-compactness*, studied in Chapters 4 and 6, is preserved under projection.

Theorem 3.1.2 (BinO2010b, Th. 1) *For a shift-compact set in $\mathbb{R}^2$ its projection is shift-compact in $\mathbb{R}$.*

3.1.1 Analyticity and K-Analyticity

As before, we denote by $\mathcal{K} = \mathcal{K}(X)$ the family of compact subsets of a topological space X, by $\mathcal{G} = \mathcal{G}(X)$ the open sets, and by $\mathcal{F} = \mathcal{F}(X)$ the closed sets. As above, a set A in X is obtained from a family $\mathcal{H}$ of subsets of X by the *Souslin operation* (briefly, is *Souslin-$\mathcal{H}$*), if there is a *determining system* $H = \langle H(i \mid n) \rangle$ assigning to each finite sequence of positive integers $i \mid n := (i_1, i_2, \ldots, i_n)$ (taken to be empty when $n = 0$) a set $H(i \mid n) \in \mathcal{H}$ with

$$A = \bigcup_{i \in I} H(i) = \bigcup_{i \in I} \bigcap_{n \in \omega} H(i \mid n), \quad \text{where } H(i) := \bigcap_{n \in \omega} H(i \mid n).$$

Here, as in Chapter 1, we write

$$I := \mathbb{N}^{\mathbb{N}},$$

which is endowed with the product topology (treating $\mathbb{N}$ as discrete), so that for $i \in I$ its restrictions may be denoted

$$i \mid n := (i_1, \ldots, i_n).$$

The behaviour of the mapping $i \mapsto H(i)$ is of significance, but before we turn to considering this we mention two important dual results.

Theorem 3.1.3 (Nikodym's Theorem and Corollary; Nik1925)   *If the sets in $\mathcal{H}$ have the Baire property, then each Souslin-$\mathcal{H}$ set has the Baire property.*
Hence Souslin-$\mathcal{F}$ sets are Baire.

Theorem 3.1.4 (Marczewski's Theorem and Corollary)   *If the sets in $\mathcal{H}$ are measurable with respect to a σ-finite regular measure μ on X, then each Souslin-$\mathcal{H}$ set is μ-measurable.*
Hence Souslin-$\mathcal{F}$ sets are measurable.

These results will be established in a more general non-separable metrizable context in Chapter 13. They both rely on the existence of a *Baire* (resp. measurable) *hull* (see §3.4) of an arbitrary set A, a Baire/measurable expansion $M \supseteq A$, minimal in the sense that, for any other Baire/measurable M' with $M \supseteq M' \supseteq A$, the difference $M \backslash M'$ is meagre/null.

Definition ($\mathcal{K}$-Analytic Spaces)   We recall from RogJ1980 that for X a Hausdorff space, a map $K : I \to \wp(X)$ is *compact-valued* if $K(i)$ is compact for each $i \in I$, and *singleton-valued* if each $K(i)$ is a singleton. (For useful background on compact-valued maps, see Ber1963; Hil1974.) The map K is *upper semi-continuous* if, for each $i \in I$ and each open U in X with $K(i) \subseteq U$, there is n such that $K(i') \subseteq U$ for each i' with $i' \mid n = i \mid n$.

A subset in X is $\mathcal{K}$-*analytic* if it may be represented in the form $K(I)$ for some compact-valued, upper semi-continuous map $K : I \to \wp(X)$. We say that X is a $\mathcal{K}$-*analytic space* if X itself is a $\mathcal{K}$-analytic set.

Notation   We put

$$I(i \mid n) := \{i' : i' \mid n = i \mid n\}$$

and

$$I_<(i \mid n) := \{i' \in I : i'_m \le i_m \text{ for } m \le n\}.$$

So

$$I_<(i) := \bigcap_n I_<(i \mid n)$$

is compact (by Tychonoff's Theorem when viewing $\{i' \mid n : i'_m \leq i_m$ for $m \leq n\}$ as a compact set). Correspondingly, we write

$$K(i \mid n) := K(I(i \mid n)) = \bigcup \{K(i') : i' \mid n = i \mid n\},$$
$$K_<(i \mid n) := K(I_<(i \mid n)), \text{ and } K_<(i) = K(I_<(i));$$

the latter is also compact if K is compact-valued and upper semi-continuous.

Definition If $K : I \to \mathcal{K}(X)$ is upper semi-continuous, say that K is $\mathcal{H}$-*circumscribed* if there is a determining system $\langle H(i \mid n) \rangle$ assigning to each $i \mid n$ a set $H(i \mid n) \in \mathcal{H}$ with $K(i) = H(i) := \bigcap_{n \in \omega} H(i \mid n)$ such that $H(i) \subseteq U$ for U open implies that $H(i \mid n) \subseteq U$ for some n.

Remarks   1. If K is upper semi-continuous, as above, and X is Hausdorff, then

$$K(I) = \bigcup_{i \in I} \bigcap_{n \in \omega} \text{cl } K(i \mid n),$$

so K is $\mathcal{F}$-*circumscribed*. In particular a $\mathcal{K}$-analytic set is Souslin-$\mathcal{F}$. Indeed, as $K(i) \subseteq \bigcap_{n \in \omega} \text{cl } K(i \mid n)$, the inclusion of the left in the right is clear; for the other direction, if $x \in (\bigcap_{n \in \omega} \text{cl } K(i \mid n)) \backslash K(i)$ for some i, then there is U open with $x \notin \text{cl} U$ and $K(i) \subseteq U$, and so $K(i \mid n) \subseteq U$ for some n, yielding the contradiction that $x \notin \text{cl } K(i \mid n) \subseteq \text{cl } U$.

2. If further X is a metric space and $G(i \mid n) := B_{1/n}(K(i \mid n)) = \{x : d(x, y) < 1/n$ for some $y \in K(i \mid n)\}$, then $K(I) = \bigcup_{i \in I} \bigcap_{n \in \omega} G(i \mid n)$ is a Souslin-$\mathcal{G}$ representation such that:

(i) $\bigcap_{n \in \omega} G(i \mid n) = K(i)$ is compact;
(ii) $K(i) \subseteq U$, for U open, implies that $G(i \mid n) \subseteq U$ for some n.

Then K is $\mathcal{G}$-*circumscribed*. Indeed, if $K(i) \subseteq U$, then by compactness there is n such that $B_{2/n}(K(i)) \subseteq U$. But for $m > n$ large enough, $K(i \mid m) \subseteq B_{1/n}(K(i))$, and so $B_{1/m}(K(i \mid m)) \subseteq B_{1/m}(B_{1/n}(K(i))) \subseteq B_{2/n}(K(i)) \subseteq U$.

We will return to this observation in Theorem 8.2.1.

3. In a non-separable metric space a more relevant generalization has I replaced by

$$J := \kappa^{\mathbb{N}},$$

with κ an arbitrary infinite cardinal (again viewed, like $\mathbb{N}$ above, as a discrete space). See Chapter 13.

Definition (Analytic Spaces)   Call a Hausdorff space X *analytic* if X is the continuous image of a closed subset of I, or equivalently of a Polish space. So an analytic space is $\mathcal{K}$-analytic, since singletons are compact; a $\mathcal{K}$-analytic subset of an analytic set is analytic. A $\mathcal{K}$-analytic space is more general, since it is a continuous image of a Lindelöf–Čech-complete space – by Jayne's Theorem (RogJ1980, Th. 2.8.1); cf. Hansell (Hanse1992, Th. 3.1).

Note that if X is metric and $\mathcal{K}$-analytic, then without loss of generality we may arrange to have $\mathrm{diam}_X(K(i \mid n)) < 2^{-n}$. This implies that $K(i) = \{k(i)\}$ on a closed subset of I on which k is continuous. Since the closed subspace of I is analytic, this yields a space which is a continuous image of I, so again an analytic space. See RogJ1980, §2.8 for background on different definitions of analyticity, and §2.11 therein for applications to Banach spaces under the weak topology.

In a metric analytic space open sets, being $\mathcal{F}_\sigma$, are analytic. Note that an open subset of a Polish space is again a Polish space. This yields the following generalizations.

Lemma 3.1.5   (i) *In an analytic space, all open sets are analytic.*

(ii) *A regular $\mathcal{K}$-analytic space (in particular a regular analytic space) is Lindelöf, so open sets are $\mathcal{K}$-analytic if and only if they are $\mathcal{F}_\sigma$.*

Proof   (i) If X is the continuous image of I under α, then each open U in X is the continuous image under α of $\alpha^{-1}(U)$, an open subset in I.

(ii) By regularity, for U open, U has an open covering by sets V with cl $V \subseteq U$. If U is $\mathcal{K}$-analytic, it is Lindelöf, and so U has a countable covering by open sets V_n with cl $V_n \subseteq U$; so $U = \bigcup_n$ cl V_n is an $\mathcal{F}_\sigma$. For the converse, as Souslin-$\mathcal{F}_\sigma$ subsets of a $\mathcal{K}$-analytic space are $\mathcal{K}$-analytic, open sets are $\mathcal{K}$-analytic if they are $\mathcal{F}_\sigma$.   □

Remarks   1. Of course for U open, (cl U)\U is closed nowhere dense, so modulo a meagre set an open set is closed, so 'almost $\mathcal{K}$-analytic'.

2. The two parts of Lemma 3.1.5 are close since (Lev1983; HruA2005) every analytic Baire space has a dense completely metrizable subspace; in fact every regular analytic space has a finer topology in which it is metrizable (just declare the countably many sets cl $\alpha(I(i \mid n))$ to be open). Condition (i) of the lemma is related to the existence of winning strategies in certain topological games (see Ost2011, Proposition L3 in §3 and Whi1975, Th. 11).

3.1.2 Analytic Cantor Theorems

The following result is implicit in a number of situations and goes back to Frolík's characterization of Čech-complete spaces as $\mathcal{G}_\delta$ in some compactification (Fro1960; see Eng1989, §3.9); it may be used to lift theorems about Polish spaces to results on analytic metric spaces and to characterize analytic sets. For example, Frolík (Fro1970, Th. 2) characterizes analytic sets as intersections of a $\mathcal{G}_\delta$ and a set that is Souslin in its Stone–Čech compact-ification; in similar spirit Fremlin (Fre1980) develops the theory of Čech-analytic sets (cf. also HanseJR1983). In the opposite direction, Aarts et al. (AarGM1970b; AarGM1970a) use similar machinery to characterize completeness via compactness. Recall that Cantor's Theorem on the intersection of a nested sequence of closed (or compact, as appropriate) sets has two formulations: (i) referring to vanishing diameters (in a complete-space setting); and (ii) to (countable) compactness. In the spirit of these, we now give two topological versions. Refer to §3.1.1 for notation.

Theorem 3.1.6 (Analytic Cantor Theorem; Ost2011) *Let X be a Hausdorff space and $A = K(I)$ be $\mathcal{K}$-analytic in X, with K compact-valued and upper semi-continuous.*

If F_n is a decreasing sequence of (non-empty) closed sets in X such that $F_n \cap K(I(i_1,\ldots,i_n)) \neq \emptyset$, for some $i = (i_1,\ldots) \in I$ and each n, then $K(i) \cap \bigcap_n F_n \neq \emptyset$.

Equivalently, if there are open sets V_n in I with cl $V_{n+1} \subseteq V_n$ and $\operatorname{diam}_I V_n \downarrow 0$ such that $F_n \cap K(V_n) \neq \emptyset$, for each n, then:

(i) *$\bigcap_n \operatorname{cl} V_n$ is a singleton, $\{i\}$ say;*
(ii) *$K(i) \cap \bigcap_n F_n \neq \emptyset$.*

Proof If not, then $\bigcap_n K(i) \cap F_n = \emptyset$ and so, by compactness, $K(i) \cap F_p = \emptyset$ for some p, i.e. $K(i) \subseteq X \backslash F_p$. So by upper semi-continuity $F_p \cap K(I(i_1,\ldots,i_n)) = \emptyset$ for some $n \geq p$, yielding the contradiction $F_n \cap K(I(i_1,\ldots,i_n)) = \emptyset$. $\square$

Theorem 3.1.6 has the following filter base (finite intersection property, 'fip') generalization.

Theorem 3.1.7 *In the setting of Theorem 3.1.6, for $\mathcal{H}$ a filter base, if for some i and each $n \in \omega$, $H_0 \in \mathcal{H}$, there is $m > n$ and $H \in \mathcal{H}$ with $H \subseteq H_0$ meeting $K(I(i_1,\ldots,i_m))$, then*

$$K(i) \cap \bigcap \{\operatorname{cl} H : H \in \mathcal{H}\} \neq \emptyset.$$

Proof If not, and $\emptyset = K(i) \cap \bigcap \{\operatorname{cl} H : H \in \mathcal{H}\}$, then, for some finite subfamily $\mathcal{H}'$, we have $\emptyset = K(i) \cap \bigcap \{\operatorname{cl} H : H \in \mathcal{H}'\}$, and so $K(i) \subseteq \bigcup \{X \backslash \operatorname{cl} H : H \in \mathcal{H}'\}$. By upper semi-continuity, $K(I(i_1,\ldots,i_n)) \subseteq \bigcup \{X \backslash \operatorname{cl} H : H \in \mathcal{H}'\}$, for some n. As $\mathcal{H}$ is a filter base, there exists $H_0 \in \mathcal{H}$ with $H_0 \subseteq \bigcap \{H : H \in \mathcal{H}'\}$. Now for some $m > n$ and some $H_1 \in \mathcal{H}$ with $H_1 \subseteq H_0$, the set $K(I(i_1,\ldots,i_m))$ meets H_1, contradicting the fact that $\emptyset = K(i) \cap \bigcap \{H : H \in \mathcal{H}\}$. $\square$

The filter base version is usually rendered employing 'inclusion' as below – suggesting the shrinking 'diameters' of Cauchy's criterion, ultimately the inspiration of Frolík's (Fro1960); cf. again Eng1989, §3.9; Hanse1992, §3, p. 281 – rather than the 'trace on $K(i)$' property above ('intersection with $K(i)$'). This has a similar but simpler proof, given below for the sake of completeness. In fact the inclusion version implies the trace version (see the Remark following the next theorem). In §2.3 we see their duals in the Banach–Mazur 'inclusion' games and the Choquet 'trace' games.

Theorem 3.1.8 *In the setting of Theorem 3.1.6, for $\mathcal{H}$ a filter base, if, for some i, each $K(I(i_1, \ldots, i_n))$ contains a member of $\mathcal{H}$, then*

$$\emptyset \neq \bigcap \{\mathrm{cl}\, H : H \in \mathcal{H}\} \subseteq K(i).$$

Proof The inclusion is clear. If $\emptyset = K(i) \cap \bigcap\{\mathrm{cl}\, H : H \in \mathcal{H}\}$, then for some finite subfamily $\mathcal{H}'$ we have $\emptyset = K(i) \cap \bigcap\{\mathrm{cl}\, H : H \in \mathcal{H}'\}$, and so $K(i) \subseteq \bigcup\{X \backslash \mathrm{cl}\, H : H \in \mathcal{H}'\}$. By upper semi-continuity, $K(I(i_1, \ldots, i_n)) \subseteq \bigcup\{X \backslash \mathrm{cl}\, H : H \in \mathcal{H}'\}$, for some n. But $K(I(i_1, \ldots, i_n)) \supseteq H_0$ for some (non-empty!) $H_0 \in \mathcal{H}$, and so $\emptyset = H_0 \cap H'$ for each $H' \in \mathcal{H}'$, contradicting the fact that $\mathcal{H}$ is a filter sub-base, unless $\mathcal{H}' = \emptyset$. But then $K(i) = \emptyset \supseteq K(I(i_1, \ldots, i_n)) \supseteq H_0$, giving a final contradiction.   □

Remark To see why Theorem 3.1.8 implies Theorem 3.1.7, consider a filter base $\mathcal{H}$ with the intersection property of 3.1.7 relative to $i \in I$. One may pick for each $H_0 \in \mathcal{H}$ and $n \in \mathbb{N}$ an integer $m = m(n) > n$ and a set $H = H^n(H_0) \subseteq H_0$ in $\mathcal{H}$ such that $H_i^n(H_0) := H \cap K(i_1, \ldots, i_{m(n)}) \neq \emptyset$. Then $\{H_i^n(H_0) : n \in \mathbb{N}, H_0 \in \mathcal{H}\}$ is a filter sub-base satisfying the hypothesis of 3.1.8, and so

$$\emptyset \neq K(i) \cap \bigcap \{\mathrm{cl}\, H_i^n(H_0) : n \in \mathbb{N},\ H_0 \in \mathcal{H}\} \subseteq K(i) \cap \bigcap \{\mathrm{cl}\, H_0 : H_0 \in \mathcal{H}\}.$$

A similar argument can be conducted with a weaker hypothesis by exploiting the compactness of $K_<(i)$ (defined in §3.1.1).

Theorem 3.1.9 *In the setting of Theorem 3.1.6, suppose now that the nested sequence F_n satisfies $F_n \cap K(I_<(i_1, \ldots, i_n)) \neq \emptyset$, for some $i = (i_1, \ldots) \in I$ and each n. Then $K_<(i) \cap \bigcap_n F_n \neq \emptyset$.*

Equivalently, if there are open sets V_n in I with $H := \bigcap_n \mathrm{cl}\, V_n$ non-empty compact sets such that $F_n \cap K(V_n) \neq \emptyset$ for each n, then $K(H) \cap \bigcap_n F_n \neq \emptyset$.

Proof If not, then $K_<(i) \cap \bigcap_n F_n = \emptyset$. Since $K_<(i)$ is compact, $K_<(i) \cap F_p = \emptyset$, for some p. By upper semi-continuity, for each $j \in I_<(i)$ there is $n(j)$ such that $K(j \mid n(j)) \subseteq X \backslash F_p$. Since $I_<(i)$ is compact, there are $j(1), \ldots, j(t)$ in $I_<(i)$ and integers $n_s = n(j(s))$ such that $\{I(j(s) \mid n_s) : s = 1, \ldots, t\}$ is a finite open covering of $I_<(i)$. Put $q = p + \max_{s \leq t} n_s$.

For $j \in I_<(i \mid q)$, consider j' with $j' \mid q = j \mid q$ and $j' \in I_<(i)$. (For instance, take $j' = j_1, \ldots, j_q, i_{q+1}, i_{q+2}, \ldots$). Refer to the finite covering to find

s with $j' \mid n_s = j(s) \mid n_s$. Then $K(j \mid q) \subseteq K(j(s) \mid n_s) \subseteq X \backslash F_p$. So $K(I_< (i \mid q)) \subseteq X \backslash F_p$, and in particular $K(I_< (i \mid q)) \cap F_q = \emptyset$, a contradiction. $\quad\square$

The following generalization of Theorem 3.1.6 is at the heart of both the proof of the Gandy–Harrington Theorem (see §8.2) and likewise of van Mill's proof (Mil2004, Prop. 2.2) of his Analytic Baire Theorem (concerning the countable intersection of dense, heavy sets).

Theorem 3.1.10 *Let X be a Hausdorff space and $A_n = K_n(I)$ be $\mathcal{K}$-analytic in X, with K_n taking singleton or empty values and upper semi-continuous.*

If F_n is a decreasing sequence of (non-empty) closed sets in X such that

$$F_n \cap \bigcap_{m \le n} K_m(I(i_1, \dots, i_n)) \ne \emptyset,$$

for some $i = (i_1, \dots) \in I$ and each n, then $\bigcap_n F_n \cap \bigcap_n A_n(i) \ne \emptyset$.

Proof If not, write $H_n := K_n(i)$ and $K_n(i) := \{x_n\}$ (whenever $K_n(i)$ is non-empty). By compactness, since $\bigcap_n (F_n \cap H_n) = \emptyset$, there is p with $F_p \cap \bigcap_{n \le p} H_n = \emptyset$. If $x_m \notin F_p$ or $H_m = \emptyset$ for some $m \le p$, then $K_m(i) \subseteq X \backslash F_p$, and so $F_p \cap K_m(I_n(i \mid n)) = \emptyset$ for some $n > p + m$. Then $F_n \cap K_m(I_n(i \mid n)) = \emptyset$, a contradiction. So $x_m \in F_p$ for all $m \le p$. Since $F_p \cap \bigcap_{n \le p} H_n = \emptyset$, for some m, m' we have $x_m \ne x_{m'}$. As X is Hausdorff, for some disjoint U, V we have $x_m \in U$ and $x_{m'} \in V$. So for some $n > m + m' + p$ we have $K_m(i \mid n) \subseteq U$ and $K_{m'}(i \mid n) \subseteq V$. So $F_n \cap K_m(i \mid n) \cap K_{m'}(i \mid n) = \emptyset$, a contradiction. $\quad\square$

We generalize the last result beyond the singleton-valued to the compact-valued case, which needs a separation lemma. (This generalizes the well-known result that in a Hausdorff space disjoint compact sets may be separated by disjoint open sets – cf. Kel1955, Th. 5.9.) The next result is shown for regular Hausdorff spaces in DavJO1977 (in the course of a proof of Th. 1 there). We work in subspaces that are $\mathcal{K}$-analytic, so the following more intuitive proof applies. Recall from Lemma 3.1.5 that a $\mathcal{K}$-analytic space A is Lindelöf and that a regular Lindelöf space is normal (cf. Kel1955, Lemma 4.1). So a regular $\mathcal{K}$-analytic space A is normal.

Lemma 3.1.11 (Separation Lemma) *In a normal space, and so also in a locally compact Hausdorff space X, for an ordered finite sequence of compact sets $\langle K_1, \dots, K_n \rangle$ with empty intersection and $n \ge 2$, there is a corresponding ordered finite sequence $\langle U_1, \dots, U_n \rangle$ of open sets with empty intersection such that $K_i \subseteq U_i$.*

Proof Suppose given the compact sequence $\langle K_1, \dots, K_n \rangle$. First assume that X is normal. For each i the set K_i is disjoint from the set $K_{-i} := \bigcap_{j \ne i} K_j$.

As in Urysohn's Separation Lemma (Chapter 1), let $f_i \colon X \to [0,1]$ be a continuous function with zero set K_i such that $f_i(K_{-1}) = 1$. Put $f := \sum_{i \leq n} f_i$; then $f(x) = 0$ if and only if $x \in \bigcap_{i \leq n} K_i = \emptyset$, and so $f > 0$ on X. Now $U_i := \{x : f_i(x) < f(x)/n\}$ is open and $K_i \subseteq U_i$, as K_i is the zero set of f_i. It follows that $\langle U_1, \ldots, U_n \rangle$ has empty intersection; indeed, if $x \in \bigcap_{i \leq n} U_i$, for some x, then summing the relations $f_i(x) < f(x)/n$ we obtain the contradiction that $f(x) < f(x)$.

Now assume that X is locally compact and Hausdorff; we may choose U open containing $\bigcup_{i \leq n} K_i$ with $Y = \operatorname{cl} U$ compact. As Y is normal, we may find sets V_i open in Y separating the sets K_i as required, but in Y. Taking $U_i = V_i \cap U$ we obtain the desired separation in U and so in X. $\qquad\qquad\square$

Remark   For an alternative proof, using subnets, see Ost2011, §1.2.

We now give the promised generalization of Theorem 3.1.10 from singleton-valued to compact-valued representations.

Theorem 3.1.12 (Multiple $\mathcal{K}$-Analytic Targets)   *Let X be regular Hausdorff and $A_n = K_n(I)$ be $\mathcal{K}$-analytic in X, with K_n compact-valued and upper semicontinuous. If F_n is a decreasing sequence of (non-empty) closed sets in X with*

$$F_n \cap \bigcap_{m \leq n} K_m(i_1(m), \ldots, i_n(m)) \neq \emptyset,$$

for some $i(n) = (i_1(n), \ldots) \in I$ and each n, then $\bigcap_n F_n \cap \bigcap_n K_n(i(n)) \neq \emptyset$.

Proof   If not, write $H_n := K_n(i(n))$. Put $Y = \bigcup_n H_n$. As Y contains the sets $F_n \cap \bigcap_{m \leq n} K_m(I(i_1(m), \ldots, i_n(m)))$, and regularity is subspace hereditary, we may as well assume that $X = Y$. By compactness, since $\bigcap_n (F_n \cap H_n) = \emptyset$, there is p with $F_p \cap \bigcap_{n \leq p} H_n = \emptyset$. If $F_p \cap H_m = \emptyset$ for some $m \leq p$, then $K_m(i(m)) \subseteq X \backslash F_p$, and so $F_p \cap K_m(I(i(m) \mid n)) = \emptyset$ for some $n > p + m$. Then $F_n \cap K_m(I(i(m) \mid n)) = \emptyset$, a contradiction. So the compact set $H_m' := F_p \cap H_m$ is non-empty for each $m \leq p$. Since $\bigcap_{n \leq p} H_n' = \emptyset$, for some open-in-$F_p$ sets $U_m \supseteq H_m'$ we have $\bigcap_{n \leq p} U_n = \emptyset$ (by the Separation Lemma 3.1.11 – as X is now assumed Lindelöf, and so normal). So for some $n > p$ we have $K_m(i(m) \mid n) \subseteq U_m$ for each $m \leq p$, and so $F_n \cap \bigcap_{n \leq p} K_m(i(m) \mid n) = \emptyset$, a contradiction. $\qquad\qquad\square$

3.2 Beyond Analytic Sets: Projective Hierarchy

For the most part we work with the mindset of the practising analyst, that is in 'naive' set theory. As usual, we work in the standard mathematical framework of Zermelo–Fraenkel set theory (ZF), i.e. we do not make use of the Axiom of Choice (AC) unless we say so explicitly. Our interest in the complexities induced by the limsup operation points us in the direction of definability and descriptive set theory because of the question of whether certain specific sets, encountered in the course of analysis, have the Baire property. The answer depends on what further axioms one admits. For us there are two alternatives yielding the kind of decidability we seek. Firstly, there is Gödel's Axiom of Constructibility ($V = L$), as an appropriate strengthening of the Axiom of Choice, which creates definable sets without the Baire property (or without measurability). Secondly, at the opposite pole, there is the Axiom of Projective Determinacy (PD) (see MycSw1964 or Kec1995, 5.38.C), which guarantees the Baire property in the kind of definable sets we encounter. Thus to decide whether sets of the kind we encounter below have the Baire property, or are measurable, the answer is: it depends on the axioms of set theory that one adopts. See §14.3 and Chapter 16.

3.2.1 Projective/Analytical Hierarchy

To formulate our results we need the language of descriptive set theory, for which see, e.g., RogJ1980; Kec1995; Mos2009. Within such an approach we will regard a function as a set, namely its *graph*. We need the beginning of the *projective hierarchy* (or *analytical hierarchy* in Euclidean space (see Kec1995, S. 37.A), in particular the following classes:

the *analytic* sets Σ^1_1;
their complements, the *co-analytic* sets Π^1_1;
the common part of the previous two classes, the ambiguous class $\Delta^1_1 := \Sigma^1_1 \cap \Pi^1_1$, that is, by Souslin's Theorem (3.1.1; RogJ1980, p. 5, and MartK1980, p. 407 or Kec1995, 14.C) the *Borel* sets;
the *projections* (continuous images) of Π^1_1 sets, forming the class Σ^1_2;
their complements, forming the class Π^1_2;
the ambiguous class $\Delta^1_2 := \Sigma^1_2 \cap \Pi^1_2$;
and then: Σ^1_{n+1}, the projections of Π^1_n; their complements Π^1_{n+1}; and the ambiguous class $\Delta^1_{n+1} := \Sigma^1_{n+1} \cap \Pi^1_{n+1}$.

The notation (bold-faced to refer to coding, cf. page 32) reflects the fact that the canonical expression of the logical structure of their definitions, that is with the quantifiers (ranging over the reals, hence the superscript 1, as reals are type 1 objects – integers are of type 0) all at the front, is determined by a string of alternating quantifiers starting with an existential or universal quantifier (resp. Σ or Π). Here the subscript accounts for the number of alternations.

In §3.3 we will examine the projective character of sets naturally associated with asymptotic behaviour of regularly varying functions, i.e. we will verify their position in the hierarchy. The concern there will be with the cases $n = 1, 2$ or 3.

3.3 * Character Theorems

The Baire/measurable property discussed at various points in the Prologue is usually satisfied in mathematical practice. Indeed, any analytic subset of $\mathbb{R}$ possesses these properties (RogJD1980, Part 1 §2.9; Kec1995, 29.5), hence so do all the sets in the σ-algebra that they generate (the C-sets, Kec1995, §29.D, C for *crible* – see Burg1983a; Burg1983b, cf. BinO2011a). There is a broader class still. Recall first that an analytic set may be viewed as a projection of a planar Borel set P, so is definable as $\{x : \Phi(x)\}$ via the Σ_1^1 formula $\Phi(x) := (\exists y \in \mathbb{R})[(x, y) \in P]$; here the notation Σ_1^1 indicates *one* quantifier block (the subscripted value) of existential quantification, ranging over reals (type 1 objects – the superscripted value). As in §1.4 use of the *bold-face version* of the symbol indicates the need to refer to *arbitrary* coding (by reals not necessarily in an *effective* manner, for which see Gao2009, §1.5) of the various open sets needed to construct P. (An open set U is *coded* by the sequence of rational intervals contained in U.) Effective variants are rendered in light-face.

Consider a set A such that both A and $\mathbb{R} \backslash A$ may be defined by a Σ_2^1 formula, say respectively as $\{x : \Phi(x)\}$ and $\{x : \Psi(x)\}$, where $\Phi(x) := (\exists y \in \mathbb{R})(\forall z \in \mathbb{R})(x, y, z) \in P\}$ now, and similarly Ψ. This means that A is both Σ_2^1 and Π_2^1 (with Π indicating a leading universal quantifier block), and so is in the ambiguous class $\mathbf{\Delta}_2^1$. If in addition the equivalence

$$\Phi(x) \Longleftrightarrow \neg\Psi(x),$$

where $\neg$, meaning negation (see BelS1969), is read 'not', is provable in ZF, i.e. *without reference* to AC, then A is said to be *provably* $\mathbf{\Delta}_2^1$. It turns out that such sets have the Baire/measurable property – see FenN1973, where these are generalized to the *universally (= absolutely) measurable* sets (cf. BinO2017, §2); the idea is ascribed to Solovay in Kan2003, Ch. 3, Ex. 14.4. How much further this may go depends on what axioms of set theory are admitted, a matter to which we presently turn.

Our interest in such matters derives from the *Character Theorems* of regular variation, as noted in BinO2010b, §3 (revisited in BinO2016a, §11), which identify the logical complexity of the function

$$h^*(x) := \lim_{t \to \infty} \sup h(t + x) - h(t),$$

which is Δ_2^1 if the function h (more precisely, its graph) is Borel (and is Π_2^1 if h is analytic, and Π_3^1 if h is co-analytic). We argued in BinO2010b, §5 that Δ_2^1 is a natural setting in which to study regular variation.

Interest in the character of a function h is motivated by an interest within the theory of regular variation in the character of the level sets

$$H^k := \{s : |h(s)| < k\} = \{s : (\exists t)[(s, t) \in h \ \& \ |t| < k]\}$$

for $k \in \mathbb{N}$ (where as above h is identified with its graph). The set H^k is thus the projection of $h \cap (\mathbb{R} \times [0, k])$ and hence is Σ_n^1 if h is Σ_n^1, e.g. it is Σ_1^1, i.e. analytic, if h is analytic (in particular, Borel). Also

$$H^k = \{s : (\forall t)[(s, t) \in h \implies |t| \leq k]\} = \{s : (\forall t)[(s, t) \notin h \text{ or } |t| \leq k]\},$$

and so this is also Π_n^1 if h is Σ_n^1. Thus if h is Σ_n^1, then H^k is Δ_n^1. So if Δ_n^1 sets are Baire, for some k the set H^k is Baire/non-null, and hence shift-compact (Chapter 4), as

$$\mathbb{R} = \bigcup_{k \in \omega} H^k.$$

With this in mind, it suffices to consider upper limits; as before, we prefer to work with the additive formulation. Consider the definition

$$h^*(x) := \lim_{t \to \infty} \sup[h(t + x) - h(t)].$$

Thus in general h^* takes values in the extended real line. The problem is that the function h^* is in general *less well behaved* than the function h – for example, if we assume h measurable, h^* need not be measurable, and similarly if h has the Baire property, h^* need not. The problem we address here is the extent of this degradation – saying *exactly how much less regular* than h the limsup h^* may be. The nub is the set S on which h^* is finite. This set S is an additive semi-group (cf. §15.2) on which the function h^* is subadditive (see BinO2008) – or additive, if limits exist (see BinO2009b). Furthermore, if h has Borel graph, then h^* has Δ_2^1 graph (see below). But in the presence of certain axioms of set-theory (for which see below), the Δ_2^1 sets have the Baire property and are measurable; hence if S is large in either of these two senses, then in fact S contains a half-line. The extent of the degradation in passing from h to h^* is addressed in the following result (taken from BinO2016a), which we call the First Character Theorem, and then contrast it with two alternative character theorems. Undefined terms are explained below in the course of the proof; as in BGT1987, p. 17, we reserve the name Characterization Theorem (CT) for a result identifying the limit function arising there concerning limsupm, cf. §P.6.

Theorem 3.3.1 (First Character Theorem)

  (i) *If h is Borel (has Borel graph), then the graph of the function $h^*(x)$ is a difference of two analytic sets, hence is measurable and Δ_2^1. If the graph of h is $\mathcal{F}_\sigma$, then the graph of $h^*(x)$ is Borel.*

  (ii) *If h is analytic (has analytic graph), then the graph of the function $h^*(x)$ is Π_2^1.*

  (iii) *If h is co-analytic (has co-analytic graph), then the graph of the function $h^*(x)$ is Π_3^1.*

In our next theorem we assume much more:

Theorem 3.3.2 (Second Character Theorem) *Suppose $h \in \Delta_2^1$ and the following limit exists:*

$$\partial h(x) := \lim_{t \to \infty}[h(t + x) - h(t)].$$

Then the graph of ∂h is Δ_2^1.

The point of the next theorem is that it may be applied under the assumption of Gödel's Axiom $V = L$ (see Dev1973), as this axiom implies that Δ_2^1 ultrafilters on ω exist (see, for instance, Zap1999; Zap2001, where Ramsey ultrafilters are considered). Recall (Chapter 1) that sets of natural numbers are identified with real numbers (via their indicator functions) and so ultrafilters are regarded as sets of reals. For information on various types of ultrafilter on ω, see ComN1974 or HinS1998 (see KomT2008 for an introduction). In particular this means that we have a midway position between the results of the First and Second Character Theorems.

Theorem 3.3.3 (Third Character Theorem) *Suppose the following are of class Δ_2^1: the function h and an ultrafilter $\mathcal{U}$ on ω. Then the following is of class Δ_2^1:*

$$\partial_{\mathcal{U}} h(t) := \mathcal{U}\text{-}\lim_n [h(t + n) - h(n)].$$

Comment 1. In the circumstances of Theorem 3.3.3 $\partial_{\mathcal{U}} h(t)$ is an additive function, whereas in those of Theorem 3.3.1 one has only subadditivity. See BGT1987, p. 62, equation (2.0.3).

Comment 2. One may also consider replacing $h(t+n) - h(n)$ by $h(t + x(n)) - h(x(n))$, as in the Equivalence Theorem of BinO2009c, so as to take limits along a specified sequence $\mathbf{x}: \omega \to \omega^\omega$, in which case to have an 'effective' version of Theorem 3.3.3 one would need to specify the effective descriptive character of $\mathbf{x}$. (Here again ω^ω is identified with the reals via indicator functions.)

3.3.1 Proofs

Proof of the First Character Theorem

(i) Let us suppose that h is Borel (i.e. h has a Borel graph). As a first step consider the graph of the function of two variables $h(t + x) - h(t)$, namely the set

$$G = \{(x, t, y) : y = h(t + x) - h(t)\}.$$

One expects this to be a Borel set and indeed it is. For a proof, we must refer back to the set h itself, and to do this we must re-write the defining clause appropriately. This re-writing brings out explicitly an implicit use of quantifiers, a common enough occurrence in analysis (see end of BinO2010b for another important example). We have

$$y = h(t + x) - h(t) \Leftrightarrow (\exists u, v, w \in \mathbb{R})r(x, t, y, u, v, w),$$

where

$$r(x, t, y, u, v, w) = [\, y = u - v \,\&\, w = t + x \,\&\, (w, u) \in h \,\&\, (t, v) \in h\,]. \quad (*)$$

From a geometric viewpoint, the set of points

$$\{(x, t, y, u, v, w) : r(x, t, y, u, v, w)\}$$

is Borel in $\mathbb{R}^6$, hence the set

$$G = \{(x, t, y) : (\exists u, v, w \in \mathbb{R})r(x, t, y, u, v, w)\},$$

being a projection of a Borel set, is an analytic set in $\mathbb{R}^3$, and in general not Borel. However, in the particular present context the 'sections'

$$\{(u, v, w) : r(x, y, z, t)\},$$

corresponding to fixed $(x, t, y) \in G$, are single points (since u, v and w are defined uniquely by the values of x and t). In consequence, the projection here is Borel. The reason for this is that any Borel set is a continuous injective image of the irrationals (RogJ1980, §3.6, p. 69), and so a continuous injective image, as here under projection, of a Borel set is Borel. (So here the hidden quantifiers are 'innocuous', in that they do not degrade the character of G.) The current result may also be seen as the simplest instance of a more general result, the Rogers–Kunugui–Arsenin Theorem, which asserts that if the sections of a Borel set are $\mathcal{F}_\sigma$ (i.e. countable unions of closed sets), then its projection is Borel (RogJ1980, pp. 147/148).

By abuse of notation, let us put $h(t, x) := h(t + x) - h(t)$ and think of t as parameterizing a family of functions. By assumption, the family of functions

$h(t, x)$ is Borel; that is, the graph $\{(x, y, t) : y = h(t, x)\}$ is a Borel set (we will weaken this restriction appropriately below).

As a second step, we now consider the formal definition of $h^*(x)$, again written out as in predicate calculus using a semi-formal apparatus. The definition comes naturally as a conjunction of two clauses:

$$y = h^*(x) \Leftrightarrow P(x, y) \,\&\, Q(x, y),$$

where

$$P = (\forall n)(\forall q \in \mathbb{Q}^+)(\exists t \in \mathbb{R})(\exists z \in \mathbb{R})[t > n \,\&\, z = h(t, x) \,\&\, |z - y| < q],$$
$$Q = (\forall q \in \mathbb{Q}^+)(\exists m)(\forall t \in \mathbb{R})(\forall z \in \mathbb{R})[t > n \,\&\, (t, x, z) \in h \Longrightarrow z < y + q].$$

The first clause (predicate) asserts that y is a limit point of the set $\{h(t, x) : t \in \mathbb{R}\}$ and this requires an existential quantifier; the second clause asserts that, with finitely many exceptions, no member of the set exceeds y by more than q and this requires a universal quantifier.

From a geometric viewpoint, for fixed $q > 0$ the set of points

$$G_1 = \{(x, y, z, t) : p(x, y, z)\},$$

where

$$p(x, y, z, t) = [(t, x, z) \in h \,\&\, |z - y| < q],$$

is Borel in $\mathbb{R}^4$, hence again the set $\{(x, y) : (\exists z, t \in \mathbb{R})p(x, y, z)\}$, being a projection of a Borel set, is an analytic set in $\mathbb{R}^2$. Again, for fixed (x, y) we look at the section of G_1. Evidently $\{z : |z - y| < q\}$ is an open set, so $\mathcal{F}_\sigma$. However, only if we assume that h is $\mathcal{F}_\sigma$ can we deduce that

$$\{(x, y) : (\exists t \in \mathbb{R})(\exists z \in \mathbb{R})[t > n \,\&\, z = h(t, x) \,\&\, |z - y| < q]$$

is Borel. Otherwise it is merely analytic.

Since the quantifiers in $(\exists z \in \mathbb{R})(\exists t \in \mathbb{R})p(x, y, z, t)$ are at the front of the defining formula, we see that the formula is Σ^1_1. See RogJ1980 for a modern side-by-side exposition of the two viewpoints of mathematical logic and geometry.

Finally, the set $\{(x, y) : P(x, y)\}$ is seen to be obtainable from analytic sets (or Borel in the special case) by use of countable union and intersection operations. It is thus an analytic set (or Borel as the case may be).

By contrast, for $Q(x, y, z, t, q) := [[z = h(t, x)] \Longrightarrow z < y + q]$, the set $\{(x, y) : (\forall z, t \in \mathbb{R})(\forall q \in \mathbb{Q}^+)Q(x, y, z, t, q)\}$, is co-analytic, since its complement is the analytic set

$$\{(x, y) : (\exists z, t \in \mathbb{R})(\exists q \in \mathbb{Q}^+))[z = h(t, x) \,\&\, z \geq y + q]\}.$$

Again for given q and for arbitrary fixed (x, y) the section of $\{(x, y, z, t) : [z = h(t, x) \ \& \ z \geq y + q]\}$ will be $\mathcal{F}_\sigma$ if the graph of h is $\mathcal{F}_\sigma$, but is otherwise analytic. Thus $\{(x, y) : Q(x, y)\}$ is seen to be obtainable from co-analytic sets (or at best Borel sets) by use of countable union and intersection operations. It is thus co-analytic, or Borel as the case may be.

Again, as with Q above, the formula $(\forall z \in \mathbb{R}) q(x, y, z, t)$ is $\mathbf{\Pi}_1^1$, since the opening quantifier is universal of order 1.

The set $\{(x, y) : Q(x, y)\}$ is seen to be obtainable from co-analytic sets by use of countable union and intersection operations. It is thus co-analytic since such operations preserve this character. Finally, note that the sets which are differences of analytic sets are both in the classes $\mathbf{\Pi}_1^1$ and $\mathbf{\Sigma}_2^1$, and so in their intersection $\mathbf{\Delta}_2^1$. We have of course neglected the possibility that the lim sup is infinite, but for this case we need only note that

$$h^*(x) = \infty \Leftrightarrow (\forall n)(\exists t \in \mathbb{R})(\exists z \in \mathbb{R})[t > n \ \& \ z = h(t, x) \ \& \ z > n],$$
$$h^*(x) < \infty \Leftrightarrow (\exists y \in \mathbb{R})(y = h^*(x)),$$

so that this case is simultaneously $\mathbf{\Sigma}_1^1$ and $\mathbf{\Pi}_1^1$.

We have thus proved part (i).

(ii) Now assume that h has an analytic graph. Then G, being the projection of an analytic set, is now analytic. That is, we may write

$$y = h(t, x) \Leftrightarrow (\exists w \in \mathbb{R}) F(t, x, y, w),$$

where the set $\{(t, x, y, w) : F(t, x, y, w)\}$ is Borel. Then

$$\{(x, y) : (\exists z \in \mathbb{R})(\exists w \in \mathbb{R})[F(t, x, z, w) \ \& \ |z - y| < q]\}$$

is only analytic, since we have no information about special sections; however, the set

$$\{(x, y) : (\forall z \in \mathbb{R})(\exists w \in \mathbb{R})[t > n \ \& \ F(t, x, z, w) \implies z < y + q]\}$$

requires for its definition a quantifier alternation which begins with a universal quantifier, so is $\mathbf{\Pi}_2^1$. Since $\mathbf{\Sigma}_1^1$ sets are necessarily a subclass of $\mathbf{\Pi}_2^1$ sets, the graph of lim $\sup_t f(t, x)$ in this case is $\mathbf{\Pi}_2^1$, which proves (ii).

(iii) Now, suppose that the function $h(x)$ has a co-analytic graph. Then by (*) the set G is of class $\mathbf{\Sigma}_2^1$, i.e. the function $h(t, x)$ has a $\mathbf{\Sigma}_2^1$ graph. That is, we now have to write

$$y = h(t, x) \Leftrightarrow (\exists u \in \mathbb{R})(\forall w \in \mathbb{R}) F(t, x, y, u, w),$$

where as before the set $\{(t, x, y, w) : F(t, x, y, u, w)\}$ is Borel. Then

$$\{(x, y) : (\exists z, u \in \mathbb{R})(\forall w \in \mathbb{R})[F(t, x, z, u, w) \ \& \ |z - y| < q]\}$$

is now Σ_2^1. On the other hand, the set

$$\{(x, y) : (\forall z \in \mathbb{R})(\exists u \in \mathbb{R})(\forall w \in \mathbb{R})[F(t, x, z, u, w) \implies z < y + q]\}$$

is Π_3^1. Since Σ_1^1 sets are necessarily a subclass of Π_3^1 sets, the graph of $\limsup_t h(t, x)$ in this case is Π_3^1. $\square$

Proof of the Second Character Theorem   Here we have

$$y = \partial h(x) \iff (\forall q \in \mathbb{Q}^+)(\exists n \in \omega)(\forall t > n)(\forall zuvw)P,$$

where

$$P = [[z = u - v \ \& \ w = t + x \ \& \ (t, v) \in h \ \& \ (w, u) \in h] \implies |z - y| < q],$$

and

$$y \neq \partial h(x) \iff (\forall q \in \mathbb{Q}^+)(\exists n \in \omega)(\forall t > n)(\forall zuvw)Q,$$

where

$$Q = [[z = u - v \ \& \ w = t + x \ \& \ (v, t) \in h \ \& \ (u, w) \in h] \implies |z - y| \geq q]. \ \square$$

Proof of the Third Character Theorem   By (*) the function $y = h(t, x)$ is of class Σ_2^1. We show that $y = \partial_{\mathcal{U}} h(t)$ is of class Σ_2^1. The result will follow since the negation satisfies

$$y \neq \partial_{\mathcal{U}} h(t) \iff \exists z [z \neq y \ \& \ [z = h^*(t) \ or \ h^*(t) = \pm\infty]],$$

and so is of class Σ_2^1. Finally,

$$y = \partial_{\mathcal{U}} h(t) \iff (\forall \varepsilon \in \mathbb{Q}^+)(\exists U)(\forall n \in \omega)(\exists t)P,$$

where

$$P = [U \in \mathcal{U} \ \& \ n \in U \ \& \ (n, t) \in \mathbf{x} \ \& \ |t - y| < \varepsilon],$$

and

$$\partial_{\mathcal{U}} h(t) = \infty \iff (\forall M \in \mathbb{Q}^+)(\exists U \in \mathcal{U})(\forall n \in U)(\exists t)[(n, t) \in \mathbf{x} \ \& \ t > M].$$

     $\square$

3.4 * Appendix: Baire Hull

The proof below relies on Banach's Category Theorem. Here $\mathcal{M}$ denotes the meagre sets and $\mathcal{B}a$ the sets with the Baire property. A point x is a *heavy point* for an arbitrary set A if each neighbourhood of x meets A in a non-meagre set. The heavy points form a closed set. One way to see this is to consider the set of *light points* of A, namely those having a neighbourhood meeting A in a meagre set. The light points evidently form an open set.

A set A that is not Baire is of course not a meagre set. Consequently, there is a natural expansion of A less a meagre set that is Baire, namely the closed set of points on which A is 'heavy'. This set is minimal among Baire sets in that any smaller expansion of A differs by a meagre set, as in the next result.

Theorem 3.4.1 (Baire Hull) *For any A, there is $B \in \mathcal{B}a$ with $A \subseteq B$ such that for any $B' \in \mathcal{B}a$, if $A \subseteq B' \subseteq B$, then $B \backslash B' \in \mathcal{M}$.*

Proof For any A, we begin by decomposing X into a part on which A is light and one on which it is heavy, as follows. Write

$$\mathcal{L}(A) = \mathcal{L}_N(A) := \{U \in G(X) : U \cap A \in \mathcal{M}\},$$
$$L(A) := \bigcup \mathcal{L}(A), \qquad H(A) := X \backslash L(A).$$

Claim *for $A \in \mathcal{B}a$, that $H(A) \backslash A \in \mathcal{M}$.*

Choose G open agreeing with A modulo $\mathcal{M}$. Put $F = \operatorname{cl} G$; then $A \backslash F \subseteq A \backslash G$. So $A \cap (X \backslash F) \in \mathcal{M}$, and $(X \backslash F) \in \mathcal{L}(A)$. So $X \backslash F \subseteq L(A)$, giving $H(A) \subseteq F$. So, since $(\operatorname{cl} G) \backslash G$ and $(G \backslash A) \in \mathcal{M}$,

$$H(A) \backslash A \subseteq F \backslash A = (\operatorname{cl} G) \backslash A \subseteq (\operatorname{cl} G) \backslash G \cup (G \backslash A) \in \mathcal{M}.$$

We next show that $A \backslash H(A) = A \cap L(A) \in \mathcal{M}$ for *arbitrary* A, implying that A is covered by the union of $H(A)$ and a meagre set (so that the symmetric difference $A \triangle H(A) \in \mathcal{M}$).

For $\mathcal{L}'(A)$ a maximal family of disjoint open subsets in $\mathcal{L}(A)$, let

$$Q := \left(\bigcup \mathcal{L}(A) \right) \backslash \bigcup \mathcal{L}'(A).$$

We have that Q is nowhere dense. For suppose that Q is dense in an open set V. Then $V \cap U$ is non-empty for some $U \in \mathcal{L}(A)$. Since $A \cap U$ is in $\mathcal{M}$, so $A \cap U \cap V$ is in $\mathcal{M}$, i.e. $U \cap V$ is in $\mathcal{L}(A)$. Now Q is dense in the open subset $V \cap U$. As $\bigcup \mathcal{L}'(A)$ is open, Q is closed as a subset of $\bigcup \mathcal{L}(A)$. Hence Q contains $V \cap U$, and so $U \cap V$ is disjoint from $\mathcal{L}'(A)$, contradicting the maximality of $\mathcal{L}'(A)$. So after all Q is nowhere dense.

We now show that $A \cap \bigcup \mathcal{L}'(A) \in \mathcal{M}$. For each U in $\mathcal{L}'(A)$, write

$$A \cap U = \bigcup_{n \in \mathbb{N}} M_n(U),$$

with $M_n(U)$ nowhere dense sets and put

$$M_n := \bigcup \{M_n(U) : U \in \mathcal{L}'(A)\}.$$

Then M_n is nowhere dense, by the Banach Category Theorem (as $M_n(U) \subseteq U$ and $\mathcal{L}'(A)$ is disjoint). Finally,

$$A \setminus H(A) = A \cap L(A) \subseteq Q \cup \bigcup\nolimits_{n \in \mathbb{N}} M_n \in \mathcal{M}.$$

So

$$A \subseteq B := H(A) \cup H_0, \quad \text{with} \quad H_0 = \bar{Q} \cup \bigcup\nolimits_{n \in \mathbb{N}} \bar{M}_n \in \mathcal{M},$$

exhibiting an $\mathcal{F}_\sigma$ cover of A, as $H(A)$ is closed. Now consider $B' \in \mathcal{B}a$ with $A \subseteq B' \subseteq B$. Then

$$L(M) \subseteq L(B') \subseteq L(A), \quad \text{and} \quad H(A) \subseteq H(B') \subseteq H(B).$$

Now

$$B \setminus B' = H_0 \setminus B' \cup H(A) \setminus B' \subseteq H_0 \setminus B' \cup H(B') \setminus B'.$$

But both H_0 and $H(B') \setminus B' \in \mathcal{M}$, the latter by the claim at the outset. So B is a Baire hull. $\qquad\square$

4

Infinite Combinatorics in $\mathbb{R}^n$: Shift-Compactness

The results of this chapter develop a new aspect of measure–category duality. This has powerful applications.

4.1 Generic Dichotomy

We will meet many examples of situations which divide into two sharply contrasting cases – usually where behaviour is, in some appropriate sense, either very good or very bad. Such dichotomy often has a generic aspect: if something works at all, it works nearly everywhere. We address the source of this genericity below: *a property inheritable by supersets either holds generically or fails outright.*

Definition For X the reals $\mathbb{R}_+$ with the Euclidean or density topology (Chapter 7), denote by $\mathcal{B}a(X)$, or just $\mathcal{B}a$, the Baire sets of the space X, and recall these form a σ-algebra. Say that a correspondence $F \colon \mathcal{B}a \to \mathcal{B}a$ is *monotonic* if $F(S) \subseteq F(T)$ for $S \subseteq T$.

The nub is the following simple result, which we call the Generic Dichotomy Principle.

Theorem 4.1.1 (Generic Dichotomy Principle; BinO2010f) *For $F \colon \mathcal{B}a \to \mathcal{B}a$ monotonic: either*

(i) *there is a non-meagre $S \in \mathcal{B}a$ with $S \cap F(S) = \varnothing$; or,*
(ii) *for every non-meagre $T \in \mathcal{B}a$, $T \cap F(T)$ is quasi almost all of T.*

Equivalently: the existence condition, that $S \cap F(S) \neq \varnothing$ should hold for all non-meagre $S \in \mathcal{B}a$, implies the genericity condition that, for each non-meagre $T \in \mathcal{B}a$, $T \cap F(T)$ is quasi almost all of T.

Proof   Suppose that (i) fails. Then $S \cap F(S) \neq \varnothing$ for every non-meagre $S \in \mathcal{B}a$. We show that (ii) holds. Suppose otherwise; then, for some T non-meagre in $\mathcal{B}a$, the set $T \cap F(T)$ is not almost all of T. Then the set $U := T \backslash F(T) \subseteq T$ is non-meagre (it is in $\mathcal{B}a$ as T and $F(T)$ are), and so

$$\varnothing \neq U \cap F(U) \qquad (S \cap F(S) \neq \varnothing \quad \text{for every non-meagre } S)$$
$$\subseteq U \cap F(T) \qquad (U \subseteq T \quad \text{and } F \text{ monotonic}).$$

But, as $U := T \backslash F(T)$, we have $U \cap F(T) = \varnothing$, a contradiction.

The final assertion simply rephrases the dichotomy as an implication.   □

The next result transfers the onus of verifying the existence condition above to topological completeness: it suffices to test the F property on $\mathcal{G}_\delta$ sets.

Theorem 4.1.2 (Generic Completeness Principle; BinO2010f)   *For $F: \mathcal{B}a \to \mathcal{B}a$ monotonic, if $W \cap F(W) \neq \varnothing$ for all non-meagre $W \in \mathcal{G}_\delta$, then, for each non-meagre $T \in \mathcal{B}a$, $T \cap F(T)$ is quasi almost all of T.*

That is, either

(i) *there is a non-meagre $S \in \mathcal{G}_\delta$ with $S \cap F(S) = \varnothing$; or,*

(ii) *for every non-meagre $T \in \mathcal{B}a$, $T \cap F(T)$ is quasi almost all of T.*

Proof   By Baire's Theorem, for S non-meagre in $\mathcal{B}a$ there is a non-meagre $W \subseteq S$ with $W \in \mathcal{G}_\delta$. So $W \cap F(W) \neq \varnothing$ and thus $\varnothing \neq W \cap F(W) \subseteq S \cap F(S)$, by monotonicity. By Theorem 4.1.1, for every non-meagre $T \in \mathcal{B}a$, $T \cap F(T)$ is quasi almost all of T.   □

The rest of this section is concerned with examples of monotonic correspondences which play a part in the development of our subject. Each of the correspondences F to be introduced below will give rise to a correspondence $\Phi(A) := F(A) \cap A$ which is a lower or upper density (see Chapter 7, and also LukMZ1986) and so gives rise to a fine topology on the real line (see Chapter 8).

Example $(\mathbb{R}, \mathcal{D})$. Here the meagre sets are the null sets. Let $\mathcal{B}$ denote a countable basis of Euclidean open neighbourhoods. For any set T and $0 < \alpha < 1$ put

$$\mathcal{B}_\alpha(T) := \{I \in \mathcal{B} : |I \cap T| > \alpha|I|\},$$

which is countable, and

$$F(T) := \bigcap_{\alpha \in \mathbb{Q} \cap (0,1)} \bigcup \{I : I \in \mathcal{B}_\alpha(T)\}.$$

Thus F is increasing in T, $F(T)$ is measurable (even if T is not) and $x \in F(T)$ if and only if x is a density point of T. The genericity assertion is the Lebesgue Density Theorem of Chapter 7; see, e.g., Saks1964, IV.10; Bog2007a, Th.5.6.2.

Further related examples follow in §8.6.

4.2 Kestelman–Borwein–Ditor (KBD) Theorem: First Proof

The main result of this section is the following theorem establishing the fundamental infinite combinatorics which apply to both Baire and measurable sets. These dual foundations, originating in single-variable theory, were unified in two ways in BinO2009b and BinO2010h. In one they are unified structurally by their identical combinatorics. In the other, both are derived from a single source: Baire category. Both views translate immediately to $\mathbb{R}^d$.

Theorem 4.2.1 (Theorem KBD – Kestelman–Borwein–Ditor Theorem) *Let* $\{z_n\} \to 0$ *be a null sequence of reals. If* T *is Baire/Lebesgue measurable, then, for generically all* $t \in T$*, there is an infinite set* $\mathbb{M}_t$ *such that*

$$\{t + z_m : m \in \mathbb{M}_t\} \subseteq T.$$

Variants of this theorem, depending on context, permeate this book. One instance, Theorem 10.5.2, is concerned with bounded sequences, and there such sequences will have subsequences contained in the target set T, up to translation as above. It seems appropriate to call this property *shift-compactness*, borrowing an established probabilistic term, coined by Parthasarathy (Par1967), describing a very similar feature. The property here does in fact imply a finite sub-covering result: see §6.4.

Infinitely Often versus Co-finitely Often

There is some slight difference (which need not detain us here) between the combinatorics of category and measure, indeed of no significance to convergence properties, in view of the *sub-subsequence theorem*.

In the case of a non-meagre Baire target set T, the infinite combinatorics of Theorem 4.2.1 can be strengthened by replacing the *infinite* set of indices $m \in \mathbb{M}$ for which the translates $t + z_m$ are in T with a *co-finite* set of indices. For a study of this aspect in a general group setting, see MilleMO2021.[1]

---

[1] Before naming them *shift-compact*, our earlier work used *sub-universal* for sets containing, up to translation as in Theorem 4.2.1, a subsequence from every null sequence. This followed Kestelman's use in Kes1947a of *universal* to describe sets co-finitely containing all convergent sequences up to translation. For an analysis of the subtle relationship between Parthasarathy's term and ours, see BinO2024.

For examples where co-finite combinatorics fail, see BorwD1978; Mille1989 and, e.g., Komj1988.

The stronger combinatorics are of immediate use when checking the continuity of an additive function on T, say when establishing $f(t) = \lim_{n\to\infty} f(t+z_n)$, when $t \in T$ and $t + z_n$ are all but finitely many in T.

But a subsequence approach, albeit less elegant, is equally helpful. It suffices to establish that $f(t) = \lim_{m\in\mathbb{M}'} f(t + z_m)$ for some infinite $\mathbb{M}' \subseteq \mathbb{M}$ of an arbitrary sequence $\mathbb{M} \subseteq \mathbb{N}$. Indeed, if there is a neighbourhood U of $f(t)$ and an infinite subset $\mathbb{M}$ of $\mathbb{N}$ omitting $\{f(t + z_m) : m \in \mathbb{M}\}$, then $f(t) = \lim_{n\in\mathbb{M}'} f(t + z_m)$ cannot hold for any $\mathbb{M}' \subseteq \mathbb{M}$.

In summary: when studying convergence properties, the infinitely often combinatorics are adequate and available for both measure and category contexts. Hence we speak of identical combinatorics.

Our first proof of the KBD Theorem 4.2.1 follows from the two results below, both important in their own right. The first and its corollary address displacements of open sets in the density and the Euclidean topologies; it is mentioned (in passing) in a note added in proof (p. 32) in Kemperman, Kem1957, Th. 2.1, p. 30, for which we give an alternative proof. The second parallels an elegant result for the measure case treated in BergHW1997.

Theorem 4.2.2 (Displacements Lemma – Kemperman's Theorem; Kem1957 Th. 2.1 with $B_i = E$, $a_i = t$) *If E is non-null Borel, then $f(x) := |E \cap (E+x)|$ is continuous at $x = 0$, and so, for some $\varepsilon = \varepsilon(E) > 0$,*

$$E \cap (E + x) \text{ is non-null for } |x| < \varepsilon.$$

More generally, $f(x_1, \ldots, x_p) := |(E + x_1) \cap \cdots \cap (E + x_p)|$ is continuous at $x = (0, \ldots, 0)$, and so, for some $\varepsilon = \varepsilon_p(E) > 0$,

$$(E + x_1) \cap \cdots \cap (E + x_p) \text{ is non-null for } |x_i| < \varepsilon, \ (i = 1, \ldots, p).$$

First Proof (After BergHW1997; cf. e.g. BinO2010g, Th. 7.5) Let t be a density point of E. Choose $\varepsilon > 0$ such that

$$|E \cap B_\varepsilon(t)| > \frac{3}{4}|B_\varepsilon(0)|.$$

Now $|B_\varepsilon(t)\backslash B_\varepsilon(t + x)| \leq (1/4)|B_\varepsilon(t + x)|$ for $x \in B_{\varepsilon/2}(0)$, so

$$|E \cap B_\varepsilon(t + x)| > \frac{1}{2}|B_\varepsilon(0))|.$$

By invariance of Lebesgue measure we have

$$|(E + x) \cap B_\varepsilon(t + x)| > \frac{3}{4}|B_\varepsilon(0))|.$$

But, again by invariance, as $B_\varepsilon(t) + x = B_\varepsilon(0) + t + x$ this set has measure $|B_\varepsilon(0)|$. Using $|A_1 \cup A_2| = |A_1| + |A_2| - |A_1 \cap A_2|$ with $A_1 := E \cap B_\varepsilon(t + x)$ and $A_2 := (E + x) \cap B_\varepsilon(t + x)$ now yields

$$|E \cap (E + x)| \ge |E \cap (E + x) \cap (B_\varepsilon(t) + x)| > \frac{5}{4}|B_\varepsilon(0)| - |B_\varepsilon(0)| > 0.$$

Hence, for $x \in B_{\varepsilon/2}(0)$, we have $|E \cap (E + x)| > 0$.

For the p-fold form we need some notation. Let t again denote a density point of E and $x = (x_1, \ldots, x_n)$ a vector of variables. Set $A_j := B_\varepsilon(t) \cap E \cap (E + x_j)$ for $1 \le j \le n$. For each multi-index $\mathbf{i} = (i(1), \ldots, i(d))$ with $0 < d < n$, put

$$f_\mathbf{i}(x) := |A_{i(1)} \cap \cdots \cap A_{i(d)}|;$$
$$f_n(x) := |A_1 \cap \cdots \cap A_n|, \qquad f_0 = |B_\varepsilon(t) \cap E|.$$

We have already shown that the functions $f_j(x) = |B_\varepsilon(t) \cap E \cap (E + x_j)|$ are continuous at 0. Now argue inductively: suppose that, for $\mathbf{i}$ of length less than n, the functions $f_\mathbf{i}$ are continuous at $(0, \ldots, 0)$. Then for given $\varepsilon > 0$, there is $\delta > 0$ such that for $\|x\| < \delta$ and each such index $\mathbf{i}$ we have

$$-\varepsilon < f_\mathbf{i}(x) - f_0 < \varepsilon,$$

where $f_0 = |B_\varepsilon(t) \cap E|$. Noting that

$$\bigcup_{i=1}^n A_i \subset B(t) \cap E,$$

and using upper or lower approximations, according to the signs in the inclusion–exclusion identity

$$\left|\bigcup_{i=1}^n A_i\right| = \sum_i |A_i| - \sum_{i<j} |A_i \cap A_j| + \cdots + (-1)^{n-1}\left|\bigcap_i A_i\right|,$$

one may compute linear functions $L(\varepsilon), R(\varepsilon)$ such that

$$L(\varepsilon) < f_n(x) - f_0 < R(\varepsilon).$$

Indeed, taking $x_i = 0$ in the identity, both sides collapse to the value f_0. Continuity follows.   □

Second Proof   Apply instead Theorem 61.A of Hal1950, Ch. XII, p. 266 to establish the base case, and then proceed inductively as before.   □

Corollary 4.2.3   *Kemperman's Theorem* (4.2.2) *holds for non-meagre Baire sets E in place of Borel sets in the form: for each p in $\mathbb{N}$ there exists $\varepsilon = \varepsilon_p(E) > 0$ such that*

$$(E + x_1) \cap \cdots \cap (E + x_p) \text{ non-meagre,} \quad \text{for} \quad |x_i| < \varepsilon, \ (i = 1, \ldots, p).$$

Proof   A non-meagre Baire set differs from an open set by a meagre set.   □

We will now prove Theorem 4.2.1 using the Generic Completeness Principle (Theorem 4.1.1); this amounts to proceeding in two steps. To motivate the proof strategy, note that the embedding property is upward hereditary (i.e. monotonic in the sense of §4.1): if T includes a subsequence of z_n by a shift t in T, then so does any superset of T. We first consider a non-meagre $\mathcal{G}_\delta$ /non-null closed set T, just as in BergHW1997, modified to admit the consecutive format so that, for infinitely many m, z_m and all of its p consecutive terms in the null sequence, which we denote by

$$\bar{z}_{pm} := \{z_m, z_{m+1}, \ldots, z_{m+p}\},$$

translate into the target set T. We next deduce the theorem by appeal to $\mathcal{G}_\delta$ inner-regularity of category/measure and Generic Dichotomy. (The subset E of exceptional shifts can only be meagre/null.) The next result is a generalization of a lemma due to Bergelson, Hindman and Weiss proving the existence of a sequence embedding.

Theorem 4.2.4 (See BergHW1997, Lemma 2.2) *For T Baire non-meagre/ measurable non-null and a null sequence $z_n \to 0$, there exist $t \in T$ and an infinite $\mathbb{M}_t$ such that*

$$\{t + \bar{z}_{pm} : m \in \mathbb{M}_t\} \subseteq T.$$

Proof The conclusion of the theorem is inherited by supersets (is upward hereditary), so without loss of generality we may assume that T is Baire non-meagre/measurable non-null and completely metrizable, say under a metric $\rho = \rho_T$. (For T measurable non-null we may pass down to a compact non-null subset, and for T Baire non-meagre we simply take away a meagre set to leave a Baire non-meagre $\mathcal{G}_\delta$ subset.) Since this is an equivalent metric, for each $a \in T$ and $\varepsilon > 0$ there is $\delta = \delta(\varepsilon) > 0$ such that $B_\delta(a) \subseteq B_\varepsilon^\rho(a)$. Thus, by taking $\varepsilon = 2^{-n-1}$, the δ-ball $B_\delta(a)$ has ρ-diameter less than 2^{-n}.

Working inductively in steps of length p, we define subsets of T (of possible translators) B_{pm+i} of ρ-diameter less than 2^{-m} for $i = 1, \ldots, p$ as follows. With $m = 0$, we take $B_0 = T$. Given $n = pm$ and B_n open in T, choose N such that $|z_k| < \min\{\frac{1}{2}|x_n|, \varepsilon_p(B_n)\}$, for all $k > N$. For $i = 1, \ldots, p$, let $x_{n-1+i} = z_{N+i}$; then by Kemperman's Theorem 4.2.2 or its Corollary, 4.2.3, $B_n \cap (B_n - x_n) \cap \cdots \cap (B_n - x_{n+p})$ is non-empty (and open). We may now choose a non-empty subset B_{n+i} of T which is open in A with ρ-diameter less than 2^{-m-1} such that $\text{cl}_T B_{n+i} \subset B_n \cap (B_n - x_n) \cap \cdots \cap (B_n - x_{n+i}) \subseteq B_{n+i-1}$. By completeness, the intersection $\bigcap_{n \in \mathbb{N}} B_n$ is non-empty. Let

$$t \in \bigcap_{n \in \mathbb{N}} B_n \subset T.$$

Now $t + x_n \in B_n \subset T$, as $t \in B_{n+1}$, for each n. Hence $\mathbb{M}_t := \{m : z_{mp+1} = x_n$ for some $n \in \mathbb{N}\}$ is infinite. Moreover, if $z_{pm+1} = x_n$, then $z_{pm+2} = x_{n+1}, \ldots, z_{pm+p} = x_{n+p-1}$, and so

$$\{t + \bar{z}_{pm} : m \in \mathbb{M}_t\} \subseteq T. \qquad \square$$

We now apply Theorem 4.1.1 (Generic Dichotomy) to extend what we now know from an existence to a genericity statement, thus completing the proof of Theorem 4.2.1.

Theorem 4.2.5 *For T Baire/measurable and $z_n \to 0$, for generically all $t \in T$ there exists an infinite $\mathbb{M}_t$ such that*

$$\{t + \bar{z}_{pm} : m \in \mathbb{M}_t\} \subseteq T.$$

Hence, if $Z \subseteq X$ accumulates at 0 (has an accumulation point there), then for some $t \in T$ the set $Z \cap (T - t)$ accumulates at 0 (along Z). Such a t may be found in any open set with which T has non-null intersection.

Proof Working as usual in X, the correspondence

$$F(T) := \bigcap_{n \in \omega} \bigcup_{m > n} [(T - z_{pm+1}) \cap \cdots \cap (T - z_{pm+p})]$$

takes Baire sets to Baire sets and is monotonic. Here $t \in F(T)$ if and only if there exists an infinite $\mathbb{M}_t$ such that $\{t + \bar{z}_{pm} : m \in \mathbb{M}_t\} \subseteq T$. By Theorem 4.2.2 $F(T) \cap T \neq \emptyset$, for T Baire non-meagre, so we may appeal to the Generic Dichotomy Principle (Theorem 4.1.1) to deduce that $F(T) \cap T$ is quasi all of T (cf. the example of §4.1).

With the main assertion proved, let $Z \subseteq X$ accumulate at 0 and suppose that z_n in Z converges to 0. Take $p = 1$. Then, for some $t \in T$, there is an infinite $\mathbb{M}_t$ such that $\{t + z_m : n \in \mathbb{M}_t\} \subseteq T$. Thus $\{z_m : n \in \mathbb{M}_t\} \subseteq Z \cap (T - t)$ has 0 as a joint accumulation point. $\qquad \square$

5

Kingman Combinatorics and Shift-Compactness

The KBD theorem of the previous chapter is about shift-embedding of sub-sequences of a null sequence $\{z_n\}$ into a *single* set T with an assumption of *regularity* (Baire/measurable). Our generalizations below of a theorem of Kingman (Kin1963; Kin1964) have been motivated by the wish to establish 'multiple embedding' versions of KBD: we seek conditions on a sequence $\{z_n\}$ and a *family* of sets $\{T_k\}_{k \in \omega}$ which together guarantee that *one* shift embeds (different) subsequences of $\{z_n\}$ into *all* members of the family.

Evidently, if $t + z_n$ lies in several sets infinitely often, then the sets in question have a common limit point, a sense in which they are contiguous at t. Thus *contiguity conditions* are one goal, the other two being *regularity conditions* on the family, and *admissibility conditions* on the null sequences.

We view the original Kingman Theorem as studying contiguity at infinity, so that divergent sequences z_n (i.e. with $z_n \to +\infty$) there replace the null sequences of KBD. The theorem uses *openness* as a regularity condition on the family, *cofinality* at infinity (e.g. *unboundedness* on the right) as the simplest contiguity condition at infinity, and

$$z_{n+1}/z_n \to 1 \qquad \text{(multiplicative form),}$$
$$z_{n+1} - z_n \to 0 \qquad \text{(additive form)} \tag{*}$$

as the admissibility condition on the divergent sequence z_n; (*) follows from regular variation by Weissman's Lemma (BGT1987, Lemma 1.9.6). Taken together, these three guarantee multiple embedding (at infinity).

One can switch from $\pm\infty$ to 0 by an inversion $x \to 1/x$, and thence to any τ by a shift $y \to y + \tau$. *Openness* remains the *regularity* condition, a property of *density at zero* becomes the analogous *admissibility* condition on null sequences, and cofinality (or accumulation) at τ the contiguity condition. The transformed theorem then asserts that for admissible null sequences ζ_n

there exists a scalar σ such that the sequence $\sigma\zeta_n + \tau$ has subsequences in all the open sets T_k provided these all accumulate at τ.

Bitopology: The Euclidean and Density Topologies   The two main themes of this book, *category* and *measure*, have many similarities and many contrasts (as the Table of Contents will show). Each has a *topology* intimately linked with it. For category, it is the ordinary (*Euclidean*) topology, $\mathcal{E}$. For measure, it is the *density* topology, $\mathcal{D}$ (Kec1995, 17.47). We defer a full treatment of the density topology to Chapter 7.

Below, we will replace Kingman's regularity condition of openness by the Baire property, or alternatively measurability, to obtain two versions of Kingman's Theorem – one for category and one for measure. As above, we develop the regularity theme *bitopologically*, working with two topologies, so as to deduce the measure case from the Baire case by switching from the Euclidean $\mathcal{E}$ to the density topology $\mathcal{D}$ (see the end of §1.4 and Chapter 7).

5.1 Definitions and Notation

5.1.1 Essential Contiguity Conditions

We use the notation $B_r(x) := \{y : |x - y| < r\}$ and $\omega := \{0, 1, 2, \ldots\}$. Likewise for $a \in A \subseteq \mathbb{R}$ and metric $\rho = \rho_A$ on A; $B_r^\rho(a) := \{y \in A : \rho(a, y) < r\}$ and cl_A denotes closure in A. For S given, put $S^{>m} = S \backslash B_m(0)$. Denote by $\mathbb{R}_+$ the (strictly) positive reals. When we regard $\mathbb{R}_+$ as a multiplicative group, we write

$$A \cdot B := \{ab : a \in A,\ b \in B\}, \qquad A^{-1} := \{a^{-1} : a \in A\},$$

for A, B subsets of $\mathbb{R}_+$.

Call a Baire set S *essentially unbounded* if for each $m \in \mathbb{N}$ the set $S^{>m}$ is non-meagre. This may be interpreted in the sense of the metric (Euclidean) topology, or as we see later in the measure sense by recourse to the density topology. To distinguish the two, we will qualify the term by referring to the category/metric or the measure sense.

Say that a set $S \subset \mathbb{R}_+$ *accumulates essentially* at 0 if S^{-1} is essentially unbounded. (In BerghHW1997 such sets are called measurably/Baire *large* at 0.) Say that $S \subset \mathbb{R}_+$ *accumulates essentially* at t if $(S - t) \cap \mathbb{R}_+$ accumulates essentially at 0.

We begin by simplifying essential unboundedness modulo meagre/null sets.

Theorem 5.1.1   *In $\mathbb{R}_+$ with the Euclidean or density topology, and with S Baire/measurable and essentially unbounded, there exists an open/density-open unbounded G and meagre/null M with $G \backslash M \subset S$.*

Proof   Choose integers m_n inductively with $m_0 = 0$ and $m_{n+1} > m_n$ the least integer such that $(m_n, m_{n+1}) \cap S$ is non-meagre. For, given m_n, the integer m_{n+1} is well defined, as otherwise, for each $m > m_n$, we would have $(m_n, m) \cap S$ meagre, and so also

$$(m_n, \infty) \cap S = \bigcup_{m > m_n} (m_n, m) \cap S \text{ meagre,}$$

contradicting the essential unboundedness of S. Now, as $(m_n, m_{n+1}) \cap S$ is Baire/measurable, we may choose G_n open/density-open and M_n, M_n' meagre subsets of (m_n, m_{n+1}) such that

$$((m_n, m_{n+1}) \cap S) \cup M_n = G_n \cup M_n'.$$

Hence G_n is non-empty. Put $G := \bigcup_n G_n$ and $M := \bigcup_n M_n$. Then M is meagre and G is open unbounded and, since $M \cap (m_n, m_{n+1}) = M_n$ and $G \cap (m_n, m_{n+1}) = G_n$,

$$G \backslash M = \bigcup_n G_n \backslash M = \bigcup_n G_n \backslash M_n \subset \bigcup_n (m_n, m_{n+1}) \cap S = S,$$

as asserted.   □

Weakly Archimedean Property – An Admissibility Condition

Let $\mathbb{I}$ be $\mathbb{N}$ or $\mathbb{Q}$, with the ordering induced from the reals. Our purpose will be to take limits through subsets J of $\mathbb{I}$ which are *unbounded* on the right (more briefly: unbounded). According as $\mathbb{I}$ is $\mathbb{N}$ or $\mathbb{Q}$, we will write $n \to \infty$ or $q \to \infty$. Denote by X the line with either the metric or density topology and say that a family $\{h_i : i \in \mathbb{I}\}$ of self-homeomorphisms of the topological space X is *weakly Archimedean* if, for each non-empty open set V in X and any $j \in \mathbb{I}$, the open set

$$U_j(V) := \bigcup_{i \geq j} h_i(V)$$

meets every essentially unbounded set in X.

Theorem 5.1.2 (Implicit in BGT1987, Th. 1.9.1 (i))   *In the multiplicative group of positive reals $\mathbb{R}_+^*$ with the Euclidean topology, if d_n is divergent and the multiplicative form of condition (*) above holds, then the functions $h_n(x) = d_n x$ for $n = 1, 2, \dots$, are homeomorphisms and $\{h_n : n \in N\}$ is weakly Archimedean. For any interval $J = (a, b)$ with $0 < a < b$ and any m,*

$$U_m(J) := \bigcup_{n \geq m} d_n J$$

contains an infinite half-line, and so meets every unbounded open set. Similarly this is the case in the additive group of reals $\mathbb{R}$ with $h_n(x) = d_n + x$ and $U_m(J) = \bigcup_{n \geq m} (d_n + J)$.

Proof For given $\varepsilon > 0$ and all large enough n, we have $1 - \varepsilon < d_n / d_{n+1} < 1 + \varepsilon$. Write $x := (a + b)/2 \in J$. For ε small enough, $a < x(1 - \varepsilon) < x(1 + \varepsilon) < b$, and then $a < x d_n / d_{n+1} < b$, hence $x d_n \in d_{n+1} J$, and so $d_n J$ meets $d_{n+1} J$. Thus for large enough n consecutive $d_n J$ overlap; as $d_n \to \infty$, their union is thus a half-line. □

Remark Some such condition as (*) is necessary, otherwise the set $U_m(J)$ risks missing an unbounded sequence of open intervals. For an indirect example, see the remark in BGT1987 after Th. 1.9.2 and G.E.H. Reuter's elegant counterexample to a 'corollary' of Kingman's Theorem, a breakdown caused by the absence of our condition. For a direct example, note that if $d_n = r^n \log n$ with $r > 1$ and $J = (0, 1)$, then $d_n + J$ and $d_{n+1} + J$ miss the interval $(1 + r^n \log n, r^{n+1} \log(1 + n))$ and the omitted intervals have union an unbounded open set; to see that the omitted intervals are non-degenerate note that their lengths are unbounded:

$$r^{n+1} \log(1 + n) - r^n \log n - 1 \to \infty \qquad (n \to \infty).$$

Theorem 5.1.2 does not extend to the real line under the density topology; the homeomorphisms $h_n(x) = nx$ are no longer weakly Archimedean, as we demonstrate by an example in BinO2010f, Theorem 4.6. We are thus led to an alternative approach driven by the following lemma, a multiplicative version of an additive result in Sie1920.

Lemma 5.1.3 *For a, b density points of their respective measurable sets A, B in $\mathbb{R}_+$ and for $n = 1, 2, \ldots$, there exist positive rationals q_n and points a_n, b_n converging to a, b through A, B, respectively, such that $b_n = q_n a_n$.*

Proof For $n = 1, 2, \ldots$ and the consecutive values $\varepsilon = 1/n$ the sets $B_\varepsilon(a) \cap A$ and $B_\varepsilon(b) \cap B$ are measurable non-null, so by Steinhaus' Theorem (in multiplicative rather than additive form; see BGT1987, Th.1.1.1 or Chapter 15) the set $[B \cap B_\varepsilon(b)] \cdot [A \cap B_\varepsilon(a)]^{-1}$ contains interior points, and so in particular a rational point q_n. Thus for some $a_n \in B_\varepsilon(a) \cap A$ and $b_n \in B_\varepsilon(b) \cap B$ we have $q_n = b_n a_n^{-1}$, and as $|a - a_n| < 1/n$ and $|b - b_n| < 1/n$, $a_n \to a, b_n \to b$. □

Remarks 1. For the purposes of the next theorem, we observe that q_n may be selected arbitrarily large, for fixed a, by taking b sufficiently large (since $q_n \to ba^{-1}$).

2. The lemma addresses $\mathcal{D}$-open sets but also holds in the metric topology (the proof is similar but simpler), and so may be restated bitopologically as follows.

Theorem 5.1.4 *For $\mathbb{R}_+$ with either the Euclidean or the density topology, if a, b are respectively in the open sets A, B, then, for $n = 1, 2, \ldots$, there exist* *positive rationals q_n and points a_n, b_n converging metrically to a, b through A, B respectively, such that*

$$b_n = q_n a_n.$$

We now prove the density-topology analogue of Theorem 5.1.2.

Theorem 5.1.5 *In the multiplicative group of reals $\mathbb{R}_+^*$ with the density topology, the family of homeomorphisms $\{h_q : q \in \mathbb{Q}_+\}$ defined by $h_q(x) := qx$, where $\mathbb{Q}_+$ has its natural order, is weakly Archimedean. For any density-open set A and any $j \in \mathbb{Q}_+$,*

$$U_j(A) := \bigcup_{q \ge j, q \in \mathbb{Q}_+} qA$$

contains almost all of an infinite half-line, and so meets every unbounded density-open set.

Proof Let B be Baire and essentially unbounded in the $\mathcal{D}$-topology. Then B is measurable and essentially unbounded in the sense of measure. From Theorem 5.1.1, we may assume that B is density-open. Let A be non-empty density-open. Fix $a \in A$ and $j \in \mathbb{Q}_+$. Since B is unbounded, we may choose $b \in B$ such that $b > ja$. By Theorem 5.1.4 there is a $q \in \mathbb{Q}_+$ with $j < q < ba^{-1}$ such that $qa' = b'$, with $a' \in A$ and $b' \in B$. Thus

$$U_j(A) \cap B \supseteq h_q(A) \cap B = qA \cap B \ne \varnothing,$$

as required.

If $U_j(A)$ fails to contain almost all of any infinite half-line, then its complement $B := \mathbb{R}_+ \backslash U_j(A)$ is essentially unbounded in the sense of measure and so, as above, must meet $U_j(A)$, a contradiction.   □

Remarks 1. For A the set of irrationals in $(0, 1)$ the set $U_j(A)$ is again a set of irrationals which contains almost all, but not all, of an infinite half-line. Our result is thus best possible.

2. Note that (*) is relevant to the distinction between integer and rational skeletons; see the prime-divisor example on p. 53 of BGT1987. Theorem 5.1.5 holds with $\mathbb{Q}_+$ replaced by any countable *dense* subset of $\mathbb{R}_+^*$, although later we use the fact that $\mathbb{Q}_+$ is closed under multiplication. There is an affinity here with the use of a dense 'skeleton set' in the Heiberg–Seneta Theorem, Th. 1.4.3 of BGT1987, and its extension, Th. 3.2.5, therein.

Theorem 5.1.6 (Bitopological Kingman Theorem; Kin1963, Th. 1; Kin1964, where $\mathbb{I} = \mathbb{N}$) *If X is a Baire space, and*

(i) $\{h_i : i \in \mathbb{I}\}$ *is a countable, linearly ordered, weakly Archimedean family of self-homeomorphisms of X; and*

(ii) $\{S_k : k = 1, 2, \ldots\}$ *are essentially unbounded Baire sets, then, for quasi all* $\eta \in X$ *and all* $k \in \mathbb{N}$, *there exists an unbounded subset* $\mathbb{J}_\eta^k$ *of* $\mathbb{I}$ *with*

$$\{h_j(\eta) : j \in \mathbb{J}_\eta^k\} \subset S_k.$$

Equivalently, if (i) *and*

(ii′) $\{A_k : k = 1, 2, \ldots\}$ *are Baire and all accumulate essentially at 0, then for quasi all* η *and every* $k = 1, 2, \ldots$, *there exists* $\mathbb{J}_\eta^k$ *unbounded with*

$$\{h_j(\eta)^{-1} : j \in \mathbb{J}_\eta^k\} \subset A_k.$$

Proof We will apply Theorem 4.1.1 (Generic Dichotomy), so consider an arbitrary non-meagre Baire set T. We may assume without loss of generality that $T = V \backslash M$ with V non-empty open and M meagre. For each $k = 1, 2, \ldots$, choose G_k open and N_k and N'_k meagre such that $S_k \cup N_k = G_k \cup N'_k$. Put $N := M \cup \bigcup_{n,k} h_n^{-1}(N'_k)$; then N is meagre (as h_n, and so h_n^{-1}, is a homeomorphism).

As S_k is essentially unbounded, G_k is unbounded (otherwise, for some m, $G_k \subset (-m, m)$, and so $S_k \cap (m, \infty) \subset N_k$ is meagre). Define the open sets $G_{jk} := \bigcup_{i \geq j} h_i^{-1}(G_k)$. We first show that each G_{jk} is dense. Suppose, for some j, k, there is a non-empty open set V such that $V \cap G_{jk} = \varnothing$. Then for all $i \geq j$,

$$V \cap h_i^{-1}(G_k) = \varnothing; \qquad G_k \cap h_i(V) = \varnothing.$$

So $G_k \cap \bigcup_{i \geq j} h_i(V) = \varnothing$, i.e. for U^j the open set $U^j := \bigcup_{i \geq j} h_i(V)$, we have $G_k \cap U^j = \varnothing$. But as G_k is unbounded, this contradicts $\{h_i\}$ being a weakly Archimedean family.

Thus the open set G_{jk} is dense (meets every non-empty open set); so, as $\mathbb{I}$ is countable, the $\mathcal{G}_\delta$ set

$$H := \bigcap_{k=1}^{\infty} \bigcap_{j \in \mathbb{I}} G_{jk}$$

is dense (as X is a Baire space). So as V is a non-empty open subset we may choose $\eta \in (H \cap V) \backslash N$. (Otherwise $N \cup (X \backslash H)$ and hence V is meagre.) Thus $\eta \in T$ and for all $k = 1, 2, \ldots$,

$$\eta \in V \cap \bigcap_{j \in \mathbb{I}} \bigcup_{i \geq j} h_i^{-1}(G_k) \text{ and } \eta \notin N. \qquad (**)$$

For all m, as $h_m(\eta) \notin h_m(N)$ we have for all m, k, $h_m(\eta) \notin N'_k$. Using (**), for each k select an unbounded $\mathbb{J}_\eta^k$ such that for $j \in \mathbb{J}_\eta^k$, $\eta \in h_j^{-1}(G_k)$; for such j we have $\eta \in h_j^{-1}(S_k)$. That is, for some $\eta \in T$ we have

$$\{h_j(\eta) : j \in \mathbb{J}_\eta^k\} \subset S_k.$$

Now

$$F(T) := T \cap \bigcap_{k=1}^{\infty} \bigcap_{j \in \mathbb{I}} \bigcup_{i \geq j} h_i^{-1}(G_k)$$

takes Baire sets to Baire sets and is monotonic. Moreover, $\eta \in F(T)$ if and only if $\eta \in T$ and for each k there is an unbounded $\mathbb{J}_\eta^k$ with $\{h_j(\eta) : j \in \mathbb{J}_\eta^k\} \subset S_k$. We have just shown that $T \cap F(T) \neq \varnothing$ for T arbitrary non-meagre, so the Generic Dichotomy Principle, Theorem 4.1.1 implies that $X \cap F(X)$ is quasi almost all of X, i.e. for quasi all η in X and each k there is an unbounded $\mathbb{J}_\eta^k$ with $\{h_j(\eta) : j \in \mathbb{J}_\eta^k\} \subset S_k$. $\qquad\square$

Working in the density and the Euclidean topology in turn, we obtain the following conclusions.

Theorem 5.1.7 (Kingman Theorem for Category)   *If* $\{S_k : k = 1, 2, \ldots\}$ *are Baire and essentially unbounded in the category sense, then for quasi all η and each $k \in \mathbb{N}$ there exists an unbounded subset $\mathbb{J}_\eta^k$ of $\mathbb{N}$ with*

$$\{n\eta : n \in \mathbb{J}_\eta^k\} \subset S_k.$$

In particular this is so if the sets S_k are open.

Theorem 5.1.8 (Kingman Theorem for Measure)   *If* $\{S_k : k = 1, 2, \ldots\}$ *are measurable and essentially unbounded in the measure sense, then for almost all η and each $k \in \mathbb{N}$ there exists an unbounded subset $\mathbb{J}_\eta^k$ of $\mathbb{Q}_+$ with*

$$\{q\eta : q \in \mathbb{J}_\eta^k\} \subset S_k.$$

In the following corollary, $\mathbb{J}_t^k$ refers to unbounded subsets of $\mathbb{N}$ or $\mathbb{Q}_+$ according to the category/measure context. It specializes down to a KBD result for a single set T when $T_k \equiv T$, but it falls short of KBD in view of the extra admissibility assumption and the factor σ (the latter an artefact of the multiplicative setting).

Corollary 5.1.9   *For* $\{T_k : k \in \omega\}$ *Baire/measurable and $z_n \to 0$ admissible, for generically all $t \in \mathbb{R}$ there exist σ_t and unbounded $\mathbb{J}_t^k$ such that for $k = 1, 2, \ldots$,*

$$t \in T_k \implies \{t + \sigma_t z_m : m \in \mathbb{J}_t^k\} \subset T_k.$$

Proof For T Baire/measurable, let $N = N(T)$ be the set of points $t \in T$ which are not points of essential accumulation of T; then $t \in N$ if for some $n = n(t)$ the set $T \cap B_{1/n}(t)$ is meagre/null. As $\mathbb{R}$ with the Euclidean topology is (hereditarily) second countable, it is hereditarily Lindelöf (see Eng1989, Th. 3.8.1 or Dug1966, Th. 8.6.3), so for some countable $S \subset N$

$$N \subset \bigcup_{t \in S} T \cap B_{1/n(t)}(t),$$

and so N is meagre/null. Thus the set N_k of points $t \in T_k$ such that $T_k - t$ does not accumulate essentially at 0 is meagre/null, as is $N = \bigcup_k N_k$. For $t \notin N$, put $\Omega_t := \{k \in \omega : T_k - t$ accumulates essentially at 0$\}$. Applying Kingman's Theorem, 5.1.6, to the sets $\{T_k - t : k \in \Omega_t\}$ and the sequence $z_n \to 0$, there exist σ_t and unbounded $\mathbb{J}_t^k$ such that for $k \in \Omega_t$,

$$\{\sigma_t z_m : m \in \mathbb{J}_t^k\} \subset T_k - t, \quad \text{i.e.} \quad \{t + \sigma_t z_m : m \in \mathbb{J}_t^k\} \subset T_k.$$

Thus for $t \notin N$, so for generically all t, there exist σ_t and unbounded $\mathbb{J}_t^k$ such that, for $k = 1, 2, \ldots,$

$$t \in T_k \implies \{t + \sigma_t z_m : m \in \mathbb{J}_t^k\} \subset T_k. \qquad \square$$

6

Groups and Norms: Birkhoff–Kakutani Theorem

6.1 Introduction

Group-norms, which behave like the usual vector norms except that scaling is restricted to the basic scalars of group theory (the units ±1 in an abelian context and the exponents ±1 in the non-commutative context), have played a part in the early development of topological group theory, and created an early context for functional analysis as conducted by such American mathematicians as Michal (Mic1940; Mic1947) and Bartle (Bar1955). They appear naturally in the study of groups of homeomorphisms. See the long paper, BinO2010g, by the present authors for a detailed development and also ArhT2008. For some relatively recent work on group norms, see CabC2002, Sect. 2.1.1 which uses embedding of quasi-normed groups into Banach spaces in considering Ulam's problem (see Ula1960) on the global approximation of nearly additive functions by additive functions. See also BurBI2001.

The basic metrization theorem for groups, the Birkhoff–Kakutani Theorem of 1936 (Bir1936; Kak1936; see Kel1955; Kle1952; Bou1966; ArhM1998, compare also Eng1989), is usually stated as asserting that a first-countable Hausdorff topological group has a right-invariant metric. It is, properly speaking, a 'normability' theorem in the style of Kolmogorov's Theorem (Kol1934 or Rud1973; in this connection see also Jam1943, where strong forms of connectedness are used in an abelian setting to generate norms), as we shall see below (see also JamMW1947). Indeed the metric construction in Kak1936 is reminiscent of the more familiar construction of a Minkowski functional (for which see Rud1973), but is implicitly a supremum norm – as defined below; in Rudin's derivation of the metric (for a topological vector space setting, Rud1973) this norm is explicit.

We say that the group X is *normed* if it has a group-norm as defined below (cf. DezDD2006).

Definition We say that $\|.\|: X \to \mathbb{R}_+$ is a *group-norm* if the following properties hold:

(i) Subadditivity (triangle inequality): $\|xy\| \leq \|x\| + \|y\|$;
(ii) Positivity: $\|x\| > 0$ unless $x = e$;
(iii) Inversion (symmetry): $\|x^{-1}\| = \|x\|$.

If (i) holds we speak of a group *semi-norm*; if (i) and (iii) and $\|e\| = 0$ hold we speak of a *pseudo-norm* (cf. Pet1950); if (i) and (ii) hold we speak of a group *pre-norm* (see Low1997 for a full vocabulary).

We say that a group pre-norm, and so also a group-norm, is *abelian*, or more precisely *cyclically permutable*, if

(iv) Abelian norm (cyclic permutation): $\|xy\| = \|yx\|$ for all x, y.

Proposition 6.1.1 *If $\| \cdot \|$ is a group-norm on X, then*

$$d(x, y) = d_R(x, y) = d_R^X(x, y) := \|xy^{-1}\|$$

is a right-invariant metric; equivalently,

$$\tilde{d}(x, y)) = d_L(x, y) = d_L^X(x, y) := d(x^{-1}, y^{-1}) = \|x^{-1}y\|$$

is the conjugate left-invariant metric on the group.

Conversely, if $d(x, y)$ is a right-invariant metric, then

$$\|x\| := d(x, e) = \tilde{d}(x, e)$$

is a group-norm. Thus the metric $\| \cdot \|$ is bi-invariant if and only if $\|xy^{-1}\| = \|x^{-1}y\| = \|y^{-1}x\|$, i.e. if and only if the group-norm is abelian.

Furthermore, for $(X, \| \cdot \|)$ a normed group, the inversion mapping $x \mapsto x^{-1}$ from (X, d) to $(X, \tilde{d})$ is an isometry and hence a homeomorphism.

Proof Given a group-norm put $d(x, y) = \|xy^{-1}\|$. Then $\|xy^{-1}\| = 0$ if and only if $xy^{-1} = e$, i.e. if and only if $x = y$. Symmetry follows from inversion as $d(x, y) = \|(xy^{-1})^{-1}\| = \|yx^{-1}\| = d(y, x)$. Finally, d obeys the triangle inequality, since

$$\|xy^{-1}\| = \|xz^{-1}zy^{-1}\| \leq \|xz^{-1}\| + \|zy^{-1}\|.$$

As for the converse, given a right-invariant metric d, put $\|x\| := d(e, x)$. Now $\|x\| = d(e, x) = 0$ if and only if $x = e$. Next, $\|x^{-1}\| = d(e, x^{-1}) = d(x, e) = \|x\|$, and so

$$d(xy, e) = d(x, y^{-1}) \leq d(x, e) + d(e, y^{-1}) = \|x\| + \|y\|.$$

Also $d(xa, ya) = \|xaa^{-1}y^{-1}\| = d(x, y)$.

The metric d is bi-invariant if and only if $d(e, yx^{-1}) = d(x, y) = d(e, x^{-1}y)$ if and only if $\|yx^{-1}\| = \|x^{-1}y\|$. Inverting the first term yields the abelian property of the group-norm.

Finally, for $(X, \| \cdot \|)$ a normed group and with the notation $d(x, y) = \|xy^{-1}\|$ etc., the mapping $x \to x^{-1}$ from $(X, d_R^X) \to (X, d_L^X)$ is an isometry and so a homeomorphism, as $d_L(x^{-1}, y^{-1}) = d_R(x, y)$. $\qquad\square$

The following result clarifies the relationship between the conjugate metrics and the group structure. We define the *ε-swelling* of a set K in a metric space X for a given (e.g. right-invariant) metric d^X, to be

$$B_\varepsilon(K) := \{z : d^X(z, k) < \varepsilon \text{ for some } k \in K\} = \bigcup\nolimits_{k \in K} B_\varepsilon(k)$$

and for the conjugate (resp. left-invariant) case we can write similarly

$$\tilde{B}_\varepsilon(K) := \{z : \tilde{d}^X(z, k) < \varepsilon \text{ for some } k \in K\}.$$

We write $B_\varepsilon(x_0)$ for $B_\varepsilon(\{x_0\})$, so that

$$B_\varepsilon(x_0) := \{z : \|zx_0^{-1}\| < \varepsilon\} = \{wx_0 : w = zx_0^{-1}, \|w\| < \varepsilon\} = B_\varepsilon(e)x_0.$$

When $x_0 = e_X$, the ball $B_\varepsilon(e_X)$ is the same under either of the conjugate metrics, as

$$B_\varepsilon(e_X) := \{z : \|z\| < \varepsilon\}.$$

Proposition 6.1.2 (i) *In a locally compact group X, for K compact and for $\varepsilon > 0$ small enough so that the closed ε-ball $B_\varepsilon(e_X)$ is compact, the swelling $B_{\varepsilon/2}(K)$ is precompact.*

(ii) $B_\varepsilon(K) = \{wk : k \in K, \|w\|_X < \varepsilon\} = B_\varepsilon(e_X)K$, *where the notation refers to swellings for d^X a right-invariant metric; similarly for $\tilde{d}^X$, the conjugate metric,*

$$\tilde{B}_\varepsilon(K) = K B_\varepsilon(e_X).$$

Proof (i) If $x_n \in B_{\varepsilon/2}(K)$, then we may choose $k_n \in K$ with $d(k_n, x_n) < \varepsilon/2$. Without loss of generality k_n converges to k. Thus there exists N such that, for $n > N$, $d(k_n, k) < \varepsilon/2$. For such n, we have $d(x_n, k) < \varepsilon$. Thus the sequence x_n lies in the compact closed ε-ball centred at k and so has a convergent subsequence.

(ii) Let $d^X(x, y)$ be a right-invariant metric, so that $d^X(x, y) = \|xy^{-1}\|$. If $\|w\| < \varepsilon$, then $d^X(wk, k) = d^X(w, e) = \|w\| < \varepsilon$, so $wk \in B_\varepsilon(K)$. Conversely, if $\varepsilon > d^X(z, k) = d^X(zk^{-1}, e)$, then, putting $w = zk^{-1}$, we have $z = wk \in B_\varepsilon(K)$. $\qquad\square$

Theorem 6.1.3 (Invariance of Norm Theorem) (a) *The group-norm is abelian (and the metric is bi-invariant) if and only if*

$$\|xy(ab)^{-1}\| \le \|xa^{-1}\| + \|yb^{-1}\|$$

for all x, y, a, b, or equivalently,

$$\|uabv\| \le \|uv\| + \|ab\|$$

for all a, b, u, v.

(b) *Hence a metric d on the group X is bi-invariant if and only if the Klee property holds:*

$$d(ab, xy) \le d(a, x) + d(b, y). \tag{Klee}$$

In particular, this holds if the group X is itself abelian. (See Kle1952.)

(c) *The group-norm is abelian if and only if the norm is preserved under conjugacy (inner automorphisms).*

Proof (a) If the group-norm is abelian, then by the triangle inequality,

$$\|xyb^{-1} \cdot a^{-1}\| = \|a^{-1}xyb^{-1}\| \le \|a^{-1}x\| + \|yb^{-1}\|.$$

For the converse we demonstrate bi-invariance in the form $\|ba^{-1}\| = \|a^{-1}b\|$. In fact it suffices to show that $\|yx^{-1}\| \le \|x^{-1}y\|$; for then bi-invariance follows, since taking $x = a$, $y = b$, we get $\|ba^{-1}\| \le \|a^{-1}b\|$, whereas taking $x = b^{-1}$, $y = a^{-1}$ we get the reverse $\|a^{-1}b\| \le \|ba^{-1}\|$. As for the claim, we note that

$$\|yx^{-1}\| \le \|yx^{-1}yy^{-1}\| \le \|yy^{-1}\| + \|x^{-1}y\| = \|x^{-1}y\|.$$

(b) Klee's result is deduced as follows. If d is a bi-invariant metric, then $\|\cdot\|$ is abelian. Conversely, for d a metric, let $\|x\| := d(e, x)$. Then $\|.\|$ is a group-norm, as

$$d(ee, xy) \le d(e, x) + d(e, y).$$

Hence d is right-invariant and $d(u, v) = \|uv^{-1}\|$. Now we conclude that the group-norm is abelian since

$$\|xy(ab)^{-1}\| = d(xy, ab) \le d(x, a) + d(y, b) = \|xa^{-1}\| + \|yb^{-1}\|.$$

Hence d is also left-invariant.

(c) Suppose the norm is abelian. Then for any g, by the cyclic property $\|g^{-1}bg\| = \|gg^{-1}b\| = \|b\|$. Conversely, if the norm is preserved under automorphism, then we have bi-invariance, since $\|ba^{-1}\| = \|a^{-1}(ba^{-1})a\| = \|a^{-1}b\|$. □

Note that, taking $b = v = e$, we have the triangle inequality. Thus the result (a) characterizes maps $\|.\|$ with the positivity property as group pre-norms which are abelian. In regard to conjugacy, see also the Uniformity Theorem for Conjugation in BinO2010g, Th. 12.4. We now state the following classical result.

Theorem 6.1.4 (Normability Theorem for Groups – Birkhoff–Kakutani Theorem)   *Let X be a first-countable topological group and let V_n be a symmetric local base at e_X with*

$$V_{n+1}^4 \subseteq V_n.$$

Let $r = \sum_{n=1}^{\infty} c_n(r)2^{-n}$ be presented as a terminating representation of the dyadic number r, and put

$$A(r) = \prod_{c_n(r)=1} V_n.$$

Then

$$p(x) := \inf\{r : x \in A(r)\}$$

is a group-norm. If, further, X is locally compact and non-compact, then p may be arranged such that p is unbounded on X, but bounded on compact sets.

For a detailed proof see that offered in Rud1973, for Th. 1.24 (pp. 18–19), which derives a metrization of a topological vector space in the form $d(x, y) = p(x - y)$ and makes no use of commutativity or of the scalar field (so note how symmetric neighbourhoods here replace the 'balanced' ones in a topological vector space). That proof may be rewritten verbatim with xy^{-1} substituting for the additive notation $x - y$. We give a sketch account below in association with a reformulation.

Theorem 6.1.5 (Birkhoff–Kakutani Normability Theorem)   *A first-countable right topological group X is a normed group if and only if inversion and multiplication are continuous at the identity. In particular, a metrizable topological group is normed.*

Sketch Proof (Ost2013c)   A normed group under its right norm topology is a first-countable right topological group. It follows directly from the defining three axioms above that inversion and multiplication are continuous at the identity e. For the converse, consider any open neighbourhood U of e. Continuity of multiplication at e implies that there is an open neighbourhood V of e such that $V^4 \subseteq U$. Next, using continuity of inversion at e, one may choose an open neighbourhood $N \subseteq V$ of e such that $N^{-1} \subseteq V$. Then, as right shifts are homeomorphisms,

$$W := NN^{-1} = \{Ng^{-1} : g \in N\}$$

is an open neighbourhood of e with $W^{-1} = W \subseteq V^2$. So, since $W^2 \subseteq V^4 \subseteq U$, we conclude that for any open neighbourhood U of e one may choose an open neighbourhood W of e with $W^{-1} = W \subseteq W^2 \subseteq U$. One may now follow verbatim the argument in Kak1936. In broad outline: take a sequence of neighbourhoods $U_1, U_{1/2}, U_{1/4}, \ldots$ of e that is a basis at e with

$$U_{1/2^n}^{-1} = U_{1/2^n} \subseteq U_{1/2^n}^2 \subseteq U_{1/2^{n-1}}, \ldots$$

for $n \in \mathbb{N}$. Then construct a (Urysohn) function $f \colon G \to \mathbb{R}_+$ such that $f(x) \leq 2^{-n}$ if and only if $x \in U_{1/2^n}$ and define a metric by

$$d(g, h) := \sup_x |f(gx) - f(hx)|.$$

This is a right-invariant metric on G compatible with the topology, as shown by Kakutani (Kak1936). □

A group-norm defines two metrics: the right-invariant metric, which we denote, as in Proposition 6.1.1, by $d_R(x, y) := \|xy^{-1}\|$, and the conjugate left-invariant metric, here to be denoted $d_L(x, y) := d_R(x^{-1}, y^{-1}) = \|x^{-1}y\|$. There is correspondingly a right and a left metric topology, which we term the *right* or *left norm-topology*. We write $\to_R$ for convergence under d_R, etc. Both metrics give rise to the same norm, since $d_L(x, e) = d_R(x^{-1}, e) = d_R(e, x) = \|x\|$, and hence define the same balls centred at the origin e:

$$B_R^d(e, r) := \{x : d(e, x) < r\} = B_L^d(e, r).$$

Denoting this commonly determined set by $B(r)$, we have seen in Proposition 6.1.2 that

$$B_R(a, r) = \{x : x = ya \quad \text{and} \quad d_R(a, x) = d_R(e, y) < r\} = B(r)a,$$
$$B_L(a, r) = \{x : x = ay \quad \text{and} \quad d_L(a, x) = d_L(e, y) < r\} = aB(r).$$

Thus the open balls are right- or left-shifts of the norm balls at the origin. This is best viewed in the current context as saying that under d_R the right-shift $\rho_a \colon x \to xa$ is right uniformly continuous, since

$$d_R(xa, ya) = d_R(x, y),$$

and likewise that under d_L the left-shift $\lambda_a \colon x \to ax$ is left uniformly continuous, since

$$d_L(ax, ay) = d_L(x, y).$$

In particular, under d_R we have $y \to_R b$ if and only if $yb^{-1} \to_R e$, as $d_R(e, yb^{-1}) = d_R(y, b)$. Likewise, under d_L, we have $x \to_L a$ if and only if $a^{-1}x \to_L e$, as $d_L(e, a^{-1}x) = d_L(x, a)$.

Thus either topology is determined by the neighbourhoods of the identity (origin) and according to choice makes the appropriately sided shift continuous; said another way, the topology is determined by the neighbourhoods of the identity and the chosen shifts. We noted earlier that the triangle inequality implies that multiplication is jointly continuous at the identity e, as a mapping from (X, d_R) to (X, d_R). Likewise inversion is also continuous at the identity by the symmetry axiom. (See Theorem 6.1.5.) To obtain similar results other than at the identity one needs to have continuous conjugation, and this is linked to the equivalence of the two norm topologies (see the Equivalence Theorem, 6.1.9). The conjugacy map under $g \in G$ (inner automorphism) is defined by

$$\gamma_g(x) := gxg^{-1}.$$

Evidently the inverse of γ_g is given by conjugation under g^{-1} and that γ_g is a *homomorphism*. Its continuity, as a mapping from (X, d_R) to (X, d_R), is thus determined by behaviour at the identity, as we verify below. We work with the right topology (under d_R), and may sometimes leave unsaid equivalent assertions about the isometric case of (X, d_L) replacing (X, d_R).

Lemma 6.1.6 *The homomorphism γ_g is right-to-right continuous at any point if and only if it is right-to-right continuous at e.*

Proof This is immediate since $x \to_R a$ if and only if $xa^{-1} \to_R e$, and $\gamma_g(x) \to_R \gamma_g(a)$ if and only if $\gamma_g(xa^{-1}) \to_R \gamma_g(e)$ since

$$\|gxg^{-1}(gag^{-1})^{-1}\| = \|gxa^{-1}g^{-1}\|. \qquad \square$$

Working under d_R, we will relate inversion to left-shifts. We begin with the following.

Lemma 6.1.7 *If inversion is right-to-right continuous, then $x \to_R a$ if and only if $a^{-1}x \to_R e$.*

Proof For $x \to_R a$, we have $d_R(e, a^{-1}x) = d_R(x^{-1}, a^{-1}) \to 0$, assuming continuity. Conversely, for $a^{-1}x \to_R e$ we have $d_R(a^{-1}x, e) \to 0$, i.e. $d_R(x^{-1}, a^{-1}) \to 0$. So since inversion is assumed to be right-continuous and $(x^{-1})^{-1} = x$, etc., we have $d_R(x, a) \to 0$. $\qquad \square$

We now expand this.

Theorem 6.1.8 *The following are equivalent:*

(i) *inversion is right-to-right continuous;*
(ii) *left-open sets are right-open;*
(iii) *for each g, the conjugacy γ_g is right-to-right continuous at e; i.e. for every $\varepsilon > 0$ there is $\delta > 0$ such that*

$$gB(\delta)g^{-1} \subset B(\varepsilon);$$

(iv) *left-shifts are right-continuous.*

Proof We show that (i)$\Longleftrightarrow$(ii)$\Longleftrightarrow$(iii)$\Longleftrightarrow$(iv).

Assume (i). For any a and any $\varepsilon > 0$, by continuity of inversion at a, there is $\delta > 0$ such that, for x with $d_R(x, a) < \delta$, we have $d_R(x^{-1}, a^{-1}) < \varepsilon$, i.e. $d_L(x, a) < \varepsilon$. Thus

$$B(\delta)a = B_R(a, \delta) \subset B_L(a, \varepsilon) = aB(\varepsilon), \tag{incl}$$

i.e. left-open sets are right-open, giving (ii). For the converse, we just reverse the last argument. Let $\varepsilon > 0$. As $a \in B_L(a, \varepsilon)$ and $B_L(a, \varepsilon)$ is left open, it is right open and so there is $\delta > 0$ such that

$$B_R(a, \delta) \subset B_L(a, \varepsilon).$$

Thus for x with $d_R(x, a) < \delta$, we have $d_L(x, a) < \varepsilon$, i.e. $d_R(x^{-1}, a^{-1}) < \varepsilon$, i.e. inversion is right-to-right continuous, giving (i).

To show that (ii)$\Longleftrightarrow$(iii) note that the inclusion (incl) is equivalent to

$$a^{-1}B(\delta)a \subset B(\varepsilon),$$

i.e. to

$$\gamma_a^{-1}[B(\delta)] \subset B(\varepsilon);$$

that is, to the assertion that $\gamma_{a^{-1}}(x)$ is continuous at $x = e$ (and so continuous, by Lemma 6.1.6). The property (iv) is equivalent to (iii) since the right-shift is right-continuous and $\gamma_a(x)a = \lambda_a(x)$ is equivalent to $\gamma_a(x) = \lambda_a(x)a^{-1}$. $\square$

We saw in the Birkhoff–Kakutani Theorem that metrizable topological groups are normable (equivalently, have a right-invariant metric); we now formulate a converse, showing when the right-invariant metric derived from a group-norm equips its group with a topological group structure. As this is a characterization of metric topological groups, we will henceforth refer to them synonymously as *normed topological groups*. Below the condition (adm) refers to *admissible* topologies.

Theorem 6.1.9 (Equivalence Theorem) *A normed group is a topological group under the right (resp. left) norm-topology if and only if each conjugacy*

$$\gamma_g(x) := gxg^{-1}$$

is right-to-right (resp. left-to-left) continuous at $x = e$ (and so everywhere); i.e. for $z_n \to_R e$ and any g

$$gz_ng^{-1} \to_R e. \tag{adm}$$

Equivalently, it is a topological group if and only if left/right-shifts are continuous for the right/left norm-topology, or if and only if the two norm topologies are themselves equivalent.

In particular, if also the group structure is abelian, then the normed group is a topological group.

Proof Only one direction needs proving. We work with the d_R topology, the right topology. By Theorem 6.1.8 we need only show that under it multiplication is jointly right-continuous. First we note that multiplication is right-continuous if and only if

$$d_R(xy, ab) = \|xyb^{-1}a^{-1}\|, \quad \text{as} \quad (x, y) \to_R (a, b).$$

Here, we may write $Y = yb^{-1}$ so that $Y \to_R e$ if and only if $y \to_R b$, and we obtain the equivalent condition

$$d_R(xYb, ab) = d_R(xY, a) = \|xYa^{-1}\|, \quad \text{as} \quad (x, Y) \to_R (a, e).$$

Again by Theorem 6.1.8, as inversion is right-to-right continuous, the preceding Lemma 6.1.6 justifies re-writing the second convergence condition with $X = a^{-1}x$ and $X \to_R e$, yielding the equivalent condition

$$d_R(aXYb, ab) = d_R(aXY, a) = \|aXYa^{-1}\|, \text{ as } (X, Y) \to_R (e, e).$$

But, by Lemma 6.1.6 again, this is equivalent to continuity of conjugacy.   □

The final assertion is related to a result of Żelazko (Zel1960) (cf. Com1984). We will later apply the Equivalence Theorem several times in conjunction with the following result.

Lemma 6.1.10 (Weak Continuity Criterion)   *For fixed x, if, for all null sequences w_n, we have $\gamma_x(w_{n(k)}) \to e_X$ down some subsequence $w_{n(k)}$, then γ_x is continuous.*

Proof We are to show that for every $\varepsilon > 0$ there is $\delta > 0$ and N such that, for all $n > N$,

$$xB(\delta)x^{-1} \subset B(\varepsilon).$$

Suppose not. Then there is $\varepsilon > 0$ such that, for each $k = 1, 2, \ldots$ and each $\delta = 1/k$, there is $n = n(k) > k$ and w_k with $\|w_k\| < 1/k$ and $\|xw_kx^{-1}\| > \varepsilon$. So $w_k \to 0$. By assumption, down some subsequence $n(k)$ we have $\|xw_{n(k)}x^{-1}\| \to 0$; but this contradicts $\|xw_{n(k)}x^{-1}\| > \varepsilon$.   □

6.2 Analytic Shift Theorem

The main result of the section is a shift-compactness theorem which, thanks to its restriction to analytic target sets, is applicable in normed groups that need not be abelian. We will need some preliminary results. The first is a simple and immediate corollary of the Analytic Cantor Theorem (Theorem 3.1.6) and refers to the fact that $B_r(x) = B_r(e_X)x$.

A Convergence Criterion *In a normed group, for $r_n \searrow 0$ and $\alpha_n = a_n \cdots a_1$ with* cl $B_{r_{n+1}}(a_{n+1}) \subseteq B_{r_n}(e)a_n$, *if $X = K(I)$ is an analytic subset and $K(i_1, \ldots, i_n) \cap B_{r_n}(\alpha_n) \neq \emptyset$ for some $i \in I$ and all n, then the sequence $\{\alpha_n\}$ is convergent.*

Proof Indeed, $\alpha_n \to \alpha$, where $\{\alpha\} = K(i) \cap \bigcap_n F_n$ for $F_n = \text{cl}(B_{r_n}(\alpha_n))$. $\square$

Theorem 6.2.1 (Analytic Baire Theorem) *In a normed group X under d_R^X, if X contains a non-meagre analytic set, then X is Baire and in fact, up to a meagre set, X is analytic (and separable).*

Proof Let A be non-meagre and analytic with upper semi-continuous representation $A = K(I)$. Let $\{F_n^0 : n \in \omega\}$ be closed nowhere-dense subsets of X and G an arbitrary non-empty open set. Say $g \in G$. As A is analytic, by Nikodym's Theorem, A has the Baire property, so $A = U \setminus N \cup M$ for some open U and meagre M, N. As A is non-meagre, U is non-meagre and so non-empty. Say $u \in U \setminus N$. Since each mapping $\rho_t(x) = xt$ is a homeomorphism, $H := Gg^{-1}u$ is open and meets U in $u \in A$. As G is arbitrary, it suffices to show that H meets $X \setminus \bigcup_{m \in \omega} F_m^0$ in a point of A. (Here we are using the fact that a normed group which is locally Baire at some point is Baire – for a more general result along these lines, see TopH1980, Prop. 2.2.3.)

We choose inductively integers i_n, points x_n, radii $r_n > 0$ with $r_{n+1} < r_n/2$ and nowhere dense closed sets $\{F_m^n : m \in \omega\}$ such that $K(i_1, \ldots, i_n) \cap B(x_n, r_n)$ is non-meagre with

$$B(x_n, r_n) \subseteq B(x_{n-1}, r_{n-1}) \subseteq H,$$

and

$$K(i_1, \ldots, i_n) \supseteq B(x_n, r_n) \setminus \bigcup_{m \in \omega} F_m^n \quad \text{and}$$
$$B(x_n, r_n) \cap \bigcup_{k, m < n} F_m^k = \emptyset.$$

Begin by taking $x_0 = e_X$ and selecting r_0 arbitrarily so that $B(x_0, r_0) \cap K(I)$ is non-meagre. To verify the inductive step, note that

$$K(i_1, \ldots, i_n) \cap B(x_n, r_n) = \bigcup_{m \in \omega} K(i_1, \ldots, i_n, m) \cap B(x_n, r_n),$$

so there is i_{n+1} such that $K(i_1, \ldots, i_n, i_{n+1}) \cap B(x_n, r_n)$ is non-meagre. So, (again by Nikodym's Theorem) for some non-meagre open V, closed nowhere dense sets F_m^{n+1} and meagre N_{n+1}, we have

$$K(i_1, \ldots, i_n, i_{n+1}) \cap B(x_n, r_n) \backslash N_{n+1} = V \backslash \bigcup_{m \in \omega} F_m^{n+1}.$$

By analyticity, V is hereditarily separable, so we may choose x_{n+1}, r_{n+1} (with $r_{n+1} < r_n/2$) such that $x_{n+1} \in \mathrm{cl}(B(x_{n+1}, r_{n+1})) \subseteq V \cap B(x_n, r_n) \backslash \bigcup_{m < n+1} F_m^k$ and

$$K(i_1, \ldots, i_{n+1}) \cap B(x_{n+1}, r_{n+1})$$

is non-meagre. With the induction verified, by the Convergence Criterion above (or directly from the Analytic Cantor Theorem, 3.1.6), there is a such that $\{a\} = K(i) \cap \bigcap_{n \in \omega} B(x_n, r_n) \subseteq H$. So $a \in H$ (and $a = \lim_n x_n$). Furthermore, for any n, we have $B(x_{n+1}, r_{n+1}) \cap F_n^0 = \emptyset$, so $a \notin \bigcup_{m \in \omega} F_m^0$.

As for the second claim, recall that $A = U \backslash N \cup M$ with U non-empty and N, M meagre with $N \subseteq U$ and $U \cap M = \emptyset$. Suppose that $B := B_\varepsilon(a) \subseteq U$; then $B \cap A = B \backslash N$ and so $B \backslash N$ is analytic. Now $\{Bx : x \in\}$ is an open cover of X, so has a σ-discrete refinement, say $\mathcal{V} := \bigcup_{n \in \omega} \mathcal{V}_n$ with $\mathcal{V}_n := \{V_{tn} : t \in T_n\}$ discrete. Suppose that $V_{tn} \subseteq Bx_{tn}$. Put $N_{tn} := (N \cap B)x_{tn}$, which is meagre (since right-shifts are homeomorphisms); then $V_{tn} \backslash N_{tn} \subseteq (B \backslash N)x_{tn} \subseteq Ax_{tn}$. Without loss of generality N_{tn} is a meagre $\mathcal{F}_\sigma$ subset of V_{tn} and so $V_{tn} \backslash N_{tn} = (V_{tn} \cap (Ax_{tn})) \backslash N_{tn}$ is analytic and non-meagre. By Banach's Category Theorem, 2.2.1, $N' := \bigcup_{n \in \omega} \bigcup \{N_{tn} : t \in T_n\}$ is meagre, since $\mathcal{V}_n$ is discrete. By a theorem of Montgomery (Mon1935), for each n the set $\bigcup \{V_{tn} \backslash N_{tn} : t \in T_n\}$ is analytic, being locally analytic on $\bigcup \{V_{tn} : t \in T_n\}$, and is non-meagre. As $\mathcal{V}$ covers X, we have

$$X \backslash N' = \bigcup_{n \in \omega} \bigcup \{V_{tn} \backslash N_{tn} : t \in T_n\},$$

and so $X \backslash N'$ is analytic, being a countable union of analytic sets, and non-meagre.   □

Lemma 6.2.2 (Displacements Lemma – Baire Case)   *Working under* $d_{\mathbb{R}}^X$, *for X a Baire space, A Baire non-meagre and for almost all $a \in A$, there is* $\varepsilon = \varepsilon_A(a) > 0$ *such that* $A \cap Aa^{-1}xa$ *is non-meagre for all x with* $\|x\| < \varepsilon_A(a)$.

Indeed, for any x with $\|x\| < \varepsilon_A(a)$, *modulo meagre sets,* $Aa^{-1} \cap Aa^{-1}x \supseteq B(x, s(x))$ *for some $s(x) > 0$, and so $A \cap Aa^{-1}xa \supseteq B(xa, s(x))$.*

Proof   Omitting a meagre subset of A we may assume without loss of generality that $e \in Aa^{-1} = U \backslash N$ for some U open and N meagre. So, as Aa^{-1} is non-meagre, there is $\varepsilon > 0$ such that $B(e, \varepsilon) \subseteq Aa^{-1}$ modulo meagre sets. For any x with $\|x\| < \varepsilon$ and $s := \min\{\varepsilon, \frac{1}{2}(\varepsilon - \|x\|)\} > 0$, we have $B(e,\varepsilon) \supseteq B(x,s) = B(e,s)x$. Indeed, if $d_R(y,x) < s$, then $d(y,e) < d(y,x) + d(x,e) < s + \varepsilon = \frac{1}{2}(\varepsilon - \|x\|) + \|x\| < \varepsilon$. Modulo meagre sets: as $Aa^{-1}x \supseteq B(e,s)x$, we have

$$Aa^{-1} \supseteq Aa^{-1} \cap Aa^{-1}x \supseteq B(e,\varepsilon) \cap B(e,s)x \supseteq B(x,s),$$

which is non-meagre as the space X is Baire. See also Corollary 4.2.3 $\qquad \square$

Theorem 6.2.3 (Analytic Shift Theorem) *In a normed group under the topology d_R^X, for $z_n \to e_X$ null, A a $\mathcal{K}$-analytic and non-meagre subset: for a non-meagre set of $a \in A$ with co-meagre Baire envelope, there is an infinite set $\mathbb{M}_a$ and points $a_n \in A$ converging to a such that*

$$\{aa_m^{-1}z_m a_m : m \in \mathbb{M}_a\} \subseteq A.$$

In particular, if the normed group is topological, for quasi all $a \in A$, there is an infinite set $\mathbb{M}_a$ such that $\{az_m : m \in \mathbb{M}_a\} \subseteq A$.

Remark Notice that in the context above $aa_m^{-1}z_m a_m$ converges to a; indeed,

$$d_R^X(aa_m^{-1}z_m a_m, a) = \|aa_m^{-1}z_m a_m a^{-1}\| \le \|aa_m^{-1}\| + \|z_m\| + \|a_m a^{-1}\|.$$

Proof By Theorem 6.2.1 (Analytic Baire Theorem), we may assume that all non-empty open sets are Baire and so non-meagre. Next, note that any non-meagre Baire set is equal modulo meagre sets to a non-meagre $\mathcal{G}_\delta$. So if the Baire envelope of the set of $t \in A$ with the asserted property has non-meagre complement, we may assume that this complement is analytic. It thus suffices to prove that a non-meagre analytic set A contains at least one point t for which there exists an infinite set $\mathbb{M}_t$ and points $a_n \in A$ converging to t such that $\{ta_m^{-1}z_m a_m : m \in \mathbb{M}_t\} \subseteq A$.

We proceed to prove this. So write $A = K(I)$ with K upper semi-continuous and single-valued. For each $n \in \omega$ we find inductively integers i_n, points x_n, y_n, a_n with $a_n \in A$, numbers $r_n > 0$, $s_n > 0$, analytic subsets A_n of A, and closed nowhere dense sets $\{F_m^n : m \in \omega\}$ such that

$$K(i_1,\ldots,i_n) \supseteq A_n \cap (A_n a_n^{-1} x_n a_n) \supseteq B(y_n, s_n) \backslash \bigcup_{m \in \omega} F_m^n$$

and

$$y_n \in B(x_n a_n, r_n) \quad \text{and} \quad B(y_n, s_n) \cap \bigcup_{m,k<n} F_m^k = \emptyset.$$

Assuming this done for n, since $K(i_1,\ldots,i_n) = \bigcup_k K(i_1,\ldots,i_n,k)$ is non-meagre, there is i_{n+1} such that $K(i_1,\ldots,i_n,i_{n+1}) \cap A_n \cap (A_n a_n^{-1} x_n a_n)$ is non-meagre. Put

$$A_{n+1} := K(i_1,\ldots,i_n,i_{n+1}) \cap A_n \cap (A_n a_n^{-1} x_n a_n) \subseteq K(i_1,\ldots,i_{n+1}).$$

As A_n is non-meagre, we may pick $a_{n+1} \in A_{n+1}$ as in Lemma 6.2.2 (Displacements Lemma), and $m(n)$ so large that $\|z_m\| < \varepsilon(a_{n+1}, A_{n+1})$ for $m \geq m(n)$. Pick $x_{n+1} = z_{m(n)}^{-1}$. Then, since $A_{n+1} \subseteq K(i_1, \ldots, i_{n+1})$ and in view of Lemma 6.2.2, there is r_{n+1} and closed nowhere dense sets $\{F_m^{n+1} : m \in \omega\}$ such that

$$K(i_1, \ldots, i_{n+1}) \supseteq A_{n+1} \cap A_{n+1} a_{n+1}^{-1} x_{n+1} a_{n+1}$$

$$\supseteq B(x_{n+1} a_{n+1}, r_{n+1}) \setminus \bigcup\nolimits_{m \in \omega} F_m^{n+1}.$$

Since the set $\bigcup_{m, k < n+1} F_m^k$ is closed and nowhere dense, there is $y_{n+1} \in B(x_{n+1} a_{n+1}, r_{n+1})$ and $s_{n+1} > 0$ so small that $B(y_{n+1}, s_{n+1}) \subseteq B(x_{n+1} a_{n+1}, r_{n+1})$ and $B(y_{n+1}, s_{n+1}) \cap \bigcup_{m, k < n+1} F_m^k = \emptyset$. Hence

$$B(x_{n+1} a_{n+1}, r_{n+1}) \setminus \bigcup\nolimits_{m \in \omega} F_m^{n+1} \supseteq B(y_{n+1}, s_{n+1}) \setminus \bigcup\nolimits_{m \in \omega} F_m^{n+1}.$$

By Theorem 3.1.6 (Analytic Cantor Theorem), there is t with

$$\{t\} = K(i) \cap \bigcap\nolimits_n B(y_n, s_n) \subseteq \bigcap\nolimits_n A_n \cap (A_n a_n^{-1} x_n a_n).$$

So $t \in A$. Fix n. One has $t \notin \bigcup_{m \in \omega} F_m^n$ (since $B(y_{m+1}, s_{m+1}) \cap \bigcup_{k < m} F_k^n = \emptyset$ for each m), and so

$$t \in B(y_n, s_n) \setminus \bigcup\nolimits_{m \in \omega} F_m^n \subseteq A_n \cap (A_n a_n^{-1} x_n a_n) \subseteq K(i_1, \ldots, i_n) \subseteq A.$$

As $t \in A_n a_n^{-1} x_n a_n$, one has $t a_n^{-1} x_n^{-1} a_n = t a_n^{-1} z_{m(n)} a_n \in A_n \subseteq A$. So $\{t a_n^{-1} z_{m(n)} a_n : n \in \omega\} \subseteq A$. Moreover, $d_R(x_n a_n, t) = d_R(x_n, t a_n^{-1}) \to 0$, so since $x_n \to e$, we have $t a_n^{-1} \to e$, i.e. $a_n \to_R t$.   □

6.3 The 'Squared Pettis Theorem'

Theorem 6.3.1   *In a normed group X, for $T \subseteq X$ almost complete, U open with $T \cap U$ non-meagre, and $z_n \to e_X$, the set S_U of $t \in T \cap U$ for which there exist points $t_m \in T$ with $t_m \to_R t$ and an infinite $\mathbb{M}_t$ with*

$$\{t t_m^{-1} z_m t_m : m \in \mathbb{M}_t\} \subseteq T,$$

is non-meagre.

Proof   Suppose not; then there is an open set U such that S_U is meagre. Letting H be a meagre $\mathcal{F}_\sigma$ cover of S_U, the set $T' := (T \backslash H) \cap U$ is Baire and non-meagre. But then by the Analytic Shift Theorem (Theorem 6.2.3) there exists points $t, t_m \in T'$ and infinite set $\mathbb{M}_t$ such that

$$\{t t_m^{-1} z_m t_m : m \in \mathbb{M}_t\} \subseteq T' \subseteq T \cap U,$$

a contradiction.   □

Our next theorem is named after a simpler Baire category result (without the square), due to Pettis (Pet1950), true in simpler circumstances. See Chapter 15, which is devoted to interior-point theorems.

Theorem 6.3.2 (Squared Pettis Theorem) *Let X be a $\mathcal{K}$-analytic (e.g. topologically complete) normed group and A in X Baire non-meagre under the right norm-topology. Then e_X is an interior point of $(AA^{-1})^2$.*

Proof Suppose not. We may assume that A is analytic. Otherwise, since A is Baire, write $A = (U \backslash M) \cup N$ with M an $\mathcal{F}_\sigma$, and M, N both meagre. Then $U \backslash M$ is analytic and non-meagre. Now we may select $z_n \in B_{1/n}(e) \backslash (AA^{-1})^2$. As $z_n \to e$, we apply Theorem 6.2.3 (Analytic Shift Theorem) to A, to find $t \in A$, $\mathbb{M}_t$ infinite and $t_m \in A$ for $m \in \mathbb{M}_t$ such that $tt_m^{-1} z_m t_m \in A$ for all $m \in \mathbb{M}_t$. So, for $m \in \mathbb{M}_t$,

$$z_m \in AA^{-1}AA^{-1} = (AA^{-1})^2,$$

a contradiction. □

6.4 Shifted-Covering Compactness

The following two theorems assert that a 'covering property modulo shift' is satisfied by bounded (*right*) shift-compact sets (BinO2010g). It will be convenient to make the following definitions.

Definitions 1. Say that $\mathcal{D} := \{D_1, \ldots, D_h\}$ *shift-covers* a subset X of G or is a *shifted-cover* of X if, for some $d_1, \ldots, d_h$ in G

$$X = D_1 d_1 \cup \ldots \cup D_h d_h.$$

Say that X is *compactly shift-covered* if every open cover $\mathcal{U}$ of X contains a finite subfamily $\mathcal{D}$ which shift-covers X.

2. For N a neighbourhood of e_G say that $\mathcal{D} := \{D_1, \ldots, D_h\}$ *N-strongly shift-covers* a subset A of G or is an *N-strong shifted-cover* of A if, for some $d_1, \ldots, d_h$ in N

$$A \subseteq (D_1 - d_1) \cup \cdots \cup (D_h - d_h).$$

Say that A is *compactly strongly shift-covered*, or *compactly shift-covered with arbitrarily small shifts*, if every open cover $\mathcal{U}$ of A contains for each neighbourhood N of e_G a finite subfamily $\mathcal{D}$ which N-strongly shift-covers A (BinO2011a).

Theorem 6.4.1   *Let A be a shift-compact subset of a separable normed topological group G. Then A is compactly shift-covered, i.e. for any norm-open cover $\mathcal{U}$ of A, there is a finite subset $\mathcal{V}$ of $\mathcal{U}$, and for each member of $\mathcal{V}$ a translator, such that the corresponding translates of $\mathcal{V}$ cover A.*

Proof   Let $\mathcal{U}$ be an open cover of A. Since G is second-countable we may assume that $\mathcal{U}$ is a countable family. Write $\mathcal{U} = \{U_i : i \in \omega\}$. Let $Q = \{q_j : j \in \omega\}$ enumerate a dense subset of G. Suppose, contrary to the assertion, that there is no finite subset $\mathcal{V}$ of $\mathcal{U}$ such that elements of $\mathcal{V}$, translated each by a corresponding member of Q, cover A. For each n, choose $a_n \in A$ not covered by $\{U_i q_i : i, j < n\}$. As A is precompact, so we may assume, by passing to a subsequence (if necessary), that a_n converges to some point a_0, and also that, for some t, the sequence $a_n t$ lies entirely in A. Let U_i in $\mathcal{U}$ cover $a_0 t$. Without loss of generality we may assume that $a_n t \in U_i$ for all n. So $a_n \in U_i t^{-1}$ for all n. Thus we may select $V := U_i q_j$ to be a translation of U_i such that $a_n \in V = U_i q_j$ for all n. But this is a contradiction, since a_n is not covered by $\{U_{i'} q_{j'} : i', j' < n\}$ for $n > \max\{i, j\}$.   □

The above proof may be improved to strong shift-covering, with only a minor modification (replacing Q with a set $Q^\varepsilon = \{q_j^\varepsilon : j \in \omega\}$ which enumerates, for given $\varepsilon > 0$, a dense subset of the ε ball about e), yielding the following.

Theorem 6.4.2   *Let A be a strongly shift-compact subset of a separable normed topological group G. Then A is compactly strongly shift-covered, i.e. for any norm-open cover $\mathcal{U}$ of A, and any neighbourhood of e_G, there is a finite subset $\mathcal{V}$ of $\mathcal{U}$, and, for each member of $\mathcal{V}$, a translator in N such that the corresponding translates of $\mathcal{V}$ cover A.*

7

Density Topology

The first result below is the Lebesgue Density Theorem, 7.1.1. Closely related to this is the *density topology*, $\mathcal{D}$, the subject of this chapter. As we shall see, $\mathcal{D}$ and the Euclidean topology $\mathcal{E}$ enable us to work *bitopologically*, with $\mathcal{D}$ playing for the measure case the role played by $\mathcal{E}$ for the category case. This enables us to align as closely as possible qualitative measure properties with their Baire analogues. This is the theme of Chapter 9, on *category–measure duality*.

7.1 The Lebesgue Density Theorem

The density topology has a number of properties that will be crucially useful for us. However, it lacks some properties commonly regarded as desirable. From our point of view, the density topology is a wonderful topology (we would even say a beautiful one). We quote in this connection from page 1 of LukMZ1986: 'It was proved that the "nice" system of all approximately continuous functions is exactly the class of all continuous functions in the "bad" density topology.' We, like the authors of LukMZ1986, wish to rehabilitate the density topology.

Let $\mathcal{B}$ denote a countable basis of open neighbourhoods. For any set T put

$$\mathcal{B}_\alpha(T) := \{I \in \mathcal{B} : |I \cap T|^* > \alpha |I|\}$$

(with $|.|^*$ for outer Lebesgue measure), which is countable, and

$$F(T) := \bigcap_{\alpha \in \mathbb{Q} \cap (0,1)} \bigcup \{I : I \in \mathcal{B}_\alpha(T)\}.$$

Thus F is increasing in T, $F(T)$ is measurable (even if T is not), and $x \in F(T)$ if and only if x is a density point of T.

The following is one form of the *Lebesgue Density Theorem*.

Theorem 7.1.1 (Density Theorem)   *If K is compact and non-null, then K has a density point. Hence almost all points of a measurable set are density points.*

Before beginning the proof, we give a definition, a lemma and a corollary. Working in $\mathbb{R}$, for given K compact, put

$$\alpha(x) = \alpha_K(x) := \limsup_{x \in I \in \mathcal{B}, |I| \to 0} |I \cap K| / |I| \qquad (x \in K),$$

where K, if omitted, is to be understood from context.

Lemma 7.1.2   *For $I \in \mathcal{B}$, if $\alpha(.)$ is bounded away from 1 on $I \cap K$, then $|I \cap K| = 0$.*

Proof   Suppose not, and that $\alpha(x) \le A < 1$ on $I \cap K$. Take $k > |I \cap K|$. By the outer regularity of $|.|$, there is a sequence $I_n \in \mathcal{B}$ with $I_n \subset I$ such that $I \cap K \subset \bigcup_n I_n$ and $\sum_n |I_n| < k$. We may assume that each $I_n \cap K \ne \varnothing$ (discarding any others). Thus if $x \in I_n \cap K$, we have $\alpha(x) \le A$, and so

$$|I_n \cap K| \le A |I_n|.$$

So

$$|I \cap K| \le \sum_n |I_n \cap K| \le A \sum_n |I_n| \le Ak.$$

Letting $k \searrow |I \cap K|$ gives $|I \cap K| \le A|I \cap K|$. As $A < 1$, this gives $|I \cap K| = 0$. $\square$

Corollary 7.1.3   *If K is compact and $|K| > 0$, then there is $I_0 \in \mathcal{B}$ such that $\alpha(.)$ is not bounded away from 1 on $K \cap I_0$ and $|K \cap I_0| > 0$.*

Proof   Suppose otherwise; then for each $I \in \mathcal{B}$ either $|K \cap I| = 0$ or $\alpha(.)$ is bounded away from 1 on $K \cap I$ for each $I \in \mathcal{B}$. For each $k \in K$ choose $I_k \in \mathcal{B}$ with $k \in I_k$ (possible as $\mathcal{B}$ is a basis). By assumption, either $|K \cap I_k| = 0$ or $\alpha(.)$ is bounded away from 1 on $K \cap I_k$. Either way $|K \cap I_k| = 0$ (by the lemma). By compactness, there is a finite set $F \subseteq K$ such that $\{I_k : k \in F\}$ covers K. Then $|K| \le \sum_{k \in F} |K \cap I_k| = 0$, so $|K| = 0$, a contradiction. $\square$

Proof of Theorem 7.1.1   Suppose otherwise. Then for each $x \in K$ there is $\alpha = \alpha(x)$ with $0 < \alpha < 1$ such that for $I \in \mathcal{B}$, if $x \in I$, then

$$|I \cap K| \le \alpha |I|.$$

For K with $|K| > 0$, we construct a nested sequence $\langle I_n \rangle$ of non-empty sets in $\mathcal{B}$ whose intersection contains a density point of K. Pick $I_0 \in \mathcal{B}$ such that $\alpha(.)$ is not bounded away from 1 on $K \cap I_0$ and $|K \cap I_0| > 0$. By passing to a subinterval J with $|K \cap J| > 0$ on which $\alpha(.)$ is not bounded away from 1, we claim that we may without loss of generality assume also that

$$|K \cap I_0| \geq \frac{1}{2}|I_0|.$$

Otherwise, for each $I' \subset I_0$ with $I' \in \mathcal{B}$ and $|K \cap I'| > 0$, we would have

$$|K \cap I'| < \frac{1}{2}|I'|,$$

implying that $\alpha(x) < 1/2$ on $K \cap I_0$, and so, by the lemma, that $K \cap I_0$ is null.

As $K \cap I_0$ is precompact, we may choose $I_1 \in \mathcal{B}$ with $\bar{I}_1 \subset I_0$ and diam$(I_1) <$ diam$(I_0)/2$ such that $\alpha(.)$ is not bounded away from 1 on $K \cap I_1$ and such that

$$|K \cap I_1| \geq \frac{3}{4}|I_1|.$$

Continue inductively. Thus, if

$$\{x\} = \bigcap_{n \in \omega} K \cap \bar{I}_n,$$

then $x \in I_n$ for $n = 1, 2, \ldots$ and

$$|K \cap I_n| \geq (1 - 2^{-n})|I_n|.$$

Thus $x \in F(K)$, a contradiction. Thus after all K has a density point.

We now prove the final assertion. Let T be a measurable set of positive measure. In the notation above, we are to prove that $S := T \backslash F(T) \subseteq T$ has measure zero. Suppose otherwise. Then we may choose $K \subseteq S$ compact with $|K| > 0$ (by inner regularity of Lebesgue measure). Note that S is disjoint from $F(T)$. We have just shown that K has a density point k, i.e. for all $0 < \alpha < 1$, there is I with $k \in I \in \mathcal{B}$ such that

$$|I \cap K| > \alpha|I|.$$

But $k \in K \subseteq S \subseteq T$, so for all $0 < \alpha < 1$, there is I with $k \in I \in \mathcal{B}$ such that

$$|I \cap T| \geq |I \cap S| \geq |I \cap K| > \alpha|I|,$$

i.e. k is a density point of T, equivalently $k \in F(T)$. This contradicts disjointness of S from $F(T)$. $\qquad\square$

The last step is an instance of the Generic Dichotomy Principle of Chapter 4.

Lebesgue proved his Density Theorem as the special case for indicator functions $f = I_A$ of his version of the fundamental theorem of calculus for Lebesgue integrals: that if one takes the indefinite integral of a Lebesgue-measurable function and differentiates it, one recovers the original function almost everywhere. For this extra degree of generality, the Vitali covering theorem and

Hardy–Littlewood maximal function are relevant; for textbook expositions, see Bog2007a, Ch. 5; Saks1964, IV.10; Ste1970, pp. 5, 12. See also BinO2018b; Bru1971.

Corollary 7.1.4 *For A measurable, write $\phi(A)$ for the set of all density points of A. Then the symmetric difference of A and $\phi(A)$ is (Lebesgue-)null:*

$$|A\Delta\phi(A)| = 0.$$

Proof First, $A \setminus \phi(A)$ is (Lebesgue-)null, by the Lebesgue Density Theorem. Also $\phi(A) \setminus A \subseteq A^c \setminus \phi(A^c)$, which as A^c is measurable is similarly null. Combining, the symmetric difference is null. □

For A, B measurable, write $A \sim B$ if and only if $|A\Delta B| = 0$. Then $\sim$ is an equivalence relation on the measurable sets. We note some properties (Oxt1980, Th. 3.21).

(i) (Quasi-identity) $\phi(A) \sim A$.   This is the previous corollary.
(ii) (Neglecting property) $A \sim B$ implies $\phi(A) \sim \phi(B)$; $\phi(\emptyset) \sim \emptyset$, $\phi(\mathbb{R}) \sim \mathbb{R}$.
    This follows from the definition of $\phi(.)$.
(iii) (Multiplicative property) $\phi(A \cap B) = \phi(A) \cap \phi(B)$.
    For, if I is an interval, $I \setminus (A \cap B) = (I \setminus A) \cup (I \setminus B)$. So $|I| - |I \cap A \cap B| \leq |I| - |I \cap A| + |I| - |I \cap B|$, or $|I \cap A| + |I \cap B| - |I| \leq |I \cap A \cap B|$. Take $I = [x - h, x + h]$, divide by h and let $h \downarrow 0$: if x is a density point of both A and B, the limit of the left is 1, whence x is a density point of $A \cap B$.
(iv) (Monotonicity) If $A \subseteq B$, then $\phi(A) \subseteq \phi(B)$.
    This follows from (iii).

One can define a mapping ϕ with these properties in more general measure spaces. Such a mapping is called a *lower density*; see Chapter 8. The existence of lower densities was proved by von Neumann and Maharam (Mah1958) and is connected with the Ionescu Tulcea theory of *lifting* (IonT1961; IonT1969); see Oxt1980, Th. 22.4.

In what follows $(X, \mathcal{A}, m)$ is a measure space and m^* is an outer measure, that is a non-negative countably subadditive set function defined on all subsets of X (see, e.g., Bog2007a, §1.11: the value $+\infty$ for m^* is allowed). Typically, m^* will arise from the measure m via

$$m^*(A) := \inf\left\{ \sum_{n=1}^{\infty} m(A_n) : A_n \in \mathcal{A}, A \subseteq \bigcup_{n=1}^{\infty} A_n \right\}.$$

One calls a set A *m-measurable* if, for all $\epsilon > 0$, there is a set $A_\epsilon \in \mathcal{A}$ with the symmetric difference $A\Delta A_\epsilon$ having $m^*(A\Delta A_\epsilon) < \epsilon$. Equivalently, the measurable sets A are those that can be approximated from within and without by sets A_i, $A_o \in \mathcal{A}$:

$$A_i \subseteq A \subseteq A_o, \qquad m^*(A_o \setminus A_i) = 0.$$

Then A_i can be taken in $\mathcal{A}_{\sigma\delta}$ and A_o in $\mathcal{A}_{\delta\sigma}$. We call A_o the *measurable cover* (or measurable envelope) of A; A_i the *measurable core* (or measurable kernel) of A.

We shall be dealing principally here with Lebesgue measure, which is infinite. However, our main interest here is in *local* aspects, for which we may by truncation restrict to some compact set, on which Lebesgue measure is finite. On such a finite measure space, matters simplify to

$$m^*(A) = \inf\{m(M) : A \subseteq M \in \mathcal{A}\}.$$

Call $\mathcal{B}$ a *density basis* (cf. *derivation basis*, HayP1970, *differentiation basis*, LukMZ1986) if, for each x, there is a sequence $\langle I_n \rangle$ with $x \in I_n \in \mathcal{B}$ such that $m(I_n) \searrow 0$. Write $\mathcal{B}(x)$ for the collection of such sequences $\langle I_n \rangle$ in $\mathcal{B}$. The corresponding upper outer density is defined by

$$\bar{D}^*(A, x) = \sup\{\limsup_n m^*(E \cap I_n)/m(I_n) : \langle I_n \rangle \in \mathcal{B}(x)\}.$$

A lower outer density is defined similarly. If they are equal we speak of a density. We note in passing that, for fixed x, the set function $\bar{D}^*(A, x)$ is monotone and subadditive (as with the outer measure).

Call x an *outer density point* of A if the outer density is 1, and an *outer dispersion point* of A if the outer density is zero. The word outer is omitted when A is measurable.

Definition For $\mathcal{B}$ a density basis and $\bar{D}^*$ the corresponding upper outer density function, put

$$\mathcal{U} := \{U : (\forall u \in U)\ \bar{D}^*(X \setminus U, u) = 0\}.$$

Theorem 7.1.5 (Marti1964, Th. 4.1) *For $\mathcal{B}$ a density basis, $\mathcal{U}$ as above is a topology on X.*

Proof This follows from subadditivity of $\bar{D}^*(A, x)$ for fixed x. In particular if U_1 and U_2 are in $\mathcal{U}$ and $U = U_1 \cap U_2$, then for $u \in U$, since $U \subseteq U_i$ one has

$$\bar{D}^*(X \setminus U, u) = \bar{D}^*((X \setminus U_1) \cup (X \setminus U_2), u)$$
$$\leq \bar{D}^*((X \setminus U_1), u) + \bar{D}^*((X \setminus U_2), u) = 0.$$

So $U \in \mathcal{U}$. If $\{U_i : i \in I\}$ is any subfamily of $\mathcal{U}$ and $u \in U = \bigcup_I U_i$, then for some $i \in I$ we have $u \in U_i$ and so

$$\bar{D}^*(X \setminus U, u) \leq \bar{D}^*(X \setminus U_i, u) = 0,$$

so $U \in \mathcal{U}$. Evidently X and $\emptyset$ are in $\mathcal{U}$. $\qquad\qquad\square$

The main result is the measurability of the sets in $\mathcal{U}$. We need a lemma and a corollary of it.

Lemma 7.1.6 (Cf. Marti1964, 3.6)   *If E is a measurable cover of A and J is measurable, then $E \cap J$ is a measurable cover of $A \cap J$.*

Proof   If not, then there is F with $A \cap J \subseteq F \subseteq E \cap J$ with $m(F) = m^*(A \cap J)$ and $m(F) < m(E \cap J)$. But $F \cup (E \backslash J)$ is measurable, covers A and satisfies

$$m^*(A) \leq m(F \cup (E \backslash J)) = m(F) + m(E \backslash J)$$
$$< m(E \cap J) + m(E \backslash J) = m(E),$$

contradicting $m^*(A) = m(E)$.   □

Corollary 7.1.7   *If E is a measurable cover of A, then every density point of E is an outer density point of A.*

Proof   Suppose x is a density point of E. For any $\alpha < 1$ there is $I \in \mathcal{B}(x)$ such that $m(E \cap I) > \alpha m(I)$. So $m^*(A \cap I) = m(E \cap I) > \alpha m(I)$. Hence x is an outer density point of A.   □

Say that *a density theorem holds for* the density basis $\mathcal{B}$ if almost all points of any set A are outer density points of A (equivalently, by the preceding corollary, almost all points of any measurable set A are density points of A).

Theorem 7.1.8 (Marti1964, 4.3)   *If a density theorem holds for $\mathcal{B}$, then a set A is measurable if and only if almost all points in $X \backslash A$ are outer dispersion points of A.*

Proof   If A is measurable, then so is $X \backslash A$, and so by the assumed density theorem almost all points of $X \backslash A$ are density points of $X \backslash A$ and hence dispersion points of A, as asserted.

For the converse, let A have the asserted property and suppose that A is not measurable. Let E be a measurable cover of A. Recall that a set $F \subseteq E$ is measurable if and only if

$$m(E) = m^*(F) + m^*(E \backslash F).$$

As $m(E) = m^*(A)$ and A is non-measurable we have $m^*(E \backslash A) > 0$. We next show that almost all points of $E \backslash A$ are points of outer density of A, contradicting the assumption that almost all points of $E \backslash A$ are outer dispersion points of A.

To this end note, by the assumed density theorem, that *almost all* points of E are density points of E. By Corollary 7.1.7 *all* points of E are outer density points of A. Hence *almost all* points of E, and a fortiori almost all points of $E \setminus A$, are outer density points of A. □

Theorem 7.1.9 (Marti1964, 4.4. and 4.5) *If a density theorem holds for $\mathcal{B}$, then each set $U \in \mathcal{U}$ is measurable. So in particular all points of U are density points of U.*

Proof By definition, if $U \in \mathcal{U}$ then all points in $U = X \setminus (X \setminus U)$ are outer dispersion points of $X \setminus U$. By the preceding theorem, $X \setminus U$ is measurable and so also is U. □

The motivating example in the material above is the class of intervals $(x - h_n, x + h_n)$ containing a point x, with $h_n \downarrow 0$ rational (e.g. $h_n = 1/n$). This class is a density basis; that a density theorem holds for it is the content of the Lebesgue Density Theorem. Call a set *density open* if all its points are points of density. Then by Theorem 7.1.9, the class of such density open sets forms a topology, the *density topology* $\mathcal{D}$. The density topology is due to Haupt and Pauc (HauP1952; Gof1962) (see also GofNN1961; GofW1961). However, it is implicit in the work of Denjoy in 1914 (Den1915).

We shall need a number of properties of $\mathcal{D}$.

(1) The density topology $\mathcal{D}$ is finer than the Euclidean topology $\mathcal{E}$.

For, every point of a Euclidean neighbourhood is a density point. Thus every Euclidean neighbourhood is a density neighbourhood.

The density topology is thus an example of a *fine topology* – a topology on Euclidean space finer than the Euclidean topology – see Chapter 8.

(2) The $\mathcal{D}$-interior of a measurable set M is $\phi(M)$, the set of density points of M contained in M.

For, by the Lebesgue Density Theorem $M \setminus \phi(M)$ is null, so $\phi(M)$ is measurable as M is, and each of its points is a density point of M by definition of ϕ. So $\phi(M)$ is $\mathcal{D}$-open. But any $\mathcal{D}$-open subset of M is contained in $\phi(M)$, by definition of $\mathcal{D}$. So:

(2′) If $x \in M$ and M is measurable, M is a $\mathcal{D}$-neighbourhood of x if and only if x is a density point of M, i.e. $x \in \phi(M)$.

(3) The $\mathcal{D}$-nowhere dense sets are the Lebesgue-null sets.

For, if A is Lebesgue-null, all points of its complement A^c are density points of A^c, so A^c is $\mathcal{D}$-open, so A is $\mathcal{D}$-closed. Being null, $A = \overline{A}$ has empty interior. So A is nowhere dense. Conversely, if A is $\mathcal{D}$-nowhere dense, its closure $\overline{A}$ has empty interior, so contains no density point of A. So A is null by the Lebesgue Density Theorem.

(4) The *$\mathcal{D}$-meagre sets are the Lebesgue-null sets.*

For, meagre sets are countable unions of nowhere dense sets (i.e. of null sets by above) and so are null.

(5) $(\mathbb{R}, \mathcal{D})$ is a Baire space.

For, a countable union of $\mathcal{D}$-nowhere dense sets, that is of null sets by above, is null and so $\mathcal{D}$-meagre. More is true: $(\mathbb{R}, \mathcal{D})$ is *hereditarily Baire*: any subset of the real line is Baire under the induced topology (the proof is as above: subsets of null sets are null).

(6) The $\mathcal{D}$-Borel sets are the (Lebesgue-)measurable sets. In fact these are all density-$\mathcal{G}_\delta$.

For, by outer regularity of Lebesgue measure, any measurable set is a (Euclidean, so density) $\mathcal{G}_\delta$ set less a null set. The null sets are closed in the $\mathcal{D}$ topology (by the density theorem) and so any Lebesgue-measurable set is $\mathcal{D}_\delta$ and so $\mathcal{D}$-Borel. Conversely, any $\mathcal{D}$ set is measurable, and so the $\mathcal{D}$-Borel sets are measurable (since the measurable sets are a σ-algebra).

This result is the $\mathcal{D}$-analogue of regularity of Lebesgue measure, and so may be regarded as the density analogue of Littlewood's First Principle (§1.2).

Definition   For a family $\mathcal{H}$ of subsets of X, a topology $\mathcal{T}$ on X is said to have the $\mathcal{H}$-*insertion property* if, for $\mathcal{T}$-open U and $\mathcal{T}$-closed $F \supseteq U$, there is $H \in \mathcal{H}$ with

$$U \subseteq H \subseteq F.$$

Thus $\mathcal{D}$ has the 'density-$\mathcal{G}_\delta$-insertion property', or '$\mathcal{G}_\delta(\mathcal{D})$-insertion property'. Since a topology is the class of its open sets, we can abbreviate this to:

(6') $\mathcal{D}$ has the $\mathcal{D}_\delta$-insertion property.

The close relationship between Euclidean $\mathcal{G}_\delta$-subsets and the density topology clarifies the connection between the algebraic and the two topological structures in play here. We remind the reader that a *topological group* is a topological space endowed with a group structure such that the group operations $(x, y) \to xy$ and $x \to x^{-1}$ are continuous. Where there may be separate but not joint continuity of the group operation $(x, y) \to xy$, we speak of a *paratopological group*.

(7) $(\mathbb{R}, \mathcal{D})$ under addition is a paratopological group but not a topological group.

Separate continuity of addition follows from the commutativity of addition and shift-invariance of the density topology. However, joint continuity fails. Suppose otherwise – i.e. that $f(x, y) = x + y$ is jointly continuous as a map from $(\mathbb{R}, \mathcal{D})^2 \to (\mathbb{R}, \mathcal{D})$. For any G non-empty $\mathcal{D}$-open, e.g. an interval, the set $G\backslash\mathbb{Q}$ is non-empty $\mathcal{D}$-open with no rational elements. Then $f^{-1}(G\backslash\mathbb{Q})$ is non-empty $\mathcal{D}$-open and so contains a non-empty set of the form $U \times V$ with U, V $\mathcal{D}$-open and of positive measure. By inner regularity of Lebesgue measure, these two sets have measurable kernels that are Euclidean $\mathcal{G}_\delta$-subsets, say H_U and H_V. By Steinhaus's Theorem (see Chapter 15), $H_U + H_V$ contains an interval and so rational points. But $f(H_U \times H_V) \subseteq G\backslash\mathbb{Q}$, a contradiction. (Compare Scheinberg, Sch1971, Arhangelskii and Reznichenko, ArhR2005, or Heath and Poerio.[1])

When, as here, our interest is on local aspects, one may localize onto a compact set on which Lebesgue measure is finite. One says that a complete, finite measure μ is a *category measure* if the μ-null sets are the meagre sets. The term is due to Oxtoby (Oxt1980, Ch. 23).

(8) Lebesgue measure on, say, $([0, 1], \mathcal{D})$ is a category measure.

Call a function f defined on a Euclidean neighbourhood of x *approximately continuous* if there is a measurable set M with x a density point of M and $f(y) \to f(x)$ as $y \to x$ with $y \in M$ (the term is due to A. Denjoy in 1915).

(9) The function f is approximately continuous at x if and only if f is $\mathcal{D}$-continuous at x.

For, approximate continuity at x implies $\mathcal{D}$-continuity at x, by (3). The converse, due to GofNN1961, uses the Lusin–Menchoff Theorem (see LukMZ1986, Ch. 3 and §6A and also Theorem 7.1.11).

(9') The density topology is thus the coarsest topology on the line making the approximately continuous functions (in Denjoy's sense above) continuous. In consequence, the density topology is completely regular (cf. LukMZ1986, Th. 2.3).

Thus the density topology plays for the approximately continuous functions the role played by the fine topology for the superharmonic functions (see Chapter 8).

(10) The function f is $\mathcal{D}$-continuous a.e. if and only if it has the Baire property with respect to $\mathcal{D}$ if and only if it is measurable. This is the Denjoy–Stepanoff theorem (Denjoy in 1915, W. Stepanoff in 1942; see, e.g., LukMZ1986, 6.20). The last two properties correspond roughly to Littlewood's Second Principle.

Recall (§1.2) that if $\mathcal{T}$ is submetrizable, then $\mathcal{T}$ is finer than (i.e. refines) a metrizable topology, $\mathcal{T}_\rho$ say, with ρ a metric generating this, so that $\mathcal{T} \supseteq \mathcal{T}_\rho$. This has two consequences:

---

[1] Paper at Conference on Topology and Theoretical Computer Science in honour of Peter Collins and Mike Reed, 2006, unpublished.

(i) any $\mathcal{T}_\rho$-closed set (being the complement of a $\mathcal{T}_\rho$-open set which is also $\mathcal{T}$-open) is $\mathcal{T}$-closed;

(ii) $\mathrm{cl}_\mathcal{T}\, A \subseteq \mathrm{cl}_\rho\, A$, as there are more open sets.

Definition   A submetrizable topology $\mathcal{T}$, say, refining a metrizable topology $\mathcal{T}_\rho$ from a metric ρ, is *cometrizable* relative to $\mathcal{T}_\rho$, if for each x and every $\mathcal{T}$-neighbourhood U there is a $\mathcal{T}$-neighbourhood V such that

$$x \in V \subseteq \mathrm{cl}_\rho V \subseteq U.$$

That is, the $\mathcal{T}_\rho$-closures of $\mathcal{T}$-neighbourhoods form a $\mathcal{T}$-neighbourhood-base. Since $\mathrm{cl}_\mathcal{T} V \subseteq \mathrm{cl}_\rho V$ this property implies regularity.

Definition   The topology $\mathcal{T}$ is said to have the *Lusin–Menchoff property* relative to a topology τ (cf. LukMZ1986, Th. 3.11, p. 85) if for each τ-closed set H and every $\mathcal{T}$-neighbourhood U of H there is a $\mathcal{T}$-neighbourhood V of H such that

$$H \subseteq V \subseteq \mathrm{cl}_\tau V \subseteq U.$$

Thus the separation required here concerns some but not all $\mathcal{T}$-closed sets. It certainly includes all singletons as τ is Hausdorff, and so the definition is more demanding than in the cometrizable case. It is a normality-like property (termed binormality in the bitopology literature).

Equivalently, this may be restated with $\mathcal{T}'$ for τ as requiring that if F, F' disjoint are respectively $\mathcal{T}$ and $\mathcal{T}'$ closed, then for some disjoint G, G' which are respectively $\mathcal{T}$ and $\mathcal{T}'$ open

$$F' \subseteq G \quad \text{and} \quad F \subseteq G'.$$

(Take $F := X\backslash U$ disjoint from $F' := H$; then $G' := X\backslash \mathrm{cl}_{\mathcal{T}'} V$ is disjoint from $G := V \subseteq \mathrm{cl}_\tau V$, also $G \supseteq H$ and $G' \supseteq X\backslash U = F$.)

(11) $(\mathbb{R}, \mathcal{D})$ is regular (cf. Oxt1980, Th. 22.9) – in fact cometrizable (implicitly, by the proof just cited). More is true: $\mathcal{D}$ has the Lusin–Menchoff property relative to the Euclidean topology (GofNN1961). This also implies that $\mathcal{D}$ is completely regular, so has a (Hausdorff) compactification (GofNN1961).

(12) $(\mathbb{R}, \mathcal{D})$ is connected (GofW1961).

By contrast:

(13) $(\mathbb{R}, \mathcal{D})$ is not normal (GofNN1961), nor indeed pseudonormal (see Tal1978, Th. 15), nor countably paracompact (Tal1976, Cor. 3.10).

(14) $(\mathbb{R}, \mathcal{D})$ is not second countable. Nevertheless, it does satisfy the *countable chain condition* (ccc), i.e. any family of pairwise disjoint open sets is at most

countable (more is true: it has *property K* (Knaster's property) – any uncountable family of open sets has an uncountable subfamily of mutually intersecting sets).

This motivates the following.

Definitions For a topology $\mathcal{T}$, denote by $\mathcal{I}_{\mathcal{T}}$ the σ-ideal of meagre sets. (Thus if $\mathcal{T} = \mathcal{D}$ these are the null sets.)

For a σ-ideal $\mathcal{I}$ say that $\mathcal{T}$ is $\mathcal{I}$-*quasi-Lindelöf*, or quasi-Lindelöf relative to $\mathcal{I}$, if every $\mathcal{T}$-open family $\mathcal{H}$ contains a countable subfamily $\mathcal{H}'$ such that $\bigcup \mathcal{H}'$ differs from $\bigcup \mathcal{H}$ by a set in $\mathcal{I}$.

When $\mathcal{I}$ is the σ-ideal of μ-null sets for a measure μ, we shall refer to this property as μ-*almost-Lindelöf*, or almost-Lindelöf relative to μ.

(15) $(\mathbb{R}, \mathcal{D})$ is almost Lindelöf relative to Lebesgue measure. More is true.

Theorem 7.1.10 (LukMZ1986, 1.B.1a) *If $\mathcal{T}$ satisfies the countable chain condition and $\mathcal{I} = \mathcal{I}_{\mathcal{T}}$, then X is $\mathcal{I}$-quasi-Lindelöf.*

(16) $(\mathbb{R}, \mathcal{D})$ is not topologically complete, i.e. it is not a $\mathcal{G}_{\delta}$ in its Stone–Čech compactification, nor indeed even a Borel subset thereof (see FroN1990, generalizing Gos1985). Nonetheless, it is pseudocomplete, hence Baire (see §2.2), and in fact strongly α-favourable, for which see §2.3 (for proofs see Whi1974).

We stop to consider in detail the fact that $(\mathbb{R}, \mathcal{D})$ is Baire. There is more at work here than just the regularity of Lebesgue measure (see point (5) in the list above). Note that in turn cometrizability of $\mathcal{D}$ depends on the inner regularity of Lebesgue measure. The following theorem sheds more light on the interconnections.

Theorem 7.1.11 (LukMZ1986, Th. 4.2) *If the topology $\mathcal{T}$ on X is cometrizable relative to $\mathcal{T}_{\rho}$ (for instance, has the Lusin–Menchoff property relative to $\mathcal{T}_{\rho}$) and $\mathcal{T}_{\rho}$ is topologically complete, then any $\mathcal{G}(\mathcal{T})_{\delta}$-subset, A, of X is a Baire space under $\mathcal{T}$. In particular, X is a Baire space under $\mathcal{T}$.*

Proof Suppose otherwise, and that $A = \bigcup_i A_i$ with A_i nowhere dense in A (under the $\mathcal{T}$-subspace topology), where $A = \bigcap_i U_i$ with $U_i \in \mathcal{T}$. For X under $\mathcal{T}_{\rho}$, choose a compactification Y and open sets G_n in Y with $X = \bigcap_n G_n$. Notice that, for any $\mathcal{T}$-open set U meeting A, the set $A \cap U \setminus \mathrm{cl}_{\mathcal{T}}(A_i)$ is non-empty. We show how to choose inductively points $a_i \in A \cap U_i$, $\mathcal{T}_{\rho}$-open sets W_i and $\mathcal{T}$-open sets V_i such that

$$a_i \in V_i \subseteq \mathrm{cl}_{\rho} V_i \subseteq U_i \cap [W_i \setminus \mathrm{cl}_{\mathcal{T}}(A_i)] \quad \text{and} \quad \mathrm{cl}_Y W_i \subseteq G_i.$$

Suppose done for i; then $V_i \cap U_{i+1} \cap A$ is non-empty, and so we may choose $a_{i+1} \in (A \cap V_i \cap U_{i+1}) \backslash \mathrm{cl}_{\mathcal{T}}(A_{i+1})$. By regularity in Y, choose $\mathcal{T}_\rho$-open W_{i+1} such that $a_{i+1} \in W_{i+1} \subseteq \mathrm{cl}_Y W_{i+1} \subseteq G_{i+1}$ and such that (again by regularity but now in $\mathcal{T}_\rho$, and again passing to an open subset) $a_{i+1} \in W_{i+1} \subseteq \mathrm{cl}_\rho W_{i+1} \subseteq W_i$. As $\mathcal{T}_\rho$-open is $\mathcal{T}$-open, by cometrizability, choose $V_{i+1} \in \mathcal{T}$ with

$$a_{i+1} \in V_{i+1} \subseteq \mathrm{cl}_\rho V_{i+1} \subseteq V_i \cap U_{i+1} \cap [W_{i+1} \backslash \mathrm{cl}_{\mathcal{T}}(A_{i+1})].$$

Since the sets $\mathrm{cl}_Y(W_i)$ have the finite intersection property, by compactness $\emptyset \neq \bigcap_i \mathrm{cl}_Y(W_i) \subseteq \bigcap_n G_n = X$. So

$$X \cap \bigcap_i \mathrm{cl}_Y(W_i) = \bigcap_i \mathrm{cl}_\rho(W_i) = \bigcap_i W_i$$

is non-empty. Finally,

$$\emptyset \neq \bigcap_n W_n = \bigcap_n V_n \subseteq \bigcap_n U_n = A$$

and

$$\bigcap_n V_n \cap \mathrm{cl}_{\mathcal{T}}(A_i) = \emptyset \quad \text{for each } i.$$

So $A \backslash \bigcup_i A_i$ is non-empty, a contradiction.   □

Since by property (11) $\mathcal{D}$ has the Lusin–Menchoff property with respect to the Euclidean topology $\mathcal{E}$, and $(\mathbb{R}, \mathcal{E})$ is Baire by Baire's Theorem, this shows that $(\mathcal{R}, \mathcal{D})$ is Baire (recovering the statement of property (5)). Note however that this result does not follow from any of the several versions of Baire's Theorem in Chapter 2. We regard Theorem 7.1.11 as the 'Lusin–Menchoff form of Baire's Theorem'. For differences of non-meagre sets in a topological group, see RaoR1975.

We close by considering two contributions of Caspar Goffman taken from Gof1950 and Gof1975. These are stated so succinctly that we do not hesitate to cite them verbatim.

Theorem 7.1.12 (Goffman's Theorem)   *The set of points for which the metric density of a measurable set S exists but is not equal to 0 or 1 is of measure 0 and of first category.*

Proof   We shall suppose all sets are contained in the open interval $(0, 1)$. Let T be the set of points for which the metric density of S exists, U those points of T for which the metric density of S is 0 or 1, and $Z = T \backslash U$. By the Lebesgue density theorem, Z is of measure 0 and U is of measure 1. The metric density of S is a function of *Baire class 1* on T. For, let x be in T and let $f_n(x), n = 1, 2, \ldots,$ be the relative measure of S in the interval $(x - 1/n, x + 1/n)$. For every n, $f_n(x)$ is continuous and is, accordingly, continuous relative to T. Since the metric density of S exists at every point of T, $f(x) = \lim_{n \to \infty} f_n(x)$ exists, is equal to the metric density of S, and is a function of Baire class 1 on T relative to T. Its *points of discontinuity* must be a set of first category relative to T (by the Baire Continuity Theorem, Chapter 2). On the other hand, U, as a set of measure 1, is everywhere dense in T. Thus, every interval containing a point of Z also contains points of U; that is, points x for which $f(x)$ is either 0 or 1. Since, for every x in Z, $f(x)$ is different from 0 or 1, Z must be a subset of the set of points of discontinuity of $f(x)$. Accordingly Z is of first category relative to T and, therefore, relative to $(0, 1)$. □

Katznelson and Stromberg (KatS1974) gave a relatively simple proof of the existence of an everywhere differentiable function which is not monotonic in any interval. We show the connection between this property and the density topology.

Example of a Nowhere Monotonic Everywhere Differentiable Function, cf. point (11) above. The density topology for the reals is completely regular (GofNN1961; Zah1950). Since countable sets are closed in this topology, for each countable S and $\xi \notin S$ there is an approximately continuous f such that $0 < f(x) < 1$, $f(\xi) = 1$ and $f(x) = 0$ for each $x \in S$. Let A and B be disjoint countable sets each dense in the reals. They may be enumerated $A = \{a_n\}$, $B = \{b_n\}$. For each n, let f_n be approximately continuous, $0 < f_n(x) < 1$, $f_n(a_n) = 1$, and $f_n(x) = 0$ for each $x \in B$, and let g_n be approximately continuous, $0 < g_n(x) < 1$, $g_n(b_n) = 1$, and $g_n(x) = 0$ for each $x \in A$. The function

$$f := \sum_{n=1}^{\infty} 2^{-n} f_n - \sum_{n=1}^{\infty} 2^{-n} g_n$$

is bounded, approximately continuous, positive on A and negative on B. Let F be an indefinite integral of f. Then F is everywhere differentiable and $F' = f$. So F is not monotonic in any interval.

We close by saying that in Euclidean space of dimension 2 and above, there are multiple choices of density topologies, whose properties depend on the selected basis $\mathcal{B}$ for the topology. See the examples considered in GofNN1961. For other strange functions, see Kha2018.

8

Other Fine Topologies

8.1 The Fine Topology of Potential Theory: Polar Sets

The term *potential function* (briefly, potential) stems from the first page of the famous essay of 1828 by George Green (1793–1841) on electricity and magnetism (Gre1828). Recall from physics that for a field of force (briefly, field) which is conservative (one in which work done to arrive at a point is independent of the path taken to get there), the field F is the gradient of the potential, u:

$$F = \operatorname{grad} u.$$

See, e.g., Kellogg (Doob's favourite source) (Kell1953, III.1, pp. 48–54). There one finds discussion of, e.g., units, force, work, potential energy, and the two possible sign conventions here, depending whether one measures work done by or against the field.

To begin with some background on gravitation: following two decades of observation by Tycho Brahe (1546–1601), Johannes Kepler (1571–1630) published his *Astronomia Nova* (1609). These contain Kepler's Laws, the first of which is: *Planets move around the Sun in elliptical orbits with the Sun at one focus.* The great challenge this posed to astronomy was to explain Kepler's Laws, which had been arrived at empirically. It was suspected that an inverse square law of attraction was the key. Sir Isaac Newton (1642–1727) established this in his *Principia* of 1687, by linking it with the elliptical orbits of Kepler's First Law, by then well established experimentally (for a succinct two-page account, with references within the *Principia*, see Ram1951, pp. 167–168). The counterpart of this in electromagnetism was the work of Ch. A. de Coulomb (1736–1806), who in the period 1785–1791 established the inverse square law of electromagnetism experimentally.

In Newtonian gravitation, the potential of a unit point mass at a distance r is $1/r$; likewise for a unit point charge or magnetic pole. Differentiating this to form the gradient above gives the *inverse square law* of gravity, and also of electromagnetism. Thus already the idea of a potential has a unifying effect on two of the four fundamental forces of nature, gravitation and electromagnetism (despite the dissimilarity that in gravitation all matter attracts, while in electromagnetism unlike charges attract but like charges repel). The other two, the weak nuclear force (governing radioactivity) and the strong nuclear force (enabling stability of matter by holding together in the nucleus the positively charged protons that would repel each other electrostatically further apart), emerged much later.

We speak nowadays of *a* fine topology as one finer than (refining, having more open sets than) the usual (Euclidean) topology. But there is one such topology – the first – that deserves the definite article, as *the* fine topology – that arising in *potential theory*. As well as playing a vital role in physics and applied mathematics, as a key component of the theory of two of the four fundamental forces of nature, gravity and electromagnetism, potential theory also has a deep and rich mathematical theory. It was aptly described by Pierre Jacquinot in 1964: 'La théorie du potentiel est un véritable carrefour de la Mathématique' (quoted by Bau1975). Our sources for this are Kellogg (Kell1953) for the early theory, Helms (Hel1969) and the monumental Doob (Doo1984) for a more modern, and topological, treatment. Here we work in Euclidean space $\mathbb{R}^N$ of dimension $N \geq 2$. The case $N = 1$ is degenerate by comparison (Doo1984, Ch. 1.XIV); the case $N = 2$ gives rise to logarithmic potentials; $N = 3$ to the inverse square law above, $N \geq 3$ to power-law potentials, below.

The *fundamental kernel* of classical potential theory on $\mathbb{R}^N$ is ('G for Green') $G(x, y)$ (Doo1984, p. 6), while

$$G_y(x) := G(x, y) := \begin{cases} \log(1/|x - y|), & \text{if } N = 2, \\ 1/|x - y|^{N-2}, & \text{if } N > 2 \end{cases}$$

(with $G(x, x) = +\infty$) gives the *fundamental harmonic function* (solution of Laplace's equation) with *pole* y (it is harmonic, as its value at the centre of a ball is the average of its values over the surface of the ball). If μ is a measure on $\mathbb{R}^N$,

$$G\mu(x) := \int_{\mathbb{R}^N} G(x, y)\mu(dy) = \int_{\mathbb{R}^N} G_y(x)\mu(dy)$$

is the *potential* of μ. It is again harmonic, as it inherits the 'harmonic average property' above. This usage is that of physics, if one thinks of μ as a distribution of gravitational mass or electrostatic charge. With $N = 3$, $r = |x - y|$, one recovers the $1/r$ above.

The *fine topology* (of potential theory, understood below) is defined as the coarsest topology on $\mathbb{R}^N$ making all superharmonic functions continuous (Doo1984, 1.XI.1). One can conveniently use a superscript f to denote that it is the fine topology that is being used.

As with the Euclidean and density topologies, each with its σ-ideal of negligible ('small') sets, the meagre sets $\mathcal{M}$ and the (Lebesgue) null sets $\mathcal{N}$, so the fine topology of potential theory has its σ-ideal of negligible sets, the *polar* sets, $\mathcal{P}$, say. These are the sets each point of which has an open neighbourhood carrying a superharmonic function which has a pole (takes the value $+\infty$) at each point of the set in the neighbourhood (Doo1984, Ch. 1.V.1; Hel1969, §7.1). Equivalently, a set is polar if and only if it has no fine limit points (see (iv) below).

From the definition, a subset of a polar set is polar. Also, $\mathcal{P}$ is closed under countable unions (Doo1984, 1.V.3, p. 59). Combining, the polar sets do indeed form a σ-ideal, $\mathcal{P}$.

We note some key properties and definitions.

(i) The fine topology is strictly finer than the Euclidean topology (Doo1984, p. 166).

(ii) A set $A \subset \mathbb{R}^N$ is *thin* at a finite point ξ if and only if $\xi \notin A^f$; that is, ξ is not a fine limit point of A (Doo1984, p. 167).

(iii) The Baire property holds for polar sets (ConC1972; Doo1984, pp. 167–168). So $(\mathbb{R}^N, \mathcal{F})$ is a Baire space.

(iv) A set is polar if and only if it has no fine limit points (Doo1984, Th. 1.XI.6).

(v) The fine topology is strictly coarser than the density topology (Fug1971).

There is a hierarchy of small sets in potential theory, with the polar sets being the smallest. To summarize (Chu1982, §3.5, p. 112; see there for definitions of terms not already used):

polar = zero capacity $\subset$ very thin $\subset$ thin $\subset$ semi-polar $\subset$ zero potential.

Likewise, there is a hierarchy of topologies, both the inclusions (refinings) below being strict:

$$\text{Euclidean} \subset \text{fine} \subset \text{density:} \qquad \mathcal{E} \subset \mathcal{F} \subset \mathcal{D}.$$

The term *polar set* was introduced in 1941 by Brelot (Bre1941), and with it the view of the polar sets as the negligible sets of potential theory. The polar sets were shown to be the sets of *capacity* 0 by Cartan in 1945 (Car1945). The Choquet theory of capacities of 1955 (Cho1953) is directly motivated by potential theory, as the terminology from electromagnetism suggests. For background on the links between capacity and stochastic processes, see Dellacherie (Del1972).

The fine topology is due to Cartan (Car1946). For Cartan's letter (30 December 1940) to Brelot on its significance in potential theory, and more on the emergence of the term fine topology, see the Historical Notes in Doo1984, p. 800.

Despite the comparability between the three topologies above, and the close link each has with a σ-ideal of small sets (negligibles), these σ-ideals are not themselves comparable. For, the three concepts of smallness involved are very different. For instance, the real line $\mathbb{R}$ can be decomposed into two complementary sets, one meagre and the other (Lebesgue-)null (see Oxt1980, Th. 1.6, Th. 16.). Compare a result of Muthuvel (Mut1999): the additive group $(\mathbb{R}, +)$ is the direct sum of two subgroups (one uncountable), one meagre and the other null.

The analogies between category and measure, or between the σ-ideals $\mathcal{M}$ and $\mathcal{N}$, have attracted great attention in descriptive set theory; see, e.g., Oxt1980; Kec1995. The role of potential theory, or the σ-ideal $\mathcal{P}$, has received less attention; for a monograph treatment here, see LukMZ1986.

Notes

Regarding (iii) above: what is needed here is that the original topological space be locally compact; see, e.g., LukMZ1986, p. 55.

Regarding limits and continuity in these topologies: a function f is approximately continuous at z if and only if it is continuous at z in the density topology (LukMZ1986, p. 55; BinO2009f). Regarding links between fine and density topologies, see LukMZ1986, 7C, pp. 267–270.

Markov Processes

We refer for background here to the standard monograph account of Markov processes and potential theory by Blumenthal and Getoor (BluG1968). Recall that a Markov process is one in which only the present, and not the past, is relevant to predicting the future; equivalently, the past and future are conditionally independent given the present (for background, see, e.g., Chu1982, augmented as may be necessary by Chu1974; Chu1968).

To quote the first sentence of the introduction to Doo1984: 'Potential theory and certain aspects of probability theory are intimately related, perhaps most obviously in that the transition function determining a Markov process can be used to define the Green function of a potential theory.' Again, to quote the first sentence of Meyer's classic (Mey1966, p. 1): 'The fundamental work of Doob and Hunt has shown, during the last ten years or so, that a certain form of potential theory (the study of kernels which satisfy the "complete maximum principle") and a certain branch of probability theory (the study of Markov semi-groups and processes) in reality constitute a single theory.' (Later on the same page, Meyer continues 'Nothing of the theory of Markov processes itself will be found herein'; he refers to a forthcoming 'second volume'; this turned out to be five, with Dellacherie, over a 17-year period (DelM1975–1992).)

In the Markov context, we can also define polar sets as follows (see, e.g., Haw1975). If $X = \{X_t\}$ is a Markov process and B is a non-empty analytic set, let

$$V_B := \inf \{t > 0 : X_t \in B\}$$

be the first-entry time into B, and call *points of accessibility* of B those in the set

$$\mathrm{Ac}(B) := \{x : \mathbb{P}_x(V_B < \infty\} > 0\}.$$

Then a non-empty analytic set B is polar if and only if $\mathrm{Ac}(B)$ is empty.

Brownian Motion

The Markov process corresponding to the potential theory of classical electromagnetism as discussed above (the prime example of general potential theory) is *Brownian motion*, $B = \{B_t : t \geq 0\}$ (the prime example of a Markov process, and indeed of a stochastic process). For a monograph account of the links between Brownian motion and classical potential theory, see, e.g., PortS1978. As a diffusion (path-continuous strong Markov process), B has an *infinitesimal generator*, $\frac{1}{2}\Delta$, with Δ the *Laplacian* (or $\frac{1}{2}D^2$ in one dimension); see, e.g., Chu1982, §4.6. For more on the background here, see also the very readable, brief, and wonderfully titled book of Chung, Chu1995.

The above link between Brownian motion and the Laplacian is actually a link between Brownian motion (B_t) and *Laplace's equation*

$$\Delta u = 0$$

(elliptic). Similarly, *space-time Brownian motion* (B_t, t) corresponds to

$$\frac{1}{2}\Delta u = \partial u/\partial t,$$

the *heat equation* (parabolic). In Doo1984, Ch. 1.XV–1.XIX, Doob develops the *parabolic potential theory* corresponding to this link. See also the monograph by Constantinescu and Cornea (ConC1972), and the excellent survey by Bauer (Bau1975) (especially §3). The related subject of harmonic spaces and 'H-cones' is developed in ConC1972 and Boboc, Bucur and Cornea, BobBC1981.

Symmetric Markov Processes and Dirichlet Forms

For a Markov process X, write $P_t(x, dy)$ for the probability of going from

x to dy (the interval $(y, y + dy)$) in time t. The *Markov property* of X is then expressed by the *semi-group property*

$$P_{t+s}f(x) = P_t(P_s f)(x)$$

for $s, t \geq 0$, f bounded measurable. Then X is called *symmetric* with respect to some reference measure m if

$$\int g(x)P_t f(x)m(dx) = \int P_t g(x)f(x)m(dx)$$

for all such t, f, g.

We quote: if u is superharmonic on a set D, and has a subharmonic minorant, then it has a greatest subharmonic minorant, $GM_D u$, and this is harmonic (Doo1984, 1.III.1). Then define

$$G_D(y, .) := G(y, .) - GM_D G(y, .).$$

Then the function G_D on $D \times D$ is called the *Green function of D*, and the function $G_D(y, .)$ is called the *Green function of D with pole y*. Such a (Euclidean) open set D *has a Green function* ('is Greenian').

If D is a Greenian set in $\mathbb{R}^N$ and μ, ν are measures on D, their *mutual energy* $[\mu, \nu]$ is defined as

$$[\mu, \nu] := \int_D G_D \mu \, d\nu = \int_D \int_D G_D(x, y)\mu(dx)\nu(dy);$$

the *energy* of a measure μ is defined as

$$\|\mu\|^2 := [\mu, \mu].$$

Note that this form is symmetric,

$$[\mu, \nu] = [\nu, \mu].$$

This symmetry corresponds to that in symmetric Markov processes, which are the Markov processes most suited to this area. For background here, see Silv1974; Silv1976.

It can be extended to signed measures – *charges* (the term is visibly derived from electrostatics). Write $\mathcal{E}^+$ for the set of measures on D of finite energy. The set of charges representable as differences of positive measures in $\mathcal{E}^+$ is written $\mathcal{E}$.

A measure of finite energy vanishes on polar sets. Conversely, if A is an analytic subset of a Greenian set D and null for every measure on D of finite energy supported by A, then A is polar (Doo1984, pp. 227–228).

Forms such as $[., .]$ above derive from classical work of Dirichlet and are known as *Dirichlet forms*. The modern period in this area dates from the work of Beurling and Deny (BeuD1959) on *Dirichlet spaces*; see, inter alia, BobBC1981. For a monograph treatment of the extensive links between Dirichlet forms and Markov processes, we refer to Fukushima, Oshima and Takeda, FukOT1994.

For more on Green functions, see, e.g., Doo1984, 1.VII. For more on energy and capacity, see, e.g. Doo1984, 1.XIII.

Other σ-ideals of small (or 'exceptional') sets are those of the closed sets of uniqueness for trigonometric series in the category and measure cases, $\mathcal{MU}$ and $\mathcal{NU}$ say. (See Bary1964, Vol. II, p. 358 and §9.1.)

8.2 Analytically Heavy Topologies

Recall that a $\mathcal{K}$-analytic subset A of a topological space $(X, \mathcal{T})$ takes the form

$$A = K(I) = \bigcup_{i \in I} K(i)$$

for K a compact-valued upper semi-continuous map from $I = \mathbb{N}^{\mathbb{N}}$ to X. We denote by $\mathcal{A}(\mathcal{T})$ the family of $\mathcal{K}$-analytic subsets of $(X, \mathcal{T})$.

Definitions ($\mathcal{K}$-Analytically Heavy Topologies)

(1) $\mathcal{H}$ is a *topological base* for X if (Eng1989, §1.1) $\mathcal{H}$ covers X and, for $H_1, H_2 \in \mathcal{H}$, whenever $x \in H_1 \cap H_2$ there is $H_3 \in \mathcal{H}$ with $x \in H_3 \subseteq H_1 \cap H_2$. We write $\mathcal{G}_{\mathcal{H}}$ for the topology generated by $\mathcal{H}$.

(2) $\mathcal{B}$ is a *weak base* for a topology $\mathcal{T}$ if, for each non-empty $V \in \mathcal{T}$, there is $B \in \mathcal{B}$ with $\emptyset \neq B \subseteq V$. In fact, sometimes we need only a *very weak base*: for each non-empty $V \in \mathcal{T}$ there is $B \in \mathcal{B}$ with $\emptyset \neq B \cap V$. See Remark (2) below.

(3) Let $(X, \mathcal{T})$ be a regular Hausdorff space and $\mathcal{T}' \supseteq \mathcal{T}$ a refinement topology. We say $\mathcal{T}'$ is *analytically heavy*, or *weakly $\mathcal{K}$-analytically generated* in $\mathcal{T}$, if $\mathcal{T}'$ possesses a weak base $\mathcal{H} \subseteq \mathcal{A}(\mathcal{T})$. That is, the weak base $\mathcal{H}$ comprises sets that are $\mathcal{K}$-analytic sets in $\mathcal{T}$.

Remarks   (1) In this bitopological context we refer to $(X, \mathcal{T})$ as the *ground space* and $(X, \mathcal{T}')$ as the *refinement*.

(2) The ground space topology $\mathcal{T}$ itself is (weakly) $\mathcal{K}$-analytically generated if $\mathcal{T}$ possesses a (very weak) base $\mathcal{H}$ of sets that are $\mathcal{K}$-analytic sets in $\mathcal{T}$.

(3) If the weak base $\mathcal{H}$ in (1) is actually a base, then we say that $\mathcal{T}$ is a *generalized Gandy–Harrington topology*. (See Example (2), Th. 8.2.1 and at the end of this section.)

(4) If the base $\mathcal{H}$ in (2) is countable, then open sets are $\mathcal{K}$-analytic.

(5) A $\mathcal{K}$-analytic space, in which all open sets are $\mathcal{K}$-analytic, is $\mathcal{K}$-analytically generated – take $\mathcal{H} = \mathcal{T}$.

Examples

(1) *A complete separable metric ground space.* For $(X, \mathcal{T}_d)$ with $\mathcal{T}_d$ generated by a complete separable metric d on X, the standard basis $\mathcal{H}$ of all open (analytic) balls evidently yields $\mathcal{G}_{\mathcal{H}} = \mathcal{T}_d$.

(2) (a) *The Gandy–Harrington topology* $\mathcal{GH}$. For $\mathcal{H}$, the countable family of analytic subsets of $\mathbb{R}$ which are *effective* relative to a given real α (i.e. $\Sigma_1^1(\alpha)$), we obtain the Gandy–Harrington topology $\mathcal{GH}$. For background on the standard Gandy–Harrington case $\mathcal{GH}$ and variants, see, e.g., Lou1980, Prop. 6 or MartK1980, §9.3.

(b) Any subfamily $\mathcal{H}$ of $\mathcal{A}(\mathbb{R})$ closed under intersection, including $\mathcal{A}(\mathbb{R})$ itself, is a base for a topology in the sense of Definition (2) above. In such circumstances the topology it generates, denoted $\mathcal{G}_{\mathcal{H}}$, is of course analytically generated.

(3) *Density topology.* For $\mathcal{I} = \mathcal{N}$, we may take $\mathcal{H} = \mathcal{D} \cap \mathcal{A}(X)$ as a base for $\mathcal{D}$. Here $\mathcal{G}_{\mathcal{H}} = \mathcal{D}$. Unlike in $\mathcal{GH}$, the open sets of $\mathcal{D}$ are not analytic in the ground space, although the basic sets of $\mathcal{H}$ are.

(4) (a) *The Ellentuck topology,* $\mathcal{E}ll$. The points of this space lie in Cantor space $2^{\mathbb{N}}$, the latter equipped with the Euclidean topology. The points of $2^{\mathbb{N}}$ are interpreted as indicator functions of subsets of $\mathbb{N}$. More specifically, one considers only the points corresponding to infinite subsets of $\mathbb{N}$, denoted $[\mathbb{N}]^{\omega}$. This subspace is a $\mathcal{G}_{\delta}$ in $2^{\mathbb{N}}$, so is topologically complete; indeed, if $\langle f_n \rangle$ enumerates $[\mathbb{N}]^{<\omega}$, the family of all finite subsets of $\mathbb{N}$, then $[\mathbb{N}]^{\omega} = \bigcap_n \{1_S \in 2^{\mathbb{N}} : 1_S \neq 1_{f_n}\}$.

The refinement topology on $[\mathbb{N}]^{\omega}$, called the Ellentuck topology after one of its authors (Ell1974; see also Lou1976, and the more recent Rea1996), is generated by taking for $\mathcal{H}$ the closed subsets

$$[a, A] := [\mathbb{N}]^{\omega} \cap \{1_S \in 2^{\mathbb{N}} : a \subseteq S \subseteq a \cup A\},$$

where a is finite and where $A \subseteq \mathbb{N} \setminus \{0, 1, \ldots, \max a\}$ is infinite. Note that $A = \mathbb{N} \setminus \{0, 1, \ldots, \max a\}$ gives a set in the usual Cantor basis.

If $\mathcal{I} = \{\emptyset\}$, then (not unlike the case (2b) above) $\mathcal{H}$ is $\mathcal{I}$-heavy. Recall that a point is $\mathcal{I}$-heavy if none of its neighbourhoods lie in $\mathcal{I}$ and a set is $\mathcal{I}$-heavy if all its points are $\mathcal{I}$-heavy. The space is Choquet and so Baire (Kec1995, 8.12 and 19.13); the latter will be confirmed in Theorem 8.2.1. The topology yields a 'short-cut' for a proof of the Silver–Mathias Theorem (Sil1971) that analytic sets (in the ground space) have the Ramsey property.

(b) Unlike $\mathcal{GH}$, the Ellentuck topology is generated by a continuum of analytic (in fact $\mathcal{G}_{\delta}$) sets; a countable *effective* coarsening of significance has been studied in Avi1998. These topologies are inspired by the method of forcing used in set theory (see Chapters 14 and 16), which generate Cohen reals, Solovay reals, and, among others, Mathias reals (associated with the Ramsey property), and Hechler reals (for which see LabR1995). Relations between forcing and descriptive set theory are traced in Mill1995.

(5) *O'Malley's r-topology* (or *resolvable*-topology). To study approximate differentiability of real-valued functions, O'Malley (OMa1977) introduced the *r*-topology $\mathcal{R}$ on $\mathbb{R}$ with $\mathcal{R} \subseteq \mathcal{D}$; it is generated by taking as base $\mathcal{B} := \mathcal{D} \cap \mathcal{G}_\delta \cap \mathcal{F}_\sigma$ the sets of $\mathcal{D}$ that are ambiguously both $\mathcal{G}_\delta$ and $\mathcal{F}_\sigma$ in the real line. (For these, see also Sto1963, Th. 10. Recall that in a complete space a set that is both $\mathcal{G}_\delta$ and $\mathcal{F}_\sigma$ may be characterized as resolvable – see Kur1966, §12. III, V.) The *r*-topology is a generalized Gandy–Harrington topology, *avant la lettre*.

The argument for Theorem 8.2.1 repeatedly uses the fact that if $\bigcup_n A_n \cap B \neq \emptyset$, then $A_n \cap B \neq \emptyset$ for some n. We view this as saying that $\mathcal{I} = \{\emptyset\}$ has the localization property and $\bigcup_n A_n$ is $\mathcal{I}$-heavy on B.

Theorem 8.2.1 (Generalized Gandy–Harrington Theorem)   *In a regular Hausdorff space, if $\mathcal{T}'$ is an analytically heavy refinement topology of $\mathcal{T}$ (i.e. possessing a weak base $\mathcal{H} \subseteq \mathcal{A}(\mathcal{T}) \cap \mathcal{T}'$ whose elements are $\mathcal{T}'$-circumscribed), then $\mathcal{T}'$ is Baire.*

In particular, this applies to a Polish space, the Gandy–Harrington $\mathcal{GH}$, the density $\mathcal{D}$, the Ellentuck $\mathcal{E}ll$ and the O'Malley r-topologies.

Proof   We put $I_n := \mathbb{N}^n = \{i \mid n : i \in I\}$. For each n, let W_n be dense and open in $\mathcal{T}'$. Suppose inductively that for all $m \leq n$ there are upper semicontinuous compact-valued maps $K_m : I \to X$ which are $\mathcal{T}'$-circumscribed by $G_m = \langle G_m(i \mid n) \rangle$ say, such that $K_m(I) \subseteq W_m$ with $K_m(I) \in \mathcal{H}$, $\sigma_n(m) \in I_n$ for $m \leq n$, and

$$G_1(\sigma_n(1)) \cap \cdots \cap G_n(\sigma_n(n)) \neq \emptyset.$$

Then

$$U_n := G_1(\sigma_n(1)) \cap \cdots \cap G_n(\sigma_n(n)) \neq \emptyset \quad \text{and} \quad U_n \in \mathcal{T}'.$$

As U_n is non-empty and open in $\mathcal{T}'$ and W_{n+1} is $\mathcal{T}'$-dense, $W_{n+1} \cap U_n \neq \emptyset$. Since $\mathcal{H}$ is a weak base, there is $A_{n+1} \in \mathcal{H}$ with $\emptyset \neq A_{n+1} \subseteq (W_{n+1} \cap U_n) \subseteq W_{n+1}$ and in particular $A_{n+1} \cap U_n \neq \emptyset$. Taking $A_{n+1} = K_{n+1}(I)$ with K_{n+1} a $\mathcal{T}'$-circumscribed representation G_m and noting that $K_{n+1}(I) = \bigcup \{G_{n+1}(\sigma) : \sigma \in I_n\}$, there is $\sigma_n(n+1) \in I_n$ such that

$$G_1(\sigma_n(1)) \cap \cdots \cap G_n(\sigma_n(n)) \cap G_{n+1}(\sigma_n(n+1)) \neq \emptyset.$$

But $G_m(\sigma_n(m)) = \bigcup_k G_m(\sigma_n(m), k)$. So there are extensions $\sigma_{n+1}(m)$ of $\sigma_n(m)$ for each $m \leq n + 1$ such that

$$G_1(\sigma_{n+1}(1)) \cap \cdots \cap G_{n+1}(\sigma_{n+1}(n)) \neq \emptyset.$$

This verifies the induction step. So for each m there is $i(m) \in I$ with $i(m) \mid n = \sigma_n(m)$ for each n. Applying Theorem 3.1.12 in the ground space (taking $F_n = X$), we have

$$\emptyset \neq \bigcap_m G_m(i(m)) \subseteq \bigcap_m A_m \subseteq \bigcap_m W_m.$$

For W an arbitrary non-empty open set in $\mathcal{T}'$, as the set $W_n \cap W$ is $\mathcal{T}'$-dense on W, we conclude by the preceding argument that $\emptyset \neq W \cap \bigcap_m W_m$. So $\mathcal{T}'$ is Baire. $\qquad\square$

Remark In the case of the Gandy–Harrington topology, the members of $\mathcal{H}$ are analytic sets with representations K_m such that each of the sets $K_m(i \mid n)$ is also in $\mathcal{H}$, so open by fiat in $\mathcal{T}' = \mathcal{GH}$. That is, the representations are $\mathcal{T}'$-circumscribed.

8.3 Other Fine Topologies

The use of fine topologies in studying properties of Euclidean spaces has several precedents. The earliest seems to be the fine topology of potential theory; more recent examples, providing new and insightful proofs of important results, include the Ellentuck topology, establishing the Ramsey property of analytic sets, and the Σ_1^1-topology, establishing Silver's Theorem on Π_1^1-equivalence relations (see Kec1995; RogJ1980). Both give a Baire space. Our work, especially in the next chapter, draws heavily not only on the density topology but also on the z-topology (for $z = \langle z_n \rangle$ a null sequence) and two other topologies, which we term the *essential* topologies as they correspond to the notion of essential point of accumulation in the sense of measure or category.

The classical prototype of a fine topology, *the* fine topology, $\mathcal{F}$, arises in potential theory, which as we have seen in §8.1 is the coarsest topology under which all superharmonic functions bounded above are continuous; see, e.g., Doo1984, I.XI; LukMZ1986, Ch. 10; Fug1971. In Euclidean dimension $d = 1$, the fine topology coincides with the Euclidean topology on the line. For $d = 1$, superharmonicity reduces to midpoint convexity, and midpoint convex functions bounded above are continuous (see, e.g., BinO2008; BinO2009f; Kucz1985, §XII.3; GerK1970). For $d \geq 2$, the class of superharmonic functions becomes richer: Newtonian potential theory applies, with the logarithmic potential in the plane and the Coulomb or inverse-square potential in space.

For $\mathcal{F}$, the small sets are the *polar sets*, and $(\mathbb{R}, \mathcal{F})$ is quasi-Lindelöf relative to the σ-ideal $\mathcal{I}$ of polar sets. It is Hausdorff, completely regular and locally connected. The $\mathcal{F}$-continuous functions are the *superharmonic* functions.

8.4 * Topologies from Functions, Base Operators or Density Operators

The fine topology above was defined by specifying which family of real-valued functions should be deemed continuous under the topology. As a result the space is completely regular (see, e.g., LukMZ1986, Th. 2.3). The density topology of Chapter 7 may also be defined in this way by reference to the approximately continuous function (as was done originally by Goffman and Waterman); O'Malley (OMa1977) introduced a coarser topology, the r-topology, as the coarsest under which the smaller family of approximately derivable functions is continuous. For the topology r a base is provided by sets in $\mathcal{D}$ that are simultaneously $\mathcal{F}_\sigma$ and $\mathcal{G}_\delta$ ('ambivalent') in $\mathcal{E}$; whilst not normal, this topology nevertheless has the Lusin–Menchoff property relative to $\mathcal{D}$ (his Theorems 3.6 and 3.10). The r-open sets differ from $\mathcal{D}$-open sets by null sets (see OMa1977 for a characterization). A still coarser topology (the 'a.e.-topology') comprises those sets $U \in \mathcal{D}$ for which U differs from its $\mathcal{E}$-interior by a null set; so the a.e.-topology is that with open sets $U \cup M$ with U open in the usual topology and M a set of density points of U. Consequently, a function that is approximately continuous everywhere and continuous almost everywhere is a.e.-continuous.

With applications in mind, our initial approach here has been the measure-theoretic refinement of the usual 'limit-point property', captured by the assertion that $A \in \mathcal{D}$ if and only if $A \in \mathcal{L}$ and $A \subseteq \phi(A)$, where we recall that $\phi(A) := \{a \in A : \bar{D}^*(A, a) > 0\}$. This prompts a unified approach (as we shall make use of other kinds of limit points than in Chapter 7). The starting point is a given topology on a set X, for instance one specified by declaring a family of real-valued functions to be continuous. We then consider the following constructions.

Base Operator. Given a topology $\mathcal{T}$ on X, one may associate with any subset A either its closure $\bar{A}$ or its derived set $\mathrm{der}(A)$. Thus one may introduce (cf. LukMZ1986, 1.A) a (monotone) *base operator* on the power set $\wp(X)$, i.e. a mapping $b: \wp(X) \to \wp(X)$ with $b(\emptyset) = \emptyset$ with the *additivity* property:

$$b(A \cup B) = b(A) \cup b(B).$$

Then say that F is *b-closed* if and only if $b(F) \subseteq F$. The b-closed sets define the b-topology in which they are the closed sets, and one has $b(A) \subseteq \bar{A}$. Taking complements, the b-open sets are those sets A such that $A \subseteq b(A^c)$. Let us write $\mathcal{T}(b)$ for the topology of b-open sets. Both closure and derived set operators are base operators, and $\mathcal{T}(b) = \mathcal{T}$, i.e. both generate the original topology. But they are not the only base operators, by any means. Thus the density topology may be introduced using the base operator

$$b(A) := \{x : \bar{D}^*(A, x) > 0\},$$

so that F is b-closed if and only if $b(A) \subseteq A$. This agrees with the opening remarks as $\phi(A) = A \cap b(A)$.

Note that if a density theorem holds, then $\mathrm{cl}_{\mathcal{T}(b)}(A) = A \cup b(A)$ and is a measurable cover of A, and $\mathrm{int}_{\mathcal{T}(b)}(A) = \{a \in A : \underline{D}_*(A, x) = 1\} \cup (A \backslash bX)$ is a measurable core of A.

Lower Density. Given a measure space (X, Σ, m), and denoting by Σ_0 the σ-ideal of its null sets, write $A \sim B$ when $A \triangle B \in \Sigma_0$. A *lower density* (see also p. 104) is a mapping $S \colon \Sigma \to \Sigma$ satisfying:

(i) (quasi-identity) $S(A) \sim A$;
(ii) (neglecting property) if $A \sim B$, then $S(A) = S(B)$;
(iii) $S(\emptyset) = \emptyset$ and $S(X) = X$;
(iv) (multiplicative property) $S(A \cap B) = S(A) \cap S(B)$.

For $A \in \Sigma$, say that A is *S-open* if $A \subseteq S(A)$. This naturally reverses the inclusion used to define the b-closed sets. In summary, the topology consists of measurable sets and

$$\mathcal{T}_S := \{A \in \Sigma : A \subseteq S(A)\}.$$

Upper Density. One may take an entirely dual approach, preferring to introduce the topology via closed sets. An *upper density* is a mapping $U \colon \Sigma \to \Sigma$ satisfying:

(i) (quasi-identity) $U(A) \sim A$;
(ii) (neglecting property) if $A \sim B$, then $U(A) = U(B)$;
(iii) $U(\emptyset) = \emptyset$ and $U(X) = X$;
(iv) (additive property) $U(A \cap B) = U(A) \cap U(B)$.

Corresponding to a lower density S is the dual upper density $U(A) := S(A^c)^c$. Since complementation is self-inverse, any upper density corresponds to a dual lower density via $S(A) := U(A^c)^c$.

For an upper density U, say that $A \in \Sigma$ is a U-*closed* set if and only if $A \supseteq U(A)$. (Note that this yields the same topology as the dual inner density because $A^{cc} = A \supseteq S(A^c)^c = U(A)$ for A in Σ.)

For an upper density U and $A \in \wp(X)$, for $H(A)$ a measurable core of A (determined up to null sets), put

$$b(A) := U(H(A)).$$

Then b is a base operator and for $A \in \Sigma$

$$A \cap b(A^c) = A \backslash S(A),$$

so any $\mathcal{T}_S$ open set is b-open and conversely.

Remark   A topology introduced by a lower/upper density operation may be introduced by an abstract base operator. The corresponding closure operator then provides a way for viewing that same topology as arising from a 'derived set' operation.

Lifting.   A mapping $L: \Sigma \to \Sigma$ is a *lifting* if it is both an upper and a lower density (LukMZ1986, p. 223; see the characterization theorem on p. 224 – these are maximal density topologies, and the closure of an open set is open – and the existence theorem on p. 225).

Remark (Construction via Modifications)   O'Malley's set-wise characterization of the topology which he introduced via the family of approximately derivable functions (for which see Bog2007a, 5.8(v), pp. 370–373) motivates the treatment in LukMZ1986 in considering *modifications* as a further tool for a further refinement of any fine topology $\mathcal{T}$, say refining a topology τ. First there is the r-modification of a fine topology $\mathcal{T}$, using as base the $\mathcal{T}$-open sets that are both $\mathcal{F}_\sigma$ and $\mathcal{G}_\delta$ in τ. For the a.e.-topology, there are two possible approaches. The a.e.-topology generated by $\mathcal{T}$ has as base the sets of the form $U \cup \{x\}$ for $U \in \mathcal{T}$ and $x \in X$. Here V is a.e. open if and only if $V \backslash \mathrm{int}_{\mathcal{T}}(V)$ is $\mathcal{T}$-discrete. A related topology is the a-modification: if b is a base operator for $\mathcal{T}$, put

$$a(A) := b(\bar{A})$$

and introduce the corresponding topology. Here a set V is a-open if and only if $V = G \cup H$ with G a τ-open set $H \subseteq G^c$ and $H \subseteq (bG^c)^c$. This is in general coarser than the a.e.-topology. The a-topology inherits the Lusin–Menchoff property with respect to a completely regular τ.

8.5 * Fine Topologies from Ideals I: Base Operators

Notation We write $\mathcal{L}_+^*$ for the sets that have positive outer Lebesgue measure and $\mathcal{L}_+$ for non-null measurable sets. By analogy write $\mathcal{B}a_+^*(\mathfrak{X})$ for sets that are non-meagre in the space $\mathfrak{X}$ (since $\mathcal{B}a(\mathfrak{X})$ denotes its Baire sets) and $\mathcal{B}a_+(\mathfrak{X})$ for the Baire non-meagre sets; also we write $\mathcal{G}_+(\mathfrak{X})$ for the non-empty open sets and $\wp_+(X)$ for the non-empty subsets of X. Of course $\mathcal{B}a(\mathbb{R}, \mathcal{E})$ comprises the usual Baire sets and $\mathcal{B}a_+(\mathbb{R}, \mathcal{D}) = \mathcal{L}_+$.

We quote two general results of Martin (Marti1961), albeit not quite in his language, to introduce two fine topologies on $\mathbb{R}$ that we need to use later.

Definition Call a family $I \subseteq \wp(X)$ a proper *ideal* (with $X \notin I$) if

(i) I contains all singletons (atoms),
(ii) I is *downwards hereditary* (subsets of sets in I are in I), and
(iii) I is *multiplicative*, i.e. closed under intersection (if $I_1, I_2 \in I$, then $I_1 \cap I_2 \in I$).

Passing to negation $\mathcal{H} = I^\neg := \{H : H \notin I\}$, so that $\mathcal{H}^\neg = I$, we obtain the following dual concept used by Martin. Note that under negation (as opposed to complementation) property (iii) does not translate into closure under union (but rather the contrapositive of this, hence the 'co' below).

Definition Say that a family $\mathcal{H} \subseteq \wp_+(X)$ is *heavy* if

(i) $\mathcal{H}$ is *atomless* (no singletons in $\mathcal{H}$),
(ii) $\mathcal{H}$ is *upwards hereditary* (supersets of sets in $\mathcal{H}$ are in $\mathcal{H}$), and
(iii) $\mathcal{H}$ is 'additively maximal', i.e. *co-multiplicative* (if $H_1 \cup H_2 \in \mathcal{H}$, then at least one of H_1, H_2 is in $\mathcal{H}$).

Note that $\mathcal{F} = I^c := \{X \backslash I : I \in I\}$ is a non-principal filter. A non-principal ultrafilter $\mathcal{F}$ is heavy and $I = \mathcal{F}^c$ is a corresponding maximal ideal I. The term 'heavy' is borrowed from usage elsewhere (cf. in BraG1960). A heavy point of a set is one at which the set is locally of second category, i.e. the point is one of essential accumulation). The approach in LukMZ1986, p. 22 is slightly different – in place of heavy sets they use the family of sets not lying in a given ideal I and require that sets locally in I (i.e. every point of a set E has a neighbourhood V with $V \cap E$ in I) are in I.

Theorem 8.5.1 (Marti1961, Th. 1, cf. LukMZ1986, §1.C) *For $\mathcal{B}$ a base for $\mathfrak{X} = (X, \mathcal{T})$ and $\mathcal{H} \subseteq \wp_+(X)$ heavy, in particular for $\mathcal{H} = \mathcal{B}a_+^*$ or $\mathcal{L}_+^*$,*

$$b_{\mathcal{H}}(E) = \{x \in X : B \cap E \in \mathcal{H}\} \text{ for every } \mathcal{B} \in \mathcal{B}_x$$

is a base operator on X refining the topology $\mathcal{T}$.

Definition For $\mathcal{H} = \mathcal{B}a_+^*(\mathbb{R}, \mathcal{E})$ or $\mathcal{L}_+^*$ above, we will refer to the corresponding two refinements as the *essential topologies* in category, respectively in measure, and to the corresponding limit points of sets as *points of essential accumulation* in category, respectively in measure.

Remarks 1. The properties (i)–(iii) of the family $\mathcal{H}$ formalize a sense of largeness; thus the family $\mathcal{H}_\infty$ of infinite subsets of X satisfies them, and the corresponding operator gives rise to limit points that are points of accumulation (derived points), so here it does not refine the original topology strictly. The family $\mathcal{H}$ of uncountable subsets of X also satisfies them, and the corresponding induced finer topology gives rise to limit points that are necessarily points of condensation in the original topology.

2. Notice that $b_\mathcal{H}(E)$ is closed in X and so Baire. If y is in its closure, then for $V \in \mathcal{B}_y$ there is $x \in V \cap b_\mathcal{H}(E)$ and so there is $B \in \mathcal{B}_x$ with $B \subseteq V$. Then, as $x \in b_\mathcal{H}(E)$, one has $B \cap E \in \mathcal{H}$. By upwards closure $V \cap E \in \mathcal{H}$. The set $F(E) := E \cap b_\mathcal{H}(E)$, which is closed relative to E, may now be viewed as Stone's non-locally-$\mathcal{I}$ kernel of A, best called the local-$\mathcal{H}$ kernel (see Sto1963). The Generic Dichotomy Principle (Theorem 4.1.1) here asserts that the Stone kernel is either empty for some non-meagre $\mathcal{G}_\delta$ in X, or is almost all of A for each Baire set A.

In the refinement topology A is closed if $A \supseteq b_\mathcal{H}(A)$.

3. As $\mathbb{R}$ is a topological group under addition, if $\mathcal{H}$ is shift-invariant one may redefine the base operator by referring to a base at 0, say $\mathcal{B}_0$, and writing

$$b_\mathcal{H}(E) = \{x \in X : (\forall B \in \mathcal{B}_0)\,(B + x) \cap E \in \mathcal{H}\}$$
$$= \{x \in X : (\forall B \in \mathcal{B}_0)\,B \cap (E - x) \in \mathcal{H}\}$$

since $\mathcal{H} = \mathcal{H} - x$. Furthermore, if H is dilation-invariant (i.e. $\lambda H \in \mathcal{H}$ for $H \in \mathcal{H}$ and $\lambda > 0$), put $B = (-1/n, 1/n) = I/n$, with $I = (-1, 1)$ and $n = 1, 2, \ldots$, and write

$$b_\mathcal{H}(E) = \{x \in X : (\forall n)\,I \cap n(E - x) \in \mathcal{H}\}$$
$$= \bigcap_n \{x \in X : I \cap n(E - x) \in \mathcal{H}\}.$$

We shall later need to know that $b_\mathcal{H}$ takes values in $\mathcal{B}a(\mathbb{R}, \mathcal{E})$ or $\mathcal{B}a(\mathbb{R}, \mathcal{D})$. For this purpose, we verify the following result.

Lemma 8.5.2 *For $X = \mathbb{R}$ and $\mathcal{I} := \{S : S \notin \mathcal{H}\}$ closed under countable unions, the set $\{x : I \cap n(E - x) \in \mathcal{H}\}$ is open and so $b_\mathcal{H}(E)$ is a $\mathcal{G}_\delta$.*

Proof Put $S^+(E) := \{\langle u, v \rangle \in \mathbb{R}^2 : u < v$ and $(u, v) \cap E \in \mathcal{H}\}$. This is open. Indeed, if $\langle u, v \rangle \subseteq \langle a, b \rangle$ and $\langle u, v \rangle \in S^+(E)$, then $\langle a, b \rangle \in S^+(E)$; on the other hand, if for each rational $r > 0$ we have $\langle u + r, v - r \rangle \notin S^+(E)$, i.e. $(u + r, v - r) \cap E \notin \mathcal{H}$, then

$$(u, v) \cap E = \bigcup_{0 < r < (v-u)/2} (u + \delta, v - \delta) \cap E \in \mathcal{H}^{\neg},$$

a contradiction since $\mathcal{H}$ and $\mathcal{H}^{\neg}$ are disjoint. In turn, for $J = (a, b)$, the set $\beta_E(J) := \{y : (J + y) \cap E \in \mathcal{H}\}$ is also open since

$$\beta_E(J) = \bigcup \{(v - b, u - a) : \langle u, v \rangle \in S^+(E) \text{ and } 0 < v - u < b - a\}.$$

Indeed, if $(u, v) \subset (y + a, y + b)$ and $(u, v) \cap E \in \mathcal{H}$, then $v - b < y < u - a$. Finally,

$$b_{\mathcal{H}}(E) = \bigcap_n \left\{ x \in X : \left(\frac{1}{n} I + x\right) \cap E \in \mathcal{H} \right\} = \bigcap_n \beta_E \left(\frac{1}{n} I\right). \qquad \square$$

The lemma will enable us to show that almost all points of a set in $\mathcal{H}$ are essential points of accumulation.

4a. Suppose that $\mathcal{H}$ is heavy but not necessarily translation invariant. Let $\mathcal{B}_0$ be a countable base for the neighbourhoods of 0. Put

$$b_{\mathcal{H}}^*(E) := \{x \in X : (\forall B \in \mathcal{B}_x) \ B \cap E \in \mathcal{H} + x\}.$$

As $\mathcal{B}_0$ is countable, this defines a base operator. Indeed $B \cap E \in \mathcal{H} + x$ if and only if $(B - x) \cap (E - x) \in \mathcal{H}$, so with $B' := (B - x) \in \mathcal{B}_0$ we have that $B' \cap ((E \cup F) - x) \in \mathcal{H}$ implies that one of $B' \cap (E - x) \in \mathcal{H}$ or $B' \cap (F - x) \in \mathcal{H}$ holds, and so for infinitely many $B' \in \mathcal{B}_0$ one of $B' \cap (E - x) \in \mathcal{H}$ or $B' \cap (F - x) \in \mathcal{H}$ holds. Thus $b_{\mathcal{H}}^*(E \cup F) \subseteq b_{\mathcal{H}}^*(E) \cup b_{\mathcal{H}}^*(F)$. As $b_{\mathcal{H}}^*$ is monotone, $b_{\mathcal{H}}^*$ is additive. So $b_{\mathcal{H}}^*$ determines a translation-invariant fine topology.

4b. The *z-topology*. Let $z_n \to 0$ be a (null) sequence comprising infinitely many distinct terms. Put $Z = \{z_n : n \in \omega\}$ and take $\mathcal{H}_Z$ to consist of sets meeting Z in an infinite set; then $\mathcal{H}_Z$ is heavy. Put

$$b^z(E) = \bigcap_n \bigcup_{k > n} (E - z_k).$$

Note that in particular $b^z(E)$ is in $\mathcal{B}a(\mathbb{R}, \mathcal{E})$ or $\mathcal{B}a_+(\mathbb{R}, \mathcal{D}) = \mathcal{L}_+$ if E is. We claim that for $\mathcal{H} = \mathcal{H}_Z$ one has $x \in b^z(E)$ if and only if $x \in b_{\mathcal{H}}^*(E)$. Indeed $x \in b^z(E)$ if and only if there is an infinite set $\mathbb{M}_x$ with $\{x + z_n : n \in \mathbb{M}_x\} \subseteq E$ if and only if $(x + \{z_n : n \in \mathbb{M}_x\}) \subseteq B \cap E \in \mathcal{H} + x$ if and only if $x \in b_{\mathcal{H}}^*(E)$.

Call such an x a *translator into E*. Thus A is b-closed if and only if $A \supseteq b_{\mathcal{H}}(A)$ if and only if *every* translator into A is in A. Below we give the dual statement in terms of z-open sets, which enables us to give an illuminating example.

Theorem 8.5.3 *For any null sequence* $z = \langle z_n \rangle$, *a set S is z-open if and only if*

$$S \subseteq \bigcup_k \bigcap_{n \geq k} (S - z_n),$$

i.e. S is z-open if and only if for each $s \in S$ *the sequence* $\langle s + z_n \rangle$ *is almost contained in S.*

Proof   Writing $T = \mathbb{R} \backslash S$ note that T is z-closed if and only if $b^z(T) \subseteq T$ if and only if

$$\bigcap_k \bigcup_{n \geq k} (T - z_n) \subseteq T.$$

Passing to complements,

$$\mathbb{R} \backslash T \subseteq \bigcup_k \bigcap_{n \geq k} \mathbb{R} \backslash (T - z_n) = \bigcup_k \bigcap_{n \geq k} (S - z_n). \qquad \square$$

Example   If z is any rational null sequence, then $T := (0, 1) \backslash \mathbb{Q}$ is z-closed since $\mathbb{Q}$ is z-open, because any rational translate of the sequence z is rational. However, T is not closed in the Euclidean topology.

Remarks   1. In the measure case, the essential topology is considered briefly in Scheinberg (Sch1971, §2) as a topology in which the Borel sets coincide with $\mathcal{L}$ and it is shown also that it is connected but not regular.

2. Sch1971, §3 considers a family $\mathcal{H}$ obtained from the family $\mathcal{F}$ of measurable sets of density 1 at 0 by extending this to a maximal filter in $\mathcal{L}$ (an $\mathcal{L}$-ultrafilter). The family $\mathcal{H}$ is in particular heavy (relative to $\mathcal{L}$ – see his property (4) which implies maximal additivity) and not translation invariant; here again the same translation technique as with the z-topology may be used to generate a topology. Scheinberg shows that the Borel sets of this topology again coincide with $\mathcal{L}$ and that the topology is completely regular.

Motivated by the O'Malley characterization of his own fine topologies as 'modifications' of the Euclidean topology, we refer to the following result as a 'modification' theorem.

Theorem 8.5.4 (Modification Theorem; Marti1961, Th. 6; cf. LukMZ1986, Prop. 1.8)   *For* $\mathfrak{X} = (X, \mathcal{T})$ *second countable and* $\mathcal{H} \subseteq \wp_+(X)$ *heavy (in particular for* $\mathfrak{X} = (\mathbb{R}, \mathcal{E})$ *and* $\mathcal{H} = \mathcal{B}a_+^*$ *or* $\mathcal{L}_+^*$) *suppose that in addition:*

(iii)' $\mathcal{H}$ *is countably complete (if* $\bigcup_i H_i \in \mathcal{H}$, *then* $H_i \in \mathcal{H}$ *for some i);*

(iv) $\mathcal{H}$ *contains* $\mathcal{G}_+$.

Then F is $\mathcal{T}(b_{\mathcal{H}})$-closed if and only if $F = C \cup Z$ where C is $\mathcal{T}$-closed and $Z \notin \mathcal{H}$; dually, V is $\mathcal{T}(b_{\mathcal{H}})$-open if and only if $V = U \backslash Z$, where U is $\mathcal{T}$-open and $Z \notin \mathcal{H}$.

Hence $\mathcal{T}(b_{\mathcal{H}})$ is neither first countable nor regular.

8.6 * Fine Topologies from Ideals II: Modes of Convergence

Our lead example here is a category analogue of the density topology, obtained by first introducing a category analogue of density points. It is remarkable that such a quantitatively defined notion can have a qualitative dual, and that marks out this example as noteworthy; this was achieved by Wilczyński by altering the mode of convergence and referring to the (Riesz) Subsequence Theorem, 1.3.1 (see §1.2).

The definition of a density point of a measurable set may be reformulated in terms of indicator functions; by translation, it is enough to consider the case of 0 being a density point. For A any set on the line define its *dilation* by λ to be

$$\lambda A := \{\lambda a : a \in A\}.$$

Since

$$\frac{|A \cap [-1/n, 1/n]|}{1/(2n)} = \frac{|nA \cap [-1, 1]|}{2},$$

0 is a density point of A if and only if

$$|nA \cap [-1, 1]| \to 2.$$

Equivalently, passing to indicator functions and working in the space $L^0[-1, 1]$ of measurable function on $[-1, 1]$:

$$1_{nA \cap [-1,1]} \to 1 \quad \text{in measure (on } [-1, 1]).$$

For fixed A, writing $J_n := 1_{nA \cap [-1,1]}$, by the Subsequence Theorem, 1.3.1, this is equivalent to the assertion that every subsequence $\langle J_{n(k)} \rangle$ has a sub-subsequence $\langle J_{n(k(j))} \rangle$ converging a.e. to 1, i.e. off a null set.

The latter statement refers to the σ-ideal of null sets and to the formation of subsequences. This opens up a number of possibilities for refining the usual notion of density point, and hence the usual topology of the line, by varying both the σ-ideal and the mode of convergence demanded of the sequence J_n. Though our starting point is the base operator of the last section, modifying the mode of convergence restricts the accumulation properties to the extent of undermining the additive property of the operator; fortunately the monotone operator is multiplicative and so the fine topologies here are introduced using the *lower density* construction. We consider in detail two examples and then comment on others. The notation is as above.

1. For $\mathcal{H}$ a filter, passing to the ideal $I = \mathcal{H}^{\neg}$, the definition of $b_{\mathcal{H}}$ may be read as saying that $1_{n(E-x)\cap I}(t) \to 1$, I-a.e. on $I = [-1, 1]$ for each n. A variant may be obtained by replacing this a.e.-constancy by *ordinary* (or simple) a.e.-convergence, requiring instead that $1_{n(E-x)\cap I}(t)$ converges I-a.e. to 1 on $[-1, 1]$. This leads to

$$\Phi_{\mathcal{H}}^{\mathrm{ord}}(E) = \{x \in X : (\exists k)(\forall n \geq k)\, I \cap n(E - x) \in \mathcal{H}\}$$

$$= \bigcup_k \bigcap_{n \geq k} \left\{x \in X : \left(\frac{1}{n}I + x\right) \cap E \in \mathcal{H}\right\};$$

this time a $\mathcal{G}_{\delta\sigma}$ monotone operator. We shall see below that $\Phi_{\mathcal{H}}^{\mathrm{ord}}$ is 'multiplicative' rather than 'additive'. Hence, if applied to the σ-algebra $\mathcal{B}a_+$ is a lower-density operator (as opposed to a base operator).

2. Switching from the simple to an alternative mode of convergence for the indicator functions, which we call *sequential* (or strong), we may require now that for each subsequence $\gamma(n)$ there be a sub-subsequence $\gamma(\kappa(n))$ such that $1_{\gamma(\kappa(n))(E-x)\cap I}(t)$ converges I-a.e. to 1 on $[-1, 1]$. This leads to

$$\Phi_{\mathcal{H}}^{\mathrm{seq}}(E) = \bigcap_{\gamma \in \Gamma} \bigcap_n \bigcup_{k > n} \{x \in X : I \cap \gamma(k)(E - x) \in \mathcal{H}\},$$

where $\Gamma := \mathbb{N}^{\mathbb{N}}$ denotes the set of all subsequences. We shall presently see that $\Phi_{\mathcal{H}}^{\mathrm{seq}}$ is also 'multiplicative' (rather than 'additive'). The set here is the complement of the set

$$\Phi^c(E) := \bigcup_{\gamma \in \Gamma} \bigcup_n \bigcap_{k > n} \{x \in X : I \cap \gamma(k)(E - x) \notin \mathcal{H}\}$$

and we have

$$\Phi^c(E) = \bigcup_n \bigcup_{\gamma \in \Gamma} \bigcap_i A_i^n,$$

with

$$A_i^n := \bigcap_{n+i > k > n} \{x \in X : I \cap \gamma(k)(E - x) \notin \mathcal{H}\},$$

as

$$\bigcap_i A_i^n = \bigcap_{k > n} \{x \in X : I \cap \gamma(k)(E - x) \notin \mathcal{H}\}.$$

By Lemma 8.5.2, the sets A_i^n are closed and so $\Phi^c(E)$ is Souslin-$\mathcal{F}$. Thus, by Nikodym's theorem, $\Phi^c(E)$ and so $\Phi_{\mathcal{H}}^{\mathrm{seq}}(E)$ is Baire. Here again, if applied to the σ-algebra $\mathcal{B}a_+$, this is a lower-density operator.

We need to know now that our last two monotone operators are in fact lower densities. Here we check only that they are multiplicative. In Chapter 9, we shall see why when applied to a non-meagre/non-null set they are the identity modulo their defining ideal, equivalently why quasi all points are density points (in whichever of the two senses and the two ideals). This will follow from the facts, just established above, that they take Baire sets to Baire sets.

Lemma 8.6.1 *Both* $\Phi = \Phi_{\mathcal{H}}^{\mathrm{ord}}$ *and* $\Phi = \Phi_{\mathcal{H}}^{\mathrm{seq}}(E)$ *are multiplicative, i.e.*

$$\Phi(A \cap B) = \Phi(A) \cap \Phi(B).$$

Proof We consider first $\Phi = \Phi_{\mathcal{H}}^{\mathrm{seq}}(E)$. By monotonicity we have $\Phi(A \cap B) \subseteq \Phi(A)$ and $\Phi(A \cap B) \subseteq \Phi(B)$, and so $\Phi(A \cap B) \subseteq \Phi(A) \cap \Phi(B)$.

For the reverse inclusion, let $t_0 \in \Phi(A) \cap \Phi(B)$. So 0 is a density point of both $A' := A - t_0$ and $B' := B - t_0$.

Now consider any $\gamma \in \Gamma$, where, as before, $\Gamma := \mathbb{N}^{\mathbb{N}}$ denotes the set of all subsequences. There is $\kappa \in \Gamma$ such that $J^{A'}(t) = \lim_{\gamma(\kappa(i))} \langle J_i^{A'}(t) \rangle$ quasi all t. But there is also $\lambda \in \Gamma$ such that $J^{B'}(t) = \lim_{\gamma(\kappa(\lambda(i)))} \langle J_i^{B'}(t) \rangle$ quasi all t. The latter occurs simultaneously with $J^{A'}(t) = \lim_{\gamma(\kappa(\lambda(i)))} \langle J_i^{A'}(t) \rangle$ quasi all t (since pointwise convergence is preserved under subsequences). Hence $J^{A' \cap B'}(t) = \lim_{\gamma(\kappa(\lambda(i)))} \langle J_i^{A' \cap B'}(t) \rangle$ quasi all t down the subsequence $\kappa \circ \lambda$ of γ. So 0 is a density point of $A' \cap B' = A \cap B - t_0$, i.e. $t_0 \in \Phi(A \cap B)$.

The other case $\Phi = \Phi_I^{\mathrm{ord}}$ is similar but simpler. $\qquad\square$

In summary: by switching σ-ideals and varying modes of convergence as appropriate (ordinary-a.e., summable-a.e.) one may introduce the notion of

(i) a *simple-density point*, by demanding ordinary a.e.-convergence of the original sequence J_n relative to the null sets $\mathcal{L}_0$ (as in WilW2007) – so a simple-density point is a density point;

(ii) a summable-a.e. density point, better known as *complete-density point*, by demanding that for each $\epsilon > 0$

$$\lim_{m \to \infty} \sum_m |\{t : |J_m(t) - 1| > \epsilon\}| = 0,$$

or equivalently that

$$\sum_n |\{t : |J_n(t) - 1| > \epsilon\}| < \infty$$

(as in WilW2007); this topology is coarser than the density topology and also coarser than the simple density topology:

$$\mathcal{E} \subset \mathcal{T}_c \subset \mathcal{T}_s \subset \mathcal{D}.$$

(iii) a *category-density point* by demanding a.e.-strong (i.e. sequential) convergence relative to the σ-ideal of meagre sets (as in PorWW1985).

Correspondingly one obtains the complete density topology and the category-density topology. This is connected but not regular, and hence is not the coarsest topology making $\mathcal{I}$-*approximately* continuous functions continuous.

We are now able to complement §4.1 with the following examples.

Example 1   Our first example is a category analogue of the example in §4.1, obtained by referring to the category analogue of density points from the complete density topology above. We just saw that the topology may be introduced by a monotone operator, which we now denote by $\Phi_{\mathcal{H}}^{\mathrm{cplt}}(E)$, which takes Baire sets to Baire sets and is multiplicative. So to show that almost all points of E are in $\Phi_{\mathcal{H}}^{\mathrm{cplt}}(E)$ it is enough to demonstrate the existence of points in $\Phi_{\mathcal{H}}^{\mathrm{cplt}}(E)$ for E Baire non-meagre. This we now do.

Existence.   For E a Baire set, if E is non-meagre there is a non-empty open interval G with $G\backslash M \subseteq E$ for some meagre set M.

But every point g of G is a (simple) density point of G and so of E, as 0 is a (simple) density point of $G - g$ and so of $E - g$. That is, E has a (simple) density point. So almost all points of E are (simple) density points.

Thus in fact $\Phi_{\mathcal{H}}^{\mathrm{cplt}}(E)$ is an inner-density.

Example 2   Here we refer to the z-topology. Let $z_n \to 0$ and put $F(T) := \bigcap_{n\in\omega} \bigcup_{m>n}(T - z_m)$. Thus $F(T) \in \mathcal{B}a$ for $T \in \mathcal{B}a$ and F is monotonic. Here $t \in F(T)$ if and only if there is an infinite $\mathbb{M}_t$ such that $\{t + z_m : m \in \mathbb{M}_t\} \subseteq T$. The Generic Dichotomy Principle (Theorem 4.1.1) asserts that once we have proved (for which see Theorem KBD, 4.2.1) that an arbitrary non-meagre Baire set T contains a 'translator', i.e. an element t which shift-embeds a subsequence z_m into T, then quasi all elements of T are translators.

Example 3   For $z_n = n$ and $\{S_k\}$ a family of unbounded open sets (in the Euclidean sense), put $F(T) := T \cap \bigcap_{k\in\omega} \bigcap_{n\in\omega} \bigcup_{m>n}(S_k - z_m)$. Thus $F(T) \in \mathcal{B}a$ for $T \in \mathcal{B}a$ and F is monotonic. Here $t \in F(T)$ if and only if $t \in T$ and, for each $k \in \omega$, there is an infinite $\mathbb{M}_t^k$ such that $\{t + z_m : m \in \mathbb{M}_t^k\} \subseteq S_k$. In Kin1963, it is shown that $F(V)$ is non-empty for any non-empty open set V; but in the Bitopological Kingman Theorem, 5.1.6, we must adjust the argument to show that $F(T)$ is non-empty for arbitrary non-meagre sets $T \in \mathcal{B}a$, hence that quasi all members of T are in $F(T)$, and in particular that this is so for $T = \mathbb{R}_+$.

8.7 * Coarse versus Fine Topologies

Definitions   Let $\mathcal{B}$ be a base of open sets of a space $\mathfrak{X} = (X, \mathcal{T})$.

A *regular* filter base in $\mathcal{B}$ is a filter $\mathcal{F} \subseteq \mathcal{B}$ such that any finite intersection of sets in $\mathcal{F}$ contains the closure of an element of $\mathcal{F}$. A maximal such filter base is termed an *ultrafilter* base.

A regular filter base $\mathcal{F}$ is *pre-convergent* if $\bigcap \mathcal{F}$ is non-empty. A regular ultrafilter is called *convergent* if $\bigcap \mathcal{F}$ is a single point.

Example For $\mathcal{D}$ the density topology, $\mathcal{B} = \mathcal{D}$, $\mathcal{F} = \mathcal{F}_0$, the family of open rational intervals about 0 is a pre-convergent (countable) regular filter base.

Definitions The space $\mathfrak{X}$ is *subcompact* if for some $\mathcal{B}$ every regular filter base relative to $\mathcal{B}$ is pre-convergent.

Equivalently, the space $\mathfrak{X}$ is *subcompact* if for some $\mathcal{B}$ every ultrafilter base relative to $\mathcal{B}$ is convergent.

The space is *countably subcompact* if for some $\mathcal{B}$ every countable regular filter base relative to $\mathcal{B}$ is pre-convergent.

Theorem 8.7.1 (Gro1963) *For $\mathfrak{X}$ metrizable, (countable) subcompactness is equivalent to topological completeness.*

Theorem 8.7.2 (Gro1963) *A regular (countably) subcompact space is a Baire space.*

For studying a space $\mathfrak{X} = (X, \mathcal{T})$ by reference to coarser topologies $\mathcal{T}' \subseteq \mathcal{T}$, we need two definitions. Say that $\mathfrak{X}' = (X, \mathcal{T}')$ is a *cospace* of $\mathfrak{X}$ and $\mathcal{T}'$ a *cotopology* if for all V closed in $\mathcal{T}$ and all $x \in \text{int}_{\mathcal{T}} V$ there is a $\mathcal{T}'$-closed U (so $\mathcal{T}$-closed) such that

$$x \in \text{int}_{\mathcal{T}} U \subseteq U \subseteq V.$$

For $\mathcal{T}$ regular, this is equivalent to the existence of a neighbourhood base for $\mathcal{T}$ consisting of $\mathcal{T}'$-closed sets. When the cotopology $\mathcal{T}'$ is metrizable, we say that $\mathfrak{X}$ is *cometrizable*. Interest in coarser topologies comes from the following:

Example If C is a family of closed sets containing a (closed) neighbourhood base for each point of space, then $\{X \backslash F : F \in C\}$ is a cotopology for X.

Definition A space $\mathfrak{X}$ is *cocompact* if it possesses a compact cospace $\mathfrak{X}'$.

The next theorem requires a lemma on refinement of locally finite covers (for which see §1.2). We omit the proof of the lemma, as this would take us too far afield; see AarGM1970b (embedded within the proof of Th. 1; this is essentially Dow1947, Lemma 3.3).

Lemma 8.7.3 (A 'Centred Is Finite' Refinement) *For $\mathfrak{X}$ normal, and $\mathcal{U}$ a locally finite open covering, there is an open refinement $\mathcal{V}$ such that every subfamily of $\{\bar{V} : V \in \mathcal{V}\}$ with the finite intersection property is finite.*

Theorem 8.7.4 (AarGM1970b, Th. 1)   *A metrizable space is topologically complete if and only if it is cocompact.*

Proof   Suppose $\mathfrak{X} = \langle X, \mathcal{T} \rangle$ is metrizable and cocompact, i.e. that $\mathfrak{X}' = \langle X, \mathcal{T}' \rangle$ is compact. As X is regular, the family $\mathcal{U}$ of $\mathcal{T}'$-closed subsets of X yields a neighbourhood base for $\mathfrak{X}$. Embed X densely in a complete metric space $\langle Y, \rho \rangle$ (see Eng1989, Th. 4.3.14 and 4.3.15). In view of the Alexandroff–Hausdorff Theorem (Eng1989, Th. 4.3.23; cf. p. 35), we proceed to show that X is a $\mathcal{G}_\delta$ subset of Y. With δ denoting the ρ-diameter, put $\mathcal{U}_i := \{U \in \mathcal{U} : \delta(U) \leq 2^{-i}\}$ and

$$G_i := \{V : V \text{ open in } Y \text{ and } V \cap X \subset U \text{ for some } U \text{ with } \delta(U) \leq 2^{-i}\}.$$

So $X \subseteq \bigcap_i G_i$, since $\mathcal{U}$ is a neighbourhood base for $\mathfrak{X}$. For the reverse inclusion, consider $y \in \bigcap_i G_i$. As $y \in G_i$, pick V_i open in Y with $y \in V_i$ and U_i in $\mathcal{U}$ such that $V_i \cap X \subset U_i$ with $\delta(U_i) \leq 2^{-i}$. As X is dense in Y, and because $\{V_i : i \in \omega\}$ has the finite intersection property (witnessed by y), the family $\{V_i \cap X : i \in \omega\}$ also has the finite intersection property, and so too has $\{U_i : i \in \omega\}$. However, $\mathfrak{X}'$ is compact, so $\bigcap_i U_i \neq \emptyset$, and, since $\delta(U_i) \leq 2^{-i}$ for each i, $\bigcap_i U_i = \{x\}$, for some $x \in X$. Working in the metric space Y, we have $y \in V_i$ for all i and by density $\delta(V_i) \leq 2^{-i}$; on the other hand, $x \in U_i$ all i, so $x = y$ (otherwise for large enough i, $U_i \cap V_i = \emptyset$). Thus $X = \bigcap_i G_i$ and so is a $\mathcal{G}_\delta$ subset of Y.

For the converse, let ρ now denote a complete metric on X compatible with $\mathcal{T}$. Working in $\mathcal{T}$, by paracompactness for each $i \in \omega$ there is a locally finite open covering $\mathcal{U}_i$ refining the open balls of ρ-radius 2^{-i}. By Lemma 8.7.3, there exists for each $i \in \omega$ an open covering refinement $\mathcal{V}_i$ of $\mathcal{U}_i$ such that every subfamily of $\{\bar{V} : V \in \mathcal{V}_i\}$ with the finite intersection property is finite. Put $\mathcal{B}' := \bigcup_i \{X \setminus \bar{V} : V \in \mathcal{V}_i\} \subseteq \mathcal{T}$, and let $\mathcal{T}'$ be the (coarser) topology generated by taking $\mathcal{B}'$ as a base; then $\mathfrak{X}' = \langle X, \mathcal{T}' \rangle$ is compact by the Alexander Subbase Theorem (see Eng1989, Prob. 3.12.2 or Kel1955, Ch. 5, Th. 6). Indeed, the family $\bar{\mathcal{V}} := \bigcup_i \{\bar{V} : V \in \mathcal{V}_i\}$ is a subbase for the closed sets of $\mathcal{T}'$, and in any infinite subfamily of $\bar{\mathcal{V}}$ with the finite intersection property there is a further infinite subfamily with shrinking ρ-diameter (by the property of $\mathcal{V}_i$), and the latter subfamily has non-empty intersection (by Cantor's Theorem – §1.2). It remains to check that $\mathcal{T}'$ is a cotopology, i.e. that any point has a neighbourhood base for $\mathcal{T}$ consisting of $\mathcal{T}'$-closed sets. But this follows from the construction above, since $\mathcal{V}_i$ is an open covering refining the balls of radius 2^{-i}.   □

9

Category–Measure Duality

9.1 Introduction

The duality between measure and category emerged in the 1920s, largely in the works of Sierpiński. See the commentary by Hartman (Hart1975), in Sierpiński's selected works (Sie1975; Sie1976).

We recall first that we term subsets of $\mathbb{R}$ *non-negligible* under some topology if they have the Baire property and are non-meagre. This has the usual category meaning for the Euclidean topology, whereas in the context of the density topology this means they are measurable and non-null. This chapter is devoted to some classical results in what is known as *category–measure duality*. These are theorems in real analysis which, regarding negligibility or non-negligibility, yield identical conclusions in the category and measure cases. Thus the definition of topological completeness allows us to capture a structural feature of category–measure duality: both exhibit $\mathcal{G}_\delta$ *inner regularity*, modulo sets which we are prepared to neglect. Other instances of duality include the following:

(i) The Poincaré recurrence theorem from statistical mechanics: this occurs in both category and measure forms; see, e.g., Oxt1980, Ch. 17.

(ii) The zero–one law of probability theory, which similarly dualizes; see, e.g., Oxt1980, Ch. 21.

(iii) The Erdős–Sierpiński duality principle, which gives full duality assuming the Continuum Hypothesis (CH); see, e.g., Oxt1980, Ch. 19.

Where there are identical conclusions, the basis resides in the dual foundation unified in two ways in BinO2009b and BinO2010h. As we had occasion to say in Chapter 4, the former creates structural unification by way of identical combinatorics; the latter a common single source: Baire category. Again, both views translate immediately to $\mathbb{R}^d$.

It is as well to be aware from the outset, however, that this duality has its limits; see Chapter 14. For example, in probabilistic language, despite (ii), it does not extend as far as the strong law of large numbers (see, e.g., Oxt1980, p. 85), nor the theory of random series (see, e.g., Kah2000; Kah2001). Similarly, despite (i) it does not extend as far as the (Birkhoff–Khinchin) ergodic theorem (by above – as the ergodic theorem includes the strong law of large numbers). For other limitations of category–measure duality, see, e.g., DouF1994; Barto2000; BartoJS1993; Fre2008, §522.

9.1.1 Uniqueness Theorems for Trigonometric Series

A rather different example is the uniqueness theorems for trigonometric series. For the classical theory here, see Bary1964; Zyg1988. A set P on the torus $\mathbb{T}$ is called a *set of uniqueness* if every trigonometric series which converges to zero outside P vanishes identically (P is a *set of multiplicity* otherwise). Every (Lebesgue-)measurable set of uniqueness is null, see Bary1964, Vol. II, Ch. XIV; Zyg1988, Vol. I, 9.6; KecL1987, 1.3. Much more recent is the corresponding category result: every Baire set of uniqueness is meagre (KecL1987, VIII; DebSR1987).

9.1.2 Cauchy Functional Equation

One of the earliest dualities concerns the Cauchy functional equation:

$$f(x + y) = f(x) + f(y) \qquad (x, y \in \mathbb{R}), \tag{CFE}$$

where the unknown function f has the Baire property (meaning it is measurable when the topology on $\mathbb{R}$ is the density topology), with the conclusion that f is continuous and so linear.

The results are linked to Steinhaus' Theorem, 9.2.2, that the difference set, $S - S$, for S non-negligible, contains an interval (Oxt1980, p. 93, Supplementary notes). The category case is known as the *Piccard–Pettis Theorem*. Their generalizations are known as *Steinhaus–Weil Theorems*. We turn to these in Chapter 15.

9.1.3 Kodaira's Theorem

Yet quite another linkage between category and measure is traced in Kodaira's Theorem (Kod1941) which relies on a density topology arising from Haar measure, rather like the Lebesgue density topology of Chapter 7 (cf. p. 257). For details see BinO2010g, Th. 7.3. The theorem asserts that for X a normed locally compact group and $f \colon X \to Y$ a homomorphism into a separable normed group Y, f is Haar-measurable if and only if f is Baire under the density topology if and only if f is continuous under the norm topology.

We regard this as further evidence for the role of category taking precedence over that of measure.

Of a different nature is a pair of theorems asserting that a planar set is negligible if and only if all but a negligible set of vertical sections are negligible. In the category case, under the Euclidean topology, this is the Kuratowski–Ulam Theorem (Oxt1980, Ch. 15) and in the measure case the Fubini Theorem for Null sets (Oxt1980, Ch. 14), which we term the Fubini Null Theorem (see §9.5). Unlike the theorems of the preceding paragraph, these two results do not have a common proof. Indeed, the first of the pair does not have a generalization that is generally valid beyond separable spaces, unlike the second. We will discuss a restricted variant of the Kuratowski–Ulam Theorem, due to Fremlin, Natkaniec and Recław, following FreNR2000; examples of cases of failure are given by Pol and van Mill; see Pol1979 and MilP1986. We also include a proof of the Fubini Null Theorem.

9.2 Steinhaus Dichotomy

We deduce the two classical motivating theorems, the Piccard–Pettis and the Steinhaus theorems – which constitute the Steinhaus dichotomy. See Piccard (Pic1939; Pic1942), Pettis (Pet1950) and, for the measurable case, Steinhaus (Stei1920).

Theorem 9.2.1 (Piccard–Pettis Theorem) *For $S \subseteq \mathbb{R}$ Baire and non-meagre, the difference set $S - S$ contains an interval around the origin.*

Theorem 9.2.2 (Steinhaus Theorem) *For $S \subseteq \mathbb{R}$ (Lebesgue-)measurable and of positive measure, the difference set $S - S$ contains an interval around the origin.*

Proof of Theorem 9.2.1 Suppose otherwise. Then, for each positive integer n we may select $z_n \in (-1/n, +1/n) \setminus (S-S)$. Since $z_n \to 0$, by Theorem KBD, 4.2.1, for quasi all $s \in S$ there is an infinite $\mathbb{M}_s$ such that $\{s + z_m : m \in \mathbb{M}_s\} \subseteq S$. Then, for any $m \in \mathbb{M}_s$, we have $s + z_m \in S$, i.e. $z_m \in S - S$, a contradiction. □

Remark See Chapter 15 (and Bin02010g) for a derivation from here of the more general result that for S, T Baire and non-meagre in the Euclidean topology, the difference set $S - T$ contains an interval.

Proof of Theorem 9.2.2 Arguing as above, here again Theorem 4.2.1 applies.

□

Just as with the Pettis extension of Piccard's result, so also here, Steinhaus proved that for S, T non-null measurable $S - T$ contains an interval. For this see again Chapter 15: the result may be derived from the Category Embedding Theorem, 10.2.2; see BinO2010g.

These results extend to topological groups. See, e.g., Com1984, Th. 4.6, p. 1175, for the positive statement, and the closing remarks for a negative one.

9.3 Subgroup Dichotomy

The next result gives (with its refinement, Theorem 9.3.2) the third of the sharp dichotomies of this chapter. It concerns additive subgroups of the reals. These may be small or large in a number of senses, to be made precise below. For instance, a subgroup may be discrete (the integers, for example) or dense (the rationals); we use Kronecker's Theorem (see HarW2008, XXIII, Th. 438) to split these two cases. In analysis one needs a stronger dichotomy, in which 'large' means total – the entire real line. Theorem 9.3.1 takes small in the dual senses of category and measure, and is a direct consequence of Theorem 9.3.2 (see Remark 1 after that result).

Theorem 9.3.1 (Subgroup Theorem; cf. BGT1987, Cor. 1.1.4; Lac1998) *For an additive Baire (resp. measurable) subgroup S of $\mathbb{R}$, the following are equivalent:*

(i) $S = \mathbb{R}$;
(ii) *S is non-meagre (resp. non-null).*

Proof By Theorem 4.2.1, for some interval I containing 0, we have $I \subseteq S - S \subseteq S$, and hence $\mathbb{R} = \bigcup_n nI = S$.   □

Interest in the theorem comes typically from examples such as the one below.

Example For an additive extended real-valued function $f: \mathbb{R} \to \mathbb{R} \cup \{\infty\}$, $D_f := \{x : f(x) < \infty\}$ is a subgroup of $\mathbb{R}$ and is the domain of definition of the corresponding real-valued 'partial' function f.

For a more intricate example, where the domain of definition of a 'partial function' is a subgroup of $\mathbb{R}$, see BGT1987, Lemma 3.2.1. The task there is to give additional conditions under which the subgroup is all of $\mathbb{R}$, so that the partial function is in fact total (cf. BGT1987, Th. 3.2.5) Sometimes a sufficient additional condition is density of the subgroup (in $\mathbb{R}$), which in turn may be reduced to the existence of two rationally incommensurable elements in the subgroup (by Kronecker's Theorem). Theorem 9.3.2 uses related but stronger conditions than density guaranteeing 'totality'. See BGT1987, Th. 1.10.2, where one uses density to show that in fact the domain of definition contains a co-countable set, so that the Subgroup Theorem applies.

Here we develop a combinatorial version, in the language of Ramsey theory (TaoV2006; GraRS1990) (though this is infinite Ramsey theory, while GraRS1990 is finite Ramsey theory – see their page 185).

Definitions   1. Say that a set $S \subseteq \mathbb{R}$ has the *strong (weak) Ramsey distance property* if for any convergent sequence $\{u_n\}$ there is an infinite set (a set with two members) $\mathbb{M}$ such that

$$\{u_n - u_m : m, n \in \mathbb{M} \text{ with } m \neq n\} \subseteq S.$$

Thinking of the points of S as those having a particular colour, S has the strong Ramsey distance property if any convergent sequence has a subsequence all of whose pairwise distances have this colour.

2. Motivated by Lemma 9.3.4, say that $S \subseteq \mathbb{R}$ has the *finite covering property* if there is an interval I and finite number of points $\{x_i : i = 1, \ldots, m\}$ such that the shifts $\{S + x_i : i = 1, \ldots, m\}$ cover I. When S is a subgroup, these shifted copies of S are just S-cosets.

Theorem 9.3.2 (Combinatorial Steinhaus Theorem)   *For an additive subgroup S of $\mathbb{R}$, the following are equivalent:*

(i) $S = \mathbb{R}$;
(ii) S *contains a non-meagre Baire, or a non-null measurable set;*
(iii) S *is shift-compact;*
(iv) S *has the strong Ramsey distance property;*
(v) S *has the weak Ramsey distance property;*
(vi) S *has the finite covering property;*
(vii) S *has finite index in $\mathbb{R}$.*

The proof follows Lemmas 9.3.3 and 9.3.4; for an application see BinO2009b. Theorem 9.3.2 effects a transition from topological through combinatorial to algebraic notions bringing out different aspects of the Steinhaus Theorem. See Laczkovich (Lac1998) for a topological study of proper subgroups of $\mathbb{R}$ (cf. BinO2011a, §10, Remark 4). Taking a topological view as in §6.4, any open covering of a shift-compact subset of $\mathbb{R}$ yields a finite 'shifted-subcovering', i.e. one consisting of shifted copies of a finite number of members of the open covering (BinO2010g).

Lemma 9.3.3 (Finite Index Property)   *An additive subgroup of $\mathbb{R}$ has finite index if and only if it coincides with $\mathbb{R}$.*

Proof Suppose an additive subgroup S has finite index, n say, so that the quotient $\mathbb{R}/S$ is a finite group of order n. Then, for each $x \in \mathbb{R}$, denoting S-cosets by $[x]$, one has $n[x/n] = [0]$ by Lagrange's Theorem. That is, $x = n(x/n) \in S$ i.e. S is $\mathbb{R}$ itself. $\qquad\square$

We use the above result in combination with the next observation, which is actually an instance (by specialization to $\mathbb{R}$) of the Finite Index Lemma due to Neumann (see Neu1954a, Neu1954b; Fuc1970, Lemma 7.3): if an abelian group can be covered by a finite number of cosets of subgroups, then one of the subgroups has finite index. The dual reformulation, that a proper additive subgroup of $\mathbb{R}$ does not have the finite covering property, is actually what we need to prove that (vi) implies (i).

Lemma 9.3.4 (Finite-Covering Characterization) *For S an additive subgroup of $\mathbb{R}$, some open interval is covered by a finite union of cosets S of $\mathbb{R}$ if and only if S has finite index in $\mathbb{R}$ if and only if $S = \mathbb{R}$.*

Proof For S a subgroup of $\mathbb{R}$, note first that S is either countable or dense (or both). Indeed, if S is uncountable, then it contains two elements which are rationally incommensurable, so there are two elements s, s' of S such that $ps + qs'$ is non-zero for all non-zero integers p, q. (Otherwise there are non-zero integers p, q such that $ps + qs' = 0$ in which case $s/s' = -q/p \in \mathbb{Q}$.) But then the subgroup of S comprising the points $ps + qs'$ for p, q integers is dense in $\mathbb{R}$, since s'/s is irrational (again by Kronecker's Theorem). So S is dense.

Suppose that a finite number of cosets of S, say $\{[x_i]_S : i = 1, 2, \ldots, m\}$, covers an interval (a, b) with $a < b$, in which case S is uncountable and so dense. We will show that these cosets in fact cover all of $\mathbb{R}$. Indeed, by density of S, for any $x \in \mathbb{R}$ we may choose $s \in S \cap (x - b, x - a)$ and so $x \in s + (a, b)$. But $[s + x_i] = [x_i]$, so one of these covers x. That is, $\{[x_i] : i = 1, 2, \ldots, m\}$ cover $\mathbb{R}$, i.e. S has finite index in $\mathbb{R}$ (as in Neumann's Lemma). By the preceding lemma, $S = \mathbb{R}$. The converse is clear. $\qquad\square$

Proof of Theorem 9.3.2 It is clear that (i) implies (ii) and from Theorem 4.2.1 that (ii) implies (iii). To see that (iii) implies (iv) observe that as S is shift-compact there are t and an infinite $\mathbb{M}$ such that

$$\{t + u_n : n \in \mathbb{M}\} \subseteq S.$$

As S is a subgroup, for distinct m and n in $\mathbb{M}$

$$u_n - u_m = (t + u_n) - (t + u_m) \in S,$$

giving (iv). Clearly (iv) implies (v). To prove (v) implies (vi) we may assume that $S \neq \mathbb{R}$ (otherwise there is nothing to prove) and, by aiming for a contradiction, that S does not have the finite covering property. Suppose that $v_0, \ldots, v_{n-1}$ have been selected with $v_k < 1/(k+1)^2$ and $v_m + \cdots + v_{n-1} \notin S$ for each $m < n-1$. We want to select $v_n < 1/(n+1)^2$ such that for each $m < n$,

$$v_m + \cdots + v_n \notin S, \quad \text{i.e. } v_n \notin S - (v_m + \cdots + v_{n-1}). \qquad (*)$$

Thus we require that

$$v_n \in \bigcap_{m<n} \left(0, 1/(n+1)^2\right) \setminus (S - (v_m + \cdots + v_{n-1}))$$
$$= \left(0, 1/(n+1)^2\right) \setminus \bigcup_{m<n} (S - (v_m + \cdots + v_{n-1})).$$

If we cannot select such a v_n, then

$$\bigcup_{m<n} S - (v_m + \cdots + v_{n-1}) \supseteq \left(0, 1/(n+1)^2\right),$$

and so S does have the finite covering property, contradicting our assumptions. (The induction can be started, as S is a subgroup, so cannot contain any interval, in particular $(0, 1)$.) Thus, the induction can proceed. Put $u_n := v_1 + \cdots + v_n$; then $\{u_n\}$ is convergent. By (iv), there is a set $\mathbb{M}$ such that for m and n in $\mathbb{M}$ with $m < n$,

$$v_m + \cdots + v_{n-1} = u_n - u_m \in S.$$

This contradicts (*), so after all S does have the finite covering property, i.e. (vi) holds. Lemma 9.3.4 shows that (vi) implies (vii) and Lemma 9.3.3 that (vii) implies (i). □

Second Proof of Theorem 9.3.1 (Subgroup Theorem) Immediate from Theorem 9.3.2. □

Remarks 1. As a consequence of the KBD Theorem, 4.2.1, on shift-compactness, Steinhaus' Theorem stands between Theorems 9.3.2 (which employs shift-compactness) and 9.3.1 (which Steinhaus' Theorem implies).

2. The role of the finite covering property may be clarified by noting that, in the context of the theorem above, one may validly add the property 'S is closed' yielding amended equivalent conditions (ii)'–(vii)', and then obtain a further equivalent condition:

(ix)′ *S is closed and contains an interval around* 0.

Indeed, if S is shift-compact, then it is closed. For suppose that $s_n \to s_0$ with $s_n \in S$ for all n. Then $z_n := s_0 - s_n$ is a null sequence and, as S is shift-compact, there is $s \in S$ such that $s + (s_0 - s_n) \in S$ infinitely often. But $s_n \in S$ and $s \in S$, so $s_0 \in S$. So S is closed. In particular, if S is closed and has the finite covering property, then by Baire's Theorem some coset of S contains an interval and so S itself contains an interval, (a, b) say. So its midpoint, s say, is in S. Shifting by $-s$, the set S contains an interval about 0, say $(-\delta, \delta)$, yielding (vii)'. So, S also contains $(-n\delta, n\delta)$ for each $n \in \mathbb{N}$, so S is $\mathbb{R}$ (as in the first proof of Theorem 9.3.1).

9.4 Darboux Dichotomy

We recall the following result (Dar1875 – see Kucz1985) and give its proof, as it is short.

Theorem 9.4.1 (Darboux's Theorem) *If $f \colon \mathbb{R} \to \mathbb{R}$ is additive and locally bounded at some point, then f is linear.*

Proof By additivity we may assume that f is locally bounded at the origin. So we may choose $\delta > 0$ and M such that, for all t with $|t| < \delta$, we have $|f(t)| < M$. For $\varepsilon > 0$ arbitrary, choose any integer N with $N > M/\varepsilon$. Now provided $|t| < \delta/N$, we have

$$N|f(t)| = |f(Nt)| < M, \quad \text{or} \quad |f(t)| < M/N < \varepsilon,$$

giving continuity at 0. Linearity easily follows (see, e.g., BGT1987, Th. 1.1.7). □

We now formulate the classical Ostrowski Theorem in what we term its strong form, as it includes both its classical measure-theoretic version and the Baire analogue due to Banach (see the Remarks after Th. 9.4.2 and the two cases in Corollary 9.4.5). Below, we say that f is locally bounded (locally bounded above, or below) on a set S if any point t has a neighbourhood I such that f is bounded (resp. above, or below) on $S \cap I$.

When null sequences can be embedded in S by an arbitrary shift (not necessarily in S) we say that S is *shift precompact*: see Section 10.1.

Theorem 9.4.2 (Strong Ostrowski Theorem) *For a shift precompact set S, if $f \colon \mathbb{R} \to \mathbb{R}$ is additive and bounded (locally, above or below) on S, then f is locally bounded and hence linear.*

Proof Suppose that f is not locally bounded in any neighbourhood of some point x. Then we may choose $z_n \to 0$ such that $f(x + z_n) \geq n$, without loss of generality (otherwise replace f by $-f$). So $f(z_n) \geq n - f(x)$. Since S is shift precompact, there are $t \in \mathbb{R}$ and an infinite $\mathbb{M}_t$ such that

$$\{t + z_m : m \in \mathbb{M}_t\} \subseteq S,$$

implying that f is unbounded on S locally at t (since $f(t + z_n) = f(t) + f(z_n)$), a contradiction. So f is locally bounded, and by Darboux's Theorem, 9.4.1, f is continuous and so linear. □

Since any non-empty interval is shift precompact, Theorem 9.4.2 embraces Darboux's Theorem. A weaker result, with the condition S shift precompact strengthened to S universal (see p. 73), was given by Kestelman (Kes1947b). Theorem 9.4.2 is the basis of the Darboux dichotomy; that in turn is connected with the Steinhaus dichotomy, because an additive function bounded on a Baire non-meagre (measurable non-null) set A is bounded on the difference set $A - A$ and so on an interval contained in $A - A$. That is, the Darboux dichotomy based upon a 'thick set', an interval, may have its basis refined to a thinner set, just so long as the difference set is 'thick'. F.B. Jones (Jon1942a) refined this basis further by observing that it is enough for A to be analytic, so long as the subgroup generated by A is the reals (see BinO2010a).

The boundedness conditions above lead naturally to a consideration of the level sets of a function and their combinatorial properties. The classical measure and category contexts appeal to various forms of localization. The nub is that, when a non-negligible set is decomposed into a countable union of nice sets, one of these is non-negligible. This is captured in the combinatorics below. Here we go beyond the null sequences.

Definitions 1. For the function $h: \mathbb{R} \to \mathbb{R}$, the (symmetric) *level sets* of h are defined by

$$H^r := \{t : |h(t)| < r\}.$$

2. For $\{T_k : k \in \omega\}$ a countable family of sets of reals, we write **NT**$(\{T_k : k \in \omega\})$ to mean that, for every bounded/convergent sequence $\{u_n\}$ in $\mathbb{R}$, some T_k contains a translate of a subsequence of $\{u_n\}$, i.e. there are $k \in \omega$, $t \in \mathbb{R}$ and infinite $\mathbb{M}_t \subseteq \omega$, such that

$$\{t + u_n : n \in \mathbb{M}_t\} \subseteq T_k.$$

The **NT** notation and the term No Trumps in Theorem 9.4.3, a combinatorial principle, are used in close analogy with earlier combinatorial principles, in particular Jensen's Diamond $\diamondsuit$ (Jen1972) and Ostaszewski's Club $\clubsuit$ (Ost1976; see also Douw1992). Our proof of Theorem 9.4.3 makes explicit an argument implicit in BinGo1982a, p. 482 (and repeated in BGT1987, p. 9), itself inspired by CsiE1964 (see also BinO2009d; BinO2010h). The intuition behind our formulation may be gleaned from forcing arguments in Mill1989; Mill1995. Applied to the level sets, it is equivalent to the Uniform Convergence Theorem (UCT), as is shown in BinO2009d; it also plays a key role in the theory of subadditive functions, for which see BinO2010a.

We shall see that the **NT** property is a common generalization of both measurability and the Baire property. (Specializing to the case when $T_k = S$ for all k, we see that S is shift precompact if and only if $\mathbf{NT}(S)$ holds.) This allows a formulation of when a function may be regarded as having 'nice' level sets. Our next result shows that $\mathbf{NT}$ captures classical notions of localization. Since $\mathbb{R}$ is the union of the level sets of a function, we have as an immediate corollary of Theorem 4.2.1.

Theorem 9.4.3 (No Trumps Theorem; cf. BinO2009b) *For $h\colon \mathbb{R} \to \mathbb{R}$ measurable or Baire, $\mathbf{NT}(\{H^k : k \in \omega\})$ holds.*

As an illustration of its usefulness, we derive a common combinatorial generalization with a weaker hypothesis. See also BinO2010a for a common analysis of measurable and Baire subadditivity via $\mathbf{NT}$.

Theorem 9.4.4 (Generalized Fréchet–Banach Theorem) *If $h\colon \mathbb{R} \to \mathbb{R}$ is additive and its level sets H^k satisfy $\mathbf{NT}(\{H^k : k \in \omega\})$, then h is locally bounded and so continuous and linear.*

Proof Suppose that h is not locally bounded at the origin. Then we may choose $z_n \to 0$ such that $h(z_n) \geq n$, without loss of generality (if not replace h by $-h$). But there are $s \in \mathbb{R}$, $k \in \omega$ and an infinite $\mathbb{M}_s$ such that

$$\{s + z_m : m \in \mathbb{M}_s\} \subseteq H^k,$$

so

$$h(s + z_m) = h(s) + h(z_m) > h(s) + m,$$

so that h is unbounded on H^k, a contradiction as $|h| < k$ on H^k. Thus h is locally bounded and additive; hence by Darboux's Theorem (cf. the proof of Theorem 9.4.2) we conclude that h is continuous and so linear. $\qquad\square$

By Theorem 9.4.3 this result embraces its classical counterpart for h measurable or Baire (due to Fréchet in 1914 and Banach in 1920; see Kucz1985).

Corollary 9.4.5 (Fréchet–Banach Theorem) *If $f\colon \mathbb{R} \to \mathbb{R}$ is additive and measurable or Baire, then f is continuous and so linear.*

There is also a further combinatorial generalization of the classical Ostrowski Theorem, by reference to functions with 'nice' level sets.

Theorem 9.4.6 (Combinatorial Ostrowski Theorem; cf. BinO2009b) *For $h(x)$ an additive function, $h(x)$ is continuous and $h(x) = cx$ for some constant c if and only if $\mathbf{NT}(\{H^k : k \in \omega\})$ holds.*

Proof If $\mathbf{NT}(\{H^k : k \in \omega\})$ holds, then by Theorem 9.4.4, h is linear. Conversely, if $h(x) = cx$, then $H^k = \{t : |ct| < k\}$ is for each $k = 1, 2, \dots$ an interval and hence shift precompact. $\qquad\square$

The Subgroup Theorem may also be similarly restated. For this, we need a variant on **NT**(S) in which 'for infinitely many' is strengthened to 'for all but finitely many' ('co-finitely many') denoted **NT**$_{\text{cof}}$(S).

Theorem 9.4.7 (Combinatorial Steinhaus Theorem Restated) *For an additive subgroup S of $\mathbb{R}$, the following are equivalent:*

(i) $S = \mathbb{R}$;
(ii)′ **NT**$_{\text{cof}}$(S);
(iii)′ **NT**(S).

9.5 The Fubini Null and the Kuratowski–Ulam Theorems

Theorem 9.5.1 (Theorem FN: Fubini Theorem for Null Sets) *Suppose G is a metric group G and $A \subseteq G^2$ is measurable under $\mu \times \nu$, with $\mu, \nu \in \mathcal{M}(G)$. If the 'exceptional set' of points x for which the vertical section A_x is ν-non-null is itself μ-null, then A is $(\mu \times \nu)$-null.*

Proof   For μ-null $N \subseteq G$, the set $N \times G$ is $(\mu \times \nu)$-null, so (by passing to the complement of the null exceptional set of the theorem) we may assume without loss of generality that the exceptional set of A is empty. By inner regularity, it suffices to show that $(\mu \times \nu)(K) = 0$ for all compact $K \subseteq A$.

For K compact, denote by F the (compact) projection of K on the first axis. Let $\varepsilon > 0$. By compactness, for any $x \in F$ there is an open neighbourhood U_x of x and an open V_x with $\nu(V_x) < \varepsilon$ and

$$K \cap (U_x \times G) \subseteq R_x := U_x \times V_x.$$

By compactness of F, there are $U^j \times V^j$ for $j = 1, \ldots, n$, with U^j, V^j open and $\nu(V^j) < \varepsilon$ such that

$$F \subseteq \bigcup_j U^j : \qquad K \subseteq \bigcup_j U^j \times V^j.$$

To disjoin the sets U^j, put

$$S^j := U^j \backslash \bigcup_{j < i} U^j : \qquad \bigcup_j U^j = \bigcup_j S^j.$$

Then

$$F = \bigcup_j F \cap S^j : \qquad K \subseteq \bigcup_j S^j \times V^j.$$

So

$$\mu(K) \leq \sum_j (\mu \times \nu)(S^j \times V^j) = \sum_j \mu(S^j)\nu(V^j) \leq \sum_j \mu(S^j) \cdot \varepsilon = \varepsilon\mu(F).$$

As $\varepsilon > 0$ was arbitrary, $\mu(K) = 0$.   □

We do not need the group-theoretic assumptions above (in 9.5.1). The *converse* of Theorem 9.5.1 also holds: for a $\mu \times \nu$-null set A, the aforementioned exceptional set is μ-null. For a proof (in the more general setting of σ-finite measures), see Hal1950, §36, Th. A, p. 147.

The following is the classical category analogue in which $\mathcal{M}(X)$ denotes the family of all meagre subsets of the space X and, for $E \subseteq X \times Y$ and $x \in X$, E_x denotes the x-section of E, i.e.

$$E_x := \{y \in Y : (x, y) \in E\}.$$

In its original form the Kuratowski–Ulam Theorem of 1932 asserted as follows.

Theorem 9.5.2 (Kuratowski–Ulam Theorem; KurU1932) *For X, Y metric spaces with Y separable:*

$$E \in \mathcal{M}(X \times Y) \implies \{x \in X : E_x \notin \mathcal{M}(Y)\} \in \mathcal{M}(X).$$

In fact their proof requires only that Y have a countable *pseudo-base* (π-basis for short), i.e. a family $\mathcal{U}$ of non-empty open subsets of Y such that any non-empty open set W in Y contains a set $U \in \mathcal{U}$.

Thus generalized, Theorem 9.5.2 may be restated to refer to $\mathcal{NWD}$, the family of nowhere dense sets (Oxt1980, Ch. 15).

Theorem 9.5.3 (Kuratowski–Ulam Theorem Variant) *For X, Y topological spaces with Y having a countable pseudo-base*

$$E \in \mathcal{NWD}(X \times Y) \implies \{x : E_x \notin \mathcal{NWD}(Y)\} \in \mathcal{M}(X).$$

In the non-separable realm the Kuratowski–Ulam Theorem does not hold in general, unlike its measure counter-part. Examples of failure were by Pol in Pol1979 and later (in a vector space setting) by van Mill and Pol (MilP1986). A *converse* holds for Baire subsets $E \subseteq X \times Y$ asserting that if E_x is meagre for all but a meagre set of x, then E is meagre. For the proof, see (Oxt1980, Th. 15.4).

9.6 * Universal Kuratowski–Ulam Spaces

Since the conclusions of the Kuratowski–Ulam Theorem may fail in more general circumstances, it is of interest to determine when it may nevertheless hold. Here we follow the approach taken by Fremlin, Natkaniec and Recław (FreNR2000) as well as Fre2002. With this aim, we will call a pair of topological spaces (X, Y) a *Kuratowski–Ulam* pair (briefly, K–U pair) if the Kuratowski–Ulam Theorem holds in $X \times Y$:

K–U: If $E \in \mathcal{M}(X \times Y)$, then $\{x \in X : E_x \notin \mathcal{M}(Y)\} \in \mathcal{M}(X)$.

Definition A topological space Y is called a *universally Kuratowski–Ulam* space (uK–U space for short) if (X, Y) is a K–U pair for any topological space X.

So, according to the Kuratowski–Ulam Theorem, every space Y with a countable π-basis is a uK–U space. Note also that every space Y that is meagre in itself is uK–U. Fremlin, Natkaniec and Recław (FreNR2000) showed that there are uK–U Baire spaces without a countable π-basis, and further that every Baire uK–U space satisfies the countable chain condition. Known as 'ccc' for short, this means that every family of pairwise disjoint open sets in Y is at most countable. Examples of Baire ccc spaces exist which are not uK–U.

A strengthening of the uK–U notion is particularly important (see Theorem 9.6.1). We will say that (X, Y) is a *Kuratowski–Ulam** pair (K–U* pair, for short) if the variant formulation, as in Theorem 9.5.3, of the Kuratowski–Ulam Theorem holds for the pair (X, Y) :

$$\text{K–U}^* : \text{ If } E \in \mathcal{NWD}(X \times Y), \text{ then} \{x : E_x \notin \mathcal{NWD}(Y)\} \in \mathcal{M}(X).$$

Definition A topological space Y is called a *universally Kuratowski–Ulam** space (uK–U* space for short) if (X, Y) is a K–U* pair for any topological space X; see Fremlin (Fre2000a) and Kucharski and Plewik (KucP2007), who give an example of a uK–U space which is not uK–U*. See also KucP2007; KalK2015.

Every K–U* pair is a K–U pair, and so every uK–U* space is uK–U space. Indeed, a space Y is uK–U if and only if its Baire part, $Y \backslash L(Y)$, is uK–U*. Here the *light part* (the analogous term *heavy* appears in §8.2) is defined by $L(Y) := \bigcup \{V : V$ is open and meagre in $Y\}$. Evidently, for Y a Baire space the latter is empty, so for Y Baire: Y is K–U if and only if it is K–U*, and similarly Y is uK–U if and only if it is uK–U*.

It emerges that a finite product of uK–U spaces is also a uK–U space (see FreNR2000). For uK–U* spaces one has the stronger result.

Theorem 9.6.1 *An arbitrary product of uK–U* spaces is uK–U*. In particular 2^κ, for an arbitrary cardinal κ, is uK–U*.*

We give Fremlin's proof below, assuming the following lemma, whose proof we will omit, as that would require further developments. The result allows the proof to focus on compact spaces which are uK–U*.

Lemma 9.6.2 *A topological space is uK–U* if and only if the Stone space of its Boolean algebra of regular open sets is also uK–U*.*

For the proof see Fre2002, Prop. 4(b); for background on Stone spaces, see Joh1982.

Before beginning the proof of Theorem 9.6.1, we state and prove some preliminaries.

Fix κ, S, Z and f as in the hypothesis, and let $\Phi := \bigcup \{2^A : A \in [\kappa]^{<\omega}\}$ denote the family of indicator functions ϕ of finite subsets of κ. Any $\phi \in \Phi$ defines the following basic open subset of 2^κ:

$$U(\phi) := \{y \in 2^\kappa : \phi \subset y\}.$$

We denote by $\mathcal{U}$ the family of all such basic open sets in 2^κ. Note that $U(\psi) \subset U(\phi)$, for $\phi, \psi \in \Phi$ with $\phi \subset \psi$. A set $U \subset S$ is *basic open in S* if $U = \tilde{U} \cap S$ for some basic open set $\tilde{U} \subset 2^\kappa$. Say that a set $A \subset 2^\kappa$ is *determined* by a set of coordinates $\tau \subset \kappa$ if $A = \{y \in 2^\kappa : y \mid \tau \in A^*\}$ for some $A^* \subset 2^\tau$.

Denote the closure and interior of $A \subseteq 2^\kappa$ by $\mathrm{cl}(A)$ and $\mathrm{int}(A)$, and the closure and interior of A in S by $\mathrm{cl}_S(A)$, and $\mathrm{int}_S(A)$.

We now follow Fremlin's exposition and begin with two claims.

Claim 1   *If $Y = \prod_{i \in I} Y_i$ is a product of compact uK–U* spaces, and $E \subseteq X \times Y$ nowhere dense, then the following is meagre:*

$$A_n := \{x \in X : V \subseteq E_x \text{ and } V \in \tau_n\},$$

where τ_n is the family of non-empty open sets in Y determined by at most n coordinates (from I).

Proof of Claim 1   (We proceed by induction for fixed E.) For $n = 0$, $V = Y$. Here $A_0 \times Y \subseteq E$, so A_0 is in fact nowhere dense. For the step from $n - 1$ to n, suppose that A_n is not meagre. Then A_n contains a heavy point x, i.e. for each U open in X with $x \in U$, the set U is non-meagre. As E is nowhere dense, $U \times Y$ contains an open set $U' \times V$ disjoint from E with U' open in X and V basic in Y, i.e. determined by a finite set of coordinates J. For $x \in U' \cap A_n$ choose $J(x)$ comprising at most n members of I determining a non-empty open $V(x) \subseteq E_x$. Now $V(x) \cap V = \emptyset$ as $U' \times V$ is disjoint from E and $V(x) \subseteq E$, so $V(x)$ and V must disagree on some determining coordinate, and such a coordinate lies in $J(x) \cap J$. But J is finite, so there is $j \in J$ with the following set non-meagre:

$$A_n' = \{x \in A_n \cap U' : j \in J(x)\}.$$

Fix such a j and view E as a subset of $(X \times Y_j) \times Y^j$ with $Y^j := \prod_{i \neq j} Y_i$. Denoting by π_j projection from Y onto Y_j, consider

$$B := \{(x, u) \in X \times Y_j : x \in A_n' \text{ and } u \in \pi_j[V(x)]\}.$$

Here Y_j is Baire and uK–U and B_x contains an open set for each x in the non-meagre set A_n'. So B is non-meagre. For $(x, u) \in B$, the set $E_{(x,u)}$ contains $V(x)_u$, a non-empty open set determined by coordinates in $J(x) \setminus \{j\}$, i.e. by at most $n - 1$ coordinates. This contradicts that A_{n-1} is meagre. In turn this implies that A_n is meagre, establishing the inductive step. □

Claim 2 *Any product of compact Hausdorff uK–U* spaces Y_i is uK–U*.*

Proof To see this, proceed as in Claim 1 and note that $\bigcup_n A_n$ is meagre. But the latter union is just the set $\{x : \text{int} E_x \neq \emptyset\}$, proving the present implication. □

Proof of Theorem 9.6.1 Given uK–U* spaces $\{Y_i\}_{i \in I}$ with topologies τ_i, put $Y = \prod_{i \in I} Y_i$. Now pass to the (coarser) regular open topology σ_i on Y_i generated by ρ_i the family of regular open sets, i.e. those of the form $\text{int}(\text{cl}(U))$ for U open in τ_i. Let Z_i be the Stone space of the regular open algebra of Y_i. Then Z_i is a compact uK–U* space and so is their product $Z := \prod_{i \in I} Z_i$. Now consider the basis of the Tikhonov product corresponding to the τ_i and ρ_i topologies:

$$\mathcal{V} := \left\{ \prod_{i \in I} V_i : V_i \in \tau_i \quad \text{and} \quad \{i : V_i \neq Y_i\} \text{ finite} \right\},$$
$$\mathcal{W} := \left\{ \prod_{i \in I} V_i : V_i \in \rho_i \quad \text{and} \quad \{i : V_i \neq Y_i\} \text{ finite} \right\}.$$

The base $\mathcal{W}$ generates the product topology σ of the topologies σ_i which, being order isomorphic (for inclusion) to Z, makes (Y, σ) be uK–U*. From here, the lemma implies that (Y, τ) is uK–U*. □

Theorem 9.6.3 *If S is a dense subspace of 2^κ, Z a regular topological space and $f : S \to Z$ a continuous surjection, then Z is a uK–U space. If, further, Z is Baire, then Z is a uK–U* space.*

As a first step we prove

Lemma 9.6.4 *Any regular open set $W \subset S$ is a countable union of basic open sets in S.*

Proof The space 2^κ has the ccc property and so likewise has S, being a dense subspace. Choose a maximal sequence $\{B_n\}_{n<\omega}$ of basic open sets in 2^κ such that the complement in S of $S \cap \bigcup_{n<\omega} B_n$ is nowhere dense, so that $W \cap \bigcup_{n<\omega} B_n$ is a dense subset of W. Each B_n is determined by some $\tau_n \in [\kappa]^{<\omega}$. Put $\tau := \bigcup_{n<\omega} \tau_n \in [\kappa]^{\leq\omega}$. Then $\text{cl}(W) = \text{cl}\left(\bigcup_{n<\omega} B_n\right)$ is determined by τ. So in particular its subset $\text{int}(\text{cl}(W))$ is also determined by τ and so, being open, takes the form $\bigcup_{n<\omega} U_n$, for some basic open sets U_n in 2^κ.

But W is regular open in S, so $W = \text{int}_S(\text{cl}_S(W)) = S \cap \text{int}(\text{cl}(W))$, and so $W = S \cap \bigcup_{n<\omega} U_n$. □

Corollary 9.6.5 *If V is a non-meagre open set in Z, then there exists a basic open set $W \subset S$ such that $f[W]$ is non-meagre and $f[W] \subset V$.*

Proof Since Z is regular, there exists a non-meagre open V' with $\mathrm{cl}_Z(V') \subset V$, as follows. For each $z \in V$, by regularity, choose an open set V_z such that $z \in V_z \in \mathrm{cl}_Z(V_z) \subset V$. Now V_z for some z must be non-meagre, otherwise all the sets V_z are meagre and by the Banach Category Theorem, $V = \bigcup_x V_x$ is meagre, contradicting that V is non-meagre.

As f is continuous, $W' := f^{-1}[V']$ is non-empty open in S and so $W' \subseteq W := \mathrm{int}_S(\mathrm{cl}_S(W'))$, with W regular open in S. Now $V' = f(W') \subseteq f(W) \subseteq V$ and so also $f(W)$ is non-meagre. By Lemma 9.6.4, $W = \bigcup_{n<\omega} W_n$ with $\{W_n\}$ a sequence of basic open sets in S, so $f(W_n)$ is, for some n, non-meagre. $\qquad \square$

Proof of 9.6.3 Returning to the proof of the theorem, consider an arbitrary topological space X and let $E \subset X \times Z$ be a closed nowhere dense set. Let $\mathcal{P}$ be the set of all pairs (G, I) where G is an open set in X and $I \in [\kappa]^{<\omega}$. Define a relation $\prec$ on $\mathcal{P}$ by requiring $(H, J) \prec (G, I)$ if and only if

(i) $H \subset G$ and $J \supset I$, and,

(ii) for each basic open set $W \subset S$ determined by I, either

- $H \times f[W] \subset E$, or
- $(H \times U) \cap E = \emptyset$ for some basic open set $W_0 \subset W$ determined by J, and open set $U \subset Z$ with $f[W_0] \subset U$.

Claim *For any $(G, I) \in \mathcal{P}$ and any non-empty open set $G_0 \subset G$ there exists $(H, J) \in \mathcal{P}$ such that $(H, J) \prec (G, I)$ and $H \subset G_0$.*

Proof of the Claim Put $n = |I|$ and let $\{W_i : 0 < i \le 2^n\}$ list all the finitely many basic open sets that I can determine. Put $J_0 = I$ and define inductively open sets G_i and J_i for $0 < i \le 2^n$ according to which of the following two cases is determined by W_i.

Case 1. If $G_{i-1} \times f[W_i] \subset E$, set $G_i = G_{i-1}$ and $J_i = J_{i-1}$.

Case 2. As E is nowhere dense, there are open sets $G_i \subset G_{i-1}$ and $U_i \subset Z$ with $(G_i \times U_i) \cap E = \emptyset$ with $U_i \cap f(W_i) \ne \emptyset$, and so a basic open set W_i' in S with $W_i' \subset W_i$ with $f[W_i'] \subset U_i$. Let $J_i \in [\kappa]^{<\omega}$ be the set which determines W_i'.

Finally, set $H = G_{2^n} \subseteq G_0$ and $J = \bigcup_{i \le 2^n} J_i \supseteq J_0 = I$.

Now choose inductively a sequence $\mathcal{P}_n \subset \mathcal{P}$ such that

- $\mathcal{P}_0 = \{(X, \emptyset)\}$.
- If $(H, J), (H', J')$ are distinct members of $\mathcal{P}_n$, then $H \cap H' = \emptyset$.
- For $(H, J) \in \mathcal{P}_{n+1}$ there exists $(G, I) \in \mathcal{P}_n$ such that $(H, J) \prec (G, I)$.
- $\mathcal{P}_{n+1}$ is a maximal family which satisfies the conditions above.

Then all the sets $G_n^* = \bigcup\{H : (H, J) \in \mathcal{P}_n\}$ are open and dense, and so $\bigcap_{n<\omega} G_n^*$ has meagre complement in X.

Fix any $x \in \bigcap_{n<\omega} G_n^*$. We will show that E_x is nowhere dense in Z. For this it is suffices to consider any (non-meagre) open set $V \subset Z$ and to show that there exists a non-empty open set $V' \subset V$ with $E_x \cap V' = \emptyset$. Fix a non-meagre open set $V \subset Z$. By Corollary 9.6.5 there exists a basic open set $W_0 \subset S$ such that $f[W_0] \subset V$ is non-meagre. Let the set determining W_0 be $J \in [\kappa]^{<\omega}$. Given x, we may choose a sequence $\{(H_n, J_n)\}_n$ such that, for each n:

- $(H_n, J_n) \in \mathcal{P}_n$;

- $x \in H_n$;

- $(H_{n+1}, J_{n+1}) \prec (H_n, J_n)$, so in particular $J_{n+1} \supset J_n$.

Being the set that determines W_0, J is finite and so there exists n with $J_{n+1} \cap J = J_n \cap J$. Since $f[W_0]$ is non-meagre and J_n determines a finite partition of S, there exists an open basic set W determined by J_n such that $f[W \cap W_0]$ is not meagre. Hence $f[W]$ is not meagre.

Since $H_{n+1} \times \mathrm{cl}(f[W])) \subseteq \mathrm{cl}(H_{n+1} \times f[W])$ and has non-empty interior and by assumption E is closed nowhere dense, $H_{n+1} \times f[W] \nsubseteq E$. That is, Case 1 above does not occur.

So, by Case 2, there exists $W' \subset W$, a basic open set of S determined by J_{n+1}, and an open set $U \subset Z$ such that $(H_{n+1} \times U) \cap E = \emptyset$ and $f[W'] \subset U$.

Now $W' \cap W_0 \neq \emptyset$, since $J_{n+1} \cap J = J_n \cap J$: indeed, writing $W_0 = U(\tau_0)$, $W = U(\tau_n)$, and $W' = U(\tau_{n+1})$ with $\tau_{n+1} \supset \tau_n$, we have $\tau_{n+1}|J = \tau_n|J = \tau_0|J$ as $W \cap W_0 \neq \emptyset$; so, taking $\tau = \tau_0 \cup \tau_{n+1}$, we have $U(\tau) \subseteq W' \cap W_0$.

This implies that $\emptyset \neq f[W' \cap W_0] \subseteq U \cap V$ as $f[W_0] \subset V$ and $f[W'] \subset U$. So $U \cap V \neq \emptyset$, and also $(U \cap V) \cap E_x = \emptyset$, the latter since both $(H_{n+1} \times (U \cap V)) \cap E = \emptyset$ and $x \in H_{n+1}$ hold, as required, proving the claim and so the theorem. $\square$

Corollary 9.6.6 *There exists a uK–U* Baire space Y without a countable π-basis.*

Proof Consider $Y = 2^{\omega_1}$. By Theorem 9.6.1, Y is a uK–U* space. On the other hand, it is well known that $\pi(Y) = \omega_1$. In fact, let $\{U_n : n < \omega\}$ be a sequence of basic open sets in Y. For each n there exists $A_n \in [\omega_1]^{<\omega}$ and $\phi_n \colon A_n \to 2$ such that $U_n = U(\phi_n)$. Then $A = \bigcup_{n<\omega} A_n$ is countable so we may choose $\alpha \in \omega_1 \backslash A$ and take $V := \{y \in Y : y(\alpha) = 1\}$. Then V is open in Y; however, $U_n \subseteq V$ fails for each n since there are points $y \in U_n$ with $y(\alpha) = 0$. So $\{U_n : n \in \omega\}$ is not a π-basis for Y. $\square$

A compact space X is said to be *dyadic* if it is a continuous image of the space 2^κ for some cardinal κ (cf. Eng1989, p. 285). Theorem 9.6.3 implies

Corollary 9.6.7   *Every dyadic space is uK–U*.*

A topological space X is said to be *quasi-dyadic* if it is a continuous image of the Tikhonov product $\prod_\alpha X_\alpha$ of a family $\{X_\alpha : \alpha < \kappa\}$ of metric separable spaces (see FreG1995).

Theorem 9.6.8   *Every regular quasi-dyadic space is uK–U (and uK–U* if Baire).*

Proof   We start with the following lemma.

Lemma 9.6.9   *Every metric separable space is the continuous image of some dense subset of the space 2^ω.*

Proof of Lemma   This is a consequence of the fact that every metric separable space is homeomorphic to a subspace of the Hilbert cube I^ω (see, e.g., Kec1995, Th. 4.14, p. 22) and that I^ω is a continuous image of 2^ω. Thus every metric separable space is a continuous image of some subspace of 2^ω. On the other hand, it is easy to prove that every subset of a Cantor set is the continuous image of some dense subset of 2^ω.   □

To complete the proof of Theorem 9.6.8, assume that Y is a regular space, $X_\alpha, \alpha < \kappa$, are metric separable spaces, and $f \colon \prod_{\alpha<\kappa} X_\alpha \to Y$ is a continuous surjection. For every $\alpha < \kappa$ there exists a continuous surjection $f_\alpha \colon A_\alpha \to X_\alpha$ for A_α some dense subspace of 2^ω. Then the set $\prod_{\alpha<\kappa} A_\alpha$ is dense in $2^{\omega\kappa}$ and $f \circ \prod_{\alpha<\kappa} f_\alpha$ is a continuous surjection from $\prod_{\alpha<\kappa} A_\alpha$ onto Y. By Theorem 9.6.3, Y is a uK–U space.   □

The next result refers to the *additivity* of $\mathcal{M}(X)$, a cardinal invariant defined by:

$$\mathrm{add}(\mathcal{M}(X)) = \min\left\{|D| : D \subset \mathcal{M}(X) \text{ and } \bigcup D \notin \mathcal{M}(X)\right\}.$$

For background, see BartoJ1995, Ch. 5, cf. the Cichoń diagram of BartoJ1995, Ch. 2 and Chapter 16. A topological space X satisfies the *κ-chain condition*, in brief: is *κ-cc*, if there is no family of size κ of open, pairwise disjoint sets in X. The *countable chain condition* (briefly, *ccc*) is thus ω_1-cc.

Similarly one may define *π-weight* of X, denoted $\pi(X)$, to be the least cardinal of a π-basis for X.

Remark   The proof of the Kuratowski–Ulam Theorem yields the generalization that if $\pi(Y) < \mathrm{add}(\mathcal{M}(X))$, then (X, Y) is a K–U* pair.

Theorem 9.6.10   *Assume that X is a non-meagre space, Y a Baire space and (X, Y) a K–U pair. Then Y is $\mathrm{add}(\mathcal{M}(X))$-cc.*

Proof Suppose that $\kappa = \text{add}(\mathcal{M}X))$ and $\mathcal{B} = \{B_\alpha : \alpha < \kappa\}$ is a family of open, non-empty, pairwise disjoint sets in Y. Let $\mathcal{A} = \{A_\alpha : \alpha < \kappa\}$ be a family of nowhere dense sets in X with $\bigcup \mathcal{A} \notin \mathcal{M}(X)$. In $X \times Y$, take $W := \bigcup_{\alpha < \kappa} A_\alpha \times B_\alpha$. Note that W is nowhere dense in $X \times Y$. Indeed, for a basic open set $U \times V$ two cases may arise. First, if $V_0 := V \setminus \text{cl}_Y(\bigcup_{\alpha < \kappa} B_\alpha) \neq \emptyset$, then $U \times V_0$ is non-empty open and disjoint from W. Otherwise $V_0 = \emptyset$. In this case, as V is open, $V \cap B_\alpha \neq \emptyset$ for some $\alpha < \kappa$ and, since A_α is nowhere dense, there is an open, non-empty set $U' \subset U \setminus A_\alpha$. Thus the set $U' \times (V \cap B_\alpha)$ is non-empty, open and disjoint from W. On the other hand,

$$\{x : W_x \notin \mathcal{M}(Y)\} = \bigcup \mathcal{A} \notin \mathcal{M}(X),$$

and so (X, Y) is not a K–U pair.   □

Remark There exist completely regular spaces X non-meagre in themselves with $\text{add}(\mathcal{M}(X)) = \omega_1$. Specifically, as is well known, $X = 2^{\omega_1}$ has this property: indeed, the sets $E_\alpha = \{x \in X : x(\xi) = 0 \text{ for } \xi \geq \alpha\}$ are closed and nowhere dense in X, but $\bigcup_{\alpha < \omega_1} E_\alpha \notin \mathcal{M}(X)$. Another example is provided by the space (ω^ω, τ_d) from Example 1 .

Since a Baire space is uK–U if and only if it is uK–U*, we have the following.

Corollary 9.6.11 *Every Baire uK–U* space satisfies the ccc property.*

Now we will show that the assumption of ccc for a Baire space Y is not sufficient to make it be uK–U*.

For $s \in \omega^{<\omega}$ and $f \in \omega^\omega$ with $s \subset f$, define

$$(s, f) = \{g \in \omega^\omega : s \subset g \text{ and } f \leq g\}.$$

Note that the family of such pairs forms a basis for a ccc topology τ_d on ω^ω. It is known that (ω^ω, τ_d) is a completely regular, Baire space. Moreover, let τ denote the standard topology on ω^ω. For $f, g \in \omega^\omega$ we use the symbol $f \leq^* g$ to mean that $\{n \in \omega : f(n) > g(n)\}$ is finite.

Example 1 $((\omega^\omega, \tau), (\omega^\omega, \tau_d))$ and $((\omega^\omega, \tau_d), (\omega^\omega, \tau_d))$ are not K–U* pairs.

To see this, first define $W := \{(f, g) \in (\omega^\omega)^2 : f \leq^* g\}$. We will then prove two claims.

Claim 1 *W is meagre in the topologies $\tau_d \times \tau_d$ and $\tau \times \tau_d$.*

Proof of Claim 1 Put $W_n = \{(f, g) \in \omega^\omega \times \omega^\omega : (\forall k > n)f(k) \leq g(k)\}$. We verify that each W_n is nowhere dense in the topology $\tau_d \times \tau_d$. Let $(s, f) \times (r, h)$ be a basic set. Fix $k > n$ such that $k \notin \text{dom}(s) \cup \text{dom}(r)$. Choose $s_1, r_1 \in \omega^{<\omega}$ such that $s \subset s_1, r \subset r_1, s_1(k) > r_1(k), s_1 \geq f \mid \text{dom}(s_1)$, and $r_1 \geq h \mid \text{dom}(r_1)$.

Let f_1 be any extension of s_1 with $f_1 \geq f$ and h_1 be any extension of r_1 with $h_1 \geq h$.

Then $(s_1, f_1) \times (r_1, h_1) \subset (s, f) \times (r, h)$. Observe that $e(k) > g(k)$ for each $(e, g) \in (s_1, f_1) \times (r_1, h_1)$. Thus $(s_1, f_1) \times (r_1, h_1) \cap W_n = \emptyset$, so W_n is nowhere dense, and consequently W is meagre in the topology $\tau_d \times \tau_d$.

Similarly we can prove that W is meagre in the topology $\tau \times \tau_d$. Claim 1 is therefore proved.

Claim 2 $W_f \notin M(\tau_d)$ *for each* $f \in \omega^\omega$.

Proof of Claim 2 Note that $W_f = \{h : f \leq^* h\}$. Fix a basic set (s, g) and define $g_1 \in \omega^\omega$ such that $g_1(i) = h(i)$ if $i \in \mathrm{dom}(s)$ and $g_1(i) = \max(h(i), f(i))$ otherwise. Then $(s, g_1) \subset (s, g) \cap W_f$. Thus W_f is co-meagre in the topology τ_d, and Claim 2 is proved, and with it the statement of Example 9.6.

Corollary 9.6.12 *The space* (ω^ω, τ_d) *is not a* uK–U* *space.*

We also have another better known example of a ccc space which is not uK–U*. Let $\mathcal{D}$ denote the density topology on the real line. Recall that $(\mathbb{R}, \mathcal{D})$ is a Baire space with the ccc property, and $A \subset \mathbb{R}$ is $\mathcal{D}$-nowhere dense if and only if it is $\mathcal{D}$-meagre if and only if $m(A) = 0$. Here m denotes Lebesgue measure. (See Chapter 7 for the basic properties of the topology $\mathcal{D}$; cf. Oxt1980 and Tal1976.)

Example 2 For $X = (\mathbb{R}, \mathcal{D})$ the pair (X, X) is not a uK–U* pair.

To see this, consider

$$A = \{(x, y) : x - y \notin \mathbb{Q}\}.$$

As is easily seen, both A and its complement are $\mathcal{D} \times \mathcal{D}$-dense (this is a consequence of Steinhaus' Theorem, 9.2.2, see also AnaL1985). Moreover, A is a $\mathcal{G}_\delta$ subset of the plane with full Lebesgue measure, so it contains a Euclidean closed set E (so closed also in the $\mathcal{D} \times \mathcal{D}$ topology) with positive measure. To finish, the set E is nowhere dense in $(\mathbb{R}^2, \mathcal{D} \times \mathcal{D})$ and, by Fubini's Theorem,

$$\{x : E_x \notin M(\mathcal{D})\} = \{x : m(E_x) > 0\} \notin M(\mathcal{D}).$$

Below we present further results concerning universal Kuratowski–Ulam* (uK–U*) spaces, omitting routine proofs; the majority of them (except Property 7) holds for uK–U.

Applications. Recall that the product $X \times Y$ of Baire spaces may be non-Baire. (Some conditions for X and Y which imply that $X \times Y$ is a Baire space are described in HawoM1977.) Note that if X and Y are Baire spaces and (X, Y) is a K–U pair, then $X \times Y$ is a Baire space. Similarly, the product $X \times Y$ of a Baire space X and a uK–U* Baire space Y is a Baire space.

Property 1 *Any subspace of a uK–U* (uK–U) space is itself uK–U* (uK–U, respectively).*

Property 2 *If Y_0 is a dense subspace of a uK–U* space Y, then it is also a uK–U* space.*

Property 3 *Assume that Y_0 is a subspace of a uK–U* space Y such that $Y_0 \subset \mathrm{int}_Y(\mathrm{cl}_Y(Y_0))$. Then Y_0 is also a uK–U* space.*

Example 3 There exists a subspace Y_0 of a uK–U* space Y, which fails to be a uK–U* space.

Proof Take Y_0 to be the discrete space of size ω_1. As Y_0 has weight ω_1, it embeds into $Y = [0, 1]^{\omega_1}$ (see, e.g., Eng1989, Th. 2.3.11, p. 113). By Theorem 9.6.8, Y is uK–U*, but Y_0 is not ccc, so it is not uK–U*, by Corollary 9.6.11.   □

We say that a set $A \subset X$ is nowhere meagre in a space X if $U \cap A \notin \mathcal{M}(X)$ for every open, non-meagre set $U \subset X$.

Property 4 *Suppose that Y_0 is a uK–U* dense subspace of a space Y. If Y_0 is nowhere meagre in Y, then Y is a uK–U* space.*

The assumption about Y_0 cannot be omitted.

Example 4 There exists a non-uK–U* space Y with a dense uK–U subspace Y_0.

Proof Let Y be any complete dense-in-itself metric space which is not ccc. By Corollary 9.6.11, Y is not a uK–U* space. For every $n > 0$, choose a discrete set $Y_n \subset Y$ which forms a $1/n$-net in Y. Then $Y_0 = \bigcup_{n>0} Y_n$ is dense in Y, dense in itself, and meagre in itself. Thus Y_0 is a uK–U space.   □

Property 5 *Suppose that $\{Y_i : i < \omega\}$ is a sequence of uK–U* subspaces of a topological space Y. Then $\bigcup_i Y_i$ is also a uK–U* space.*

Corollary 9.6.13 *The topological sum of countably many uK–U* spaces is a uK–U* space.*

Example 5 The topological sum of uncountably many uK–U* spaces may fail to be a uK–U* space.

Proof Let Y be a discrete space of size ω_1. Then Y is not ccc, so is not a uK–U* space. On the other hand, every singleton is a uK–U* space.   □

Property 6 *The homeomorphic image of a uK–U* space is also a uK–U* space.*

Property 7   *The image of a uK–U* Baire space under an open continuous function is a uK–U* space.*

Note that any space Y is a continuous image of the meagre-in-itself space $Y \times \mathbb{Q}$. Thus any space Y is a continuous image of a uK–U* space.

Remark   The results above lead to the problem whether any continuous image of a uK–U* Baire space is also uK–U*. This problem has been solved by D. Fremlin (Fre2000a) in the negative.

9.7 * Baire Products

As Kuratowski and Ulam noted, the set $E = [0, 1]^N$ is nowhere dense in $[0, 2]^N$ yet its projection $[0, 1]$ is non-meagre in $[0, 2]$. (For any basic open set $U_1 \times U_2 \times \cdots \times U_n \times [0, 2]^{\{n+1, n+2, \dots\}}$ the open subset $U_1 \times U_2 \times \cdots \times U_n \times (1, 2) \times [0, 2]^{\{n+2, \dots\}}$ is disjoint from E.) In general the product of two Baire spaces need not be Baire: Oxtoby constructed an example of failure (using the Continuum Hypothesis, though absolute examples followed) and also identified an additional condition, that of having a *locally countable pseudo-base* which guarantees that $X \times Y$ is Baire if X is Baire and Y has such a pseudo-base. This condition on a space is sometimes referred to by saying that the space has a *countable-in-itself π-base*, as the defining property of the π-base is that each of its members contains countably many members of the said π-base.

We have seen that if X and Y are Baire spaces and (X, Y) is a K–U pair, then $X \times Y$ is a Baire space. Similarly, the product of a Baire space X and a uK–U Baire space Y is a Baire space. This leads to the notion of a space X being *almost locally uK–U* when the set

$$W := \{x \in X : x \text{ has an open uK–U neighbourhood}\}$$

is dense in X. Under such circumstances W is dense in X and open, i.e. its complement is closed nowhere dense. (If it contained an open subset V, this would meet W.)

It emerges that the property of being almost locally uK–U is a proper generalization of having a *locally countable pseudo-base* (countable-in-itself π-base).

Say that a space is *almost locally ccc*, provided every open set contains an open ccc subspace, i.e. if the space has *a π-base of open ccc subspaces*. This property is strictly *weaker than being almost locally uK–U*.

However, in the context of metrizable spaces all these concepts are equivalent, as we shall see.

Below we refer to the largest closed set comprising points at which the space is non-separable. Known as the nowhere-locally-separable kernel, it was introduced by A. H. Stone (Sto1963).

Theorem 9.7.1 (Pol's Product Theorem; Pol1979)   *For X, Y metrizable spaces,* (a) *and* (b) *below are equivalent:*

(a) *If A, B are non-meagre in respectively X and Y, then A × B is non-meagre in X × Y.*
(b) *At least one of X or Y has a nowhere-locally-separable kernel that is meagre.*

In particular, if (a) *fails and there are non-meagre sets A, B with A × B meagre, then both of their nowhere-locally-separable kernels are meagre.*

Since the uK–U property is inherited by open subspaces and since spaces with countable bases are uK–U (see §9.6 – cf. FreNR2000), it follows that if X is a uK–U space or has a countable-in-itself base, then X is almost locally uK–U. For X metrizable the converse also holds.

Corollary 9.7.2   *If X is metrizable, Baire and almost locally uK–U, then X has a countable-in-itself base.*

Proof   Let $U \subseteq X$ be an open uK–U subspace with $X \backslash U$ meagre. Let Y be a nowhere-locally-separable Baire metric space, so that Y is its own non-meagre nowhere-locally-separable kernel. We claim that X has a meagre nowhere-locally-separable kernel. For this it suffices to prove that (a) above holds. So suppose $A \subseteq X$ and $B \subseteq Y$ with $A \times B$ meagre in $X \times Y$ and A, B non-meagre. We will derive a contradiction, namely that A is meagre.

Now, $U \times Y \cap A \times B$ is a meagre subset of $U \times Y$, so, since U is uK–U, quasi all $b \in B$ have meagre section. But B is non-meagre, so there is $y \in B$ such that $U \cap A = \{x \in U : (x, y) \in A \times B\}$ is a section that is meagre in U and hence also meagre in X, since U is open. As $X \backslash U$ is meagre, A is meagre in X. Now the points in X without a separable neighbourhood form a closed meagre subset, by Pol's Theorem, 9.7.1. So, being a Baire space, X has a dense open locally separable subspace. Locally separable metrizable spaces can be partitioned into clopen separable subspaces, and so taking the union of their respective countable bases yields a countable-in-itself base for X. $\square$

The *Krom space* below allows the application of Pol's Theorem.

Definition   For a topological space (Y, τ) let $\tilde{\tau} := \tau \backslash \{\emptyset\}$, the non-empty open sets, which we view as the points of a discrete (metrizable) space. We give the space $\tilde{\tau}^{\omega}$ of sequences the metric of first difference (just as with $\mathbb{N}^{\mathbb{N}}$). The

Krom space is the subspace of $\tilde{\tau}^\omega$ corresponding to descending sequences of open sets with non-empty intersection:

$$K(Y) := \left\{ f \in \tilde{\tau}^\omega : \bigcap\nolimits_{n \in \omega} f(n) \neq \emptyset \text{ and } f(0) \supseteq f(1) \supseteq \cdots \supseteq f(n) \supseteq \cdots \right\}.$$

One writes $f \in\downarrow \tau^\omega$ when $f(0) \supseteq f(1) \supseteq \cdots \supseteq f(n) \supseteq \cdots$, and by analogy $f \in\downarrow \tau^n$ for some $n < \omega$ if $f \in \tilde{\tau}^n$ and $f(0) \supseteq f(1) \supseteq \cdots \supseteq f(n)$. The basic open set generated by $f \in\downarrow \tau^n$ in $K(Y)$, denoted $[f]$, comprises all infinite extensions of f.

Theorem 9.7.3 (Krom's Theorem; Kro1974)   *For X, Y Baire, $X \times Y$ is Baire if and only if $X \times K(Y)$ is Baire if and only if $K(X) \times K(Y)$ is Baire.*

For a proof see Kro1974. The result extends to arbitrary products (LiZ2017, Th. 4.1).

Theorem 9.7.4 (Zsi2004)   *For X, Y Baire spaces with Y almost locally ccc, $X \times Y$ is a Baire space.*

Proof   First note that the Krom space $K(Y)$ has a countable-in-itself π-base: indeed, let $f \in\downarrow \tau^n$ for some $n < \omega$, choose $U \subset f(n)$ which is ccc, and define the extension $f_U = f^\frown U$. Consider a pairwise disjoint open partition $\{[g] : g \in J\}$ of $[f_U]$, for an arbitrary $J \subset n < \omega \downarrow \tau^n$. For each $g \in J$ choose $n_g < \omega$ with $g \in\downarrow \tau^{n_g}$. Then $\{g(n_g) : g \in J\}$ is a pairwise disjoint open partition of U, hence countable, since Y is ccc and likewise $\{[g] : g \in J\}$. Thus, $K(Y)$ is an almost locally ccc *metric* space, and so has a countable-in-itself π-base, by Corollary 9.7.2.

By Krom's Theorem $K(Y)$ is a Baire space, so by Oxt1960, Th. 2, $X \times K(Y)$ is a Baire space, implying that $X \times Y$ is a Baire space, again by Krom's Theorem.   □

This result generalizes to arbitrary products (LiZ2017, Th. 1.2).

Theorem 9.7.5 (Zsi2004)   *For X, Y Baire spaces with Y almost locally ccc, $X \times Y$ is a Baire space.*

Theorem 9.7.6   *For $\{X_i : i \in I\}$ an arbitrary family of almost locally ccc Baire spaces, the product $\prod_i X_i$ is a Baire space.*

10

Category Embedding Theorem and Infinite Combinatorics

Motivated by the Kestelman–Borwein–Ditor Theorem, 4.2.1, we focus on the concept of *shift-compactness*, as it has a key role in unifying the Baire and measurable approaches. For convenience, we recall the term here (taken, as we have said earlier, from the probability literature – see Par1967 and BinO2024 – and adapted to our context) in tandem with two related concepts as follows, the weaker of which is important for applications.

10.1 Preliminaries

Recall that *quasi everywhere* (q.e.), or *for quasi all points*, means *for all points off a meagre set*. We will use *for generically all* to mean for quasi all in the category case, and for almost all in the measure case.

Definitions   1. Say that $S \subseteq \mathbb{R}$ is *boundedly shift-compact*, resp. *convergently shift-compact* (shift-compact for bounded sequences in $\mathbb{R}$, resp. in S), and write $S \in \mathcal{SK}_{\mathbb{R}}$, resp. $S \in \mathcal{SK}$, if, for any bounded/convergent sequence u_n, there are $t \in \mathbb{R}$ and infinite $\mathbb{M} = \mathbb{M}_t$ such that

(i) $\{t + u_m : m \in \mathbb{M}\} \subseteq S$, and
(ii) $\lim_{\mathbb{M}}(t + u_m) \in S$.

2. As earlier, say that $S \subseteq \mathbb{R}$ is *shift-compact* (shift-compact for null sequences), $S \in \mathcal{S}_*$, if, for any null sequence $z_n \to 0$, there are $t \in S$ and infinite $\mathbb{M} = \mathbb{M}_t$ such that

(i) $\{t + z_m : m \in \mathbb{M}\} \subseteq S$, and
(ii) $t = \lim_{\mathbb{M}}(t + z_m) \in S$.

3. As in §9.4, say that $S \subseteq \mathbb{R}$ is *shift precompact*, and write $S \in \mathcal{S}$, if, for any null sequence $z_n \to 0$, there are $t \in \mathbb{R}$ and infinite $\mathbb{M} = \mathbb{M}_t$ such that

(i) $\{t + z_m : m \in \mathbb{M}\} \subseteq S$.

Note that Definition 1, in contrast to Definition 2, does not require $t \in S$. The choice of the asterisk notation in Definition 2 is suggested by genericity (cf. Kec1995, 17.26, and the Near-Closure Theorem, 10.5.1). It is clear that $\mathcal{SK}_{\mathbb{R}} \subseteq S_*$ and $\mathcal{SK}_{\mathbb{R}} \subseteq \mathcal{SK}$; writing $\overline{S}$ for the family of closed sets in S, we have the following proposition.

Proposition 10.1.1   *If $A \subseteq \mathbb{R}$ is shift-compact, then A is boundedly shift-compact and so convergently shift-compact:*

$$\overline{S} \subseteq S_* = \mathcal{SK}_{\mathbb{R}} \subseteq \mathcal{SK} \quad and \quad S_* \subseteq S.$$

Proof   Let a_n be a convergent sequence (sequence in A) with limit a_0. Then $z_n := a_n - a_0$ is a null sequence, hence for some $t \in A$ and infinite $\mathbb{M}_t$ we have $t + z_n$ in A for $n \in \mathbb{M}_t$. Thus with $s := t - u_0$ we have $s + u_n = t + z_n \in A$ for $n \in \mathbb{M}_t$ and convergence through $\mathbb{M}_t$ to $s + u_0 = t \in A$. Thus A is boundedly shift-compact (convergently shift-compact). □

These are forms of compactness (for a topological analysis of this insight involving open shifted-covers, and further applications, see BinO2010g). They generalize their forerunner *universality* in relation to null sequences, introduced in a related context by Kestelman (Kes1947a), where the more demanding requirement on the set $\mathbb{M}$ in Definition 3 above is that it be co-finite. The latter concept of universality is implicit in some of Banach's work (see, e.g., Ban1932).

The Near-Closure Theorem, 10.5.1, implies that a *Baire non-meagre/measurable non-null set T is shift-compact*. Although the weakest of the three concepts, shift precompactness is the key combinatorial concept in many applications.

10.2 The Category Embedding Theorem, CET

Theorem 10.2.2 is a topological version of the Kestelman–Borwein–Ditor Theorem, 4.2.1, from which Theorem 10.3.3 is rederived. The latter is a (homeomorphic) *embedding* theorem (see, e.g., Eng1989, p. 67); Trautner uses the term covering principle in Tra1987 in relation essentially to bounded shift-compactness (see again BGT, p. 10). For other generalizations of a homotopic nature, see Mille1989 (which inspired BinO2011b). We need the following definition.

Definition (Category Convergence)   A sequence of Baire functions $h_n : X \to X$ satisfies the category condition (cc) if, for any non-empty open set U, there is a non-empty open set $V \subseteq U$ such that, for each $k \in \omega$,

$$\bigcap_{n \geq k} V \backslash h_n^{-1}(V) \quad \text{is meagre.} \tag{cc}$$

Equivalently, for each $k \in \omega$, there is a meagre set M such that, for $t \notin M$,

$$t \in V \implies (\exists n \geq k), \quad h_n(t) \in V.$$

We will see in Theorem 10.2.4 that this is a weak form of convergence to the identity and indeed Theorems 10.3.1 and 10.3.1′ verify that, for $z_n \to 0$, the homeomorphisms $h_n(x) := x + z_n$ satisfy (cc) in the Euclidean and in the density topologies. However, it is not true that $h_n(x)$ converges to the identity pointwise in the sense of the density topology; furthermore, whereas addition (a two-argument operation) is not $\mathcal{D}$-continuous (see Property 7 of $\mathcal{D}$ in Chapter 7), translation (a one-argument operation) is.

In Theorem 10.2.2, the topological space $\mathcal{X}$ may be assumed to be non-meagre (of second category) in itself, and the Baire set T to be non-meagre, as otherwise there is nothing to prove. To verify that X is non-meagre, one would typically assume that $\mathcal{X}$ is a Baire space (cf. Chapter 2). The proof makes use of the monotone operator (where 'i.o.' means 'infinitely often')

$$b^h(T) := \limsup h_n^{-1}(T) = \{x : x \in h_n^{-1}(T) \text{ i.o.}\}$$
$$= \bigcap_{k \in \omega} \bigcup_{n \geq k} h_n^{-1}(T)$$

associated with a sequence h of Baire functions. The following result has a routine proof so is omitted. Base operators referred to below are discussed in §8.5.

Lemma 10.2.1   *Suppose $h = \langle h_n \rangle$ is a sequence of Baire functions on X of a space $\mathcal{X}$. Then $b^h(T)$ is additive and is in fact a base operator.*

Thus b^h may be used to refine the topology of $\mathcal{X}$ to obtain an analogue of the z-topology on $\mathbb{R}$ defined in §8.5, Remark 4b, p. 129.

Theorem 10.2.2 (Category Embedding Theorem – CET; BinO2009b)   *Let X be a topological space and $h_n : X \to X$ be Baire functions satisfying* (cc) *with preimages of meagre sets being meagre. Then, for any Baire set T, for quasi all $t \in T$ there is an infinite set $\mathbb{M}_t$ such that*

$$\{h_m(t) : m \in \mathbb{M}_t\} \subseteq T.$$

In particular, the conclusion holds for h_n homeomorphisms satisfying (cc).

Proof Take T Baire and non-meagre. We may assume that $T = U \backslash M$ with U non-empty and open and M meagre. Let $V \subseteq U$ satisfy (cc). Since preimages of meagre sets under h_n are meagre, the set

$$M' := M \cup \bigcup_n h_n^{-1}(M)$$

is meagre. Put

$$W = \mathbf{h}(V) := \bigcap_{k \in \omega} \bigcup_{n \geq k} V \cap h_n^{-1}(V)$$
$$= \lim \sup [h_n^{-1}(V) \cap V]$$
$$= \{x : x \in h_n^{-1}(V) \cap V \text{ i.o.}\}$$
$$\subseteq V \subseteq U.$$

So for $t \in W$ we have $t \in V$ and

$$v_m := h_m(t) \in V, \qquad\qquad (*)$$

for infinitely many m – for $m \in \mathbb{M}_t$, say. Now W is co-meagre in V. Indeed

$$V \backslash W = \bigcup_{k \in \omega} \bigcap_{n \geq k} V \backslash h_n^{-1}(V),$$

which by (cc) is meagre.

Take $t \in W \backslash M' \subseteq U \backslash M = T$, as $V \subseteq U$ and $M \subseteq M'$. Thus $t \in T$. For $m \in \mathbb{M}_t$, we have $t \notin h_m^{-1}(M)$, since $t \notin M'$ and $h_m^{-1}(M) \subseteq M'$; but $v_m = h_m(t)$, so $v_m \notin M$. By (*), $v_m \in V \backslash M \subseteq U \backslash M = T$. Thus $\{h_m(t) : m \in \mathbb{M}_t\} \subseteq T$ for t in a co-meagre subset of V.

To deduce that quasi all $t \in T$ satisfy the conclusion of the theorem, put $S := T \backslash \mathbf{h}(T)$. Then S is Baire since

$$h_m^{-1}(U \mathcal{M}) = h_m^{-1}(U) \backslash h_m^{-1}(M),$$

both sets being Baire (the first since h_m is Baire and U is open, the second as it is meagre), and $S \cap \mathbf{h}(T) = \varnothing$. If S is non-meagre, then by the preceding argument there are $s \in S$ and an infinite $\mathbb{M}_s$ such that $\{h_m(s) : m \in \mathbb{M}_s\} \subseteq S$, i.e. $s \in \mathbf{h}(S) \subseteq \mathbf{h}(T)$, a contradiction. (This last step is an implicit appeal to a generic dichotomy – see Chapter 4.) $\square$

Corollary 10.2.3 (Quasi Identity) *Suppose $h = \langle h_n \rangle$ is a sequence of self-homeomorphisms of a space $\mathfrak{X}$, and T is a Baire set in $\mathfrak{X}$. Then*

$$b^h(T) = \lim \sup h_n^{-1}(T) \sim T \quad \text{(modulo the meagre sets)}.$$

Clearly the theorem relativizes to any open subset of T; that is, the embedding property is a *local* one. The following theorem sheds some light on the significance of the category convergence condition (cc). The result is capable of improvement, by reference to more general (topological) countability conditions. (Typically these lift category and measure arguments out of the classical context of separable metric spaces; see Chapter 2, or Eng1989, §§3.9, 4.4 for an account of Čech-completeness and metrization theory, and Arh1963, §7 for an account of *p*-spaces, their common generalization.) Here, for instance, a σ-discrete family could replace the countable family $\mathcal{B}$ of the theorem as the generator of the coarser topology. Such a replacement would offer a route to Bing's Metrization Theorem, given sufficient regularity assumptions – see Eng1989, Th. 4.4.8, thus making $\mathfrak{X}$ submetrizable, or even cometrizable.

Theorem 10.2.4 (Convergence to the Identity)   *Assume that the homeomorphisms* $h_n \colon X \to X$ *satisfy the category convergence condition* (cc) *and that* X *is a Baire space. Suppose there is a countable family* $\mathcal{B}$ *of open subsets of* X *which generates a (coarser) Hausdorff topology on* X. *Then, for quasi all (under the original topology)* t, *there is an infinite* $\mathbb{N}_t$ *such that*

$$\lim_{m \in \mathbb{N}_t} h_m(t) = t.$$

Proof   For U in the countable base $\mathcal{B}$ of the coarser topology and for $k \in \omega$, select open $V_k(U)$ so that $M_k(U) := \bigcap_{n \geq k} V_k(U) \backslash h_n^{-1}(V_k(U))$ is meagre. Thus

$$M := \bigcup_{k \in \omega} \bigcup_{U \in \mathcal{B}} M_k(U)$$

is meagre. Now $\mathcal{B}_t = \{U \in \mathcal{B} : t \in U\}$ is a basis for the neighbourhoods of t. But, for $t \in V_k(U) \backslash M$, we have $t \in h_m^{-1}(V_k(U))$ for some $m = m_k(t) \geq k$, i.e. $h_m(t) \in V_k(U) \subseteq U$. Thus $h_{m_k(t)}(t) \to t$, for all $t \notin M$. □

10.3 Kestelman–Borwein–Ditor Theorem: CET Proof

We now deduce the category and measure cases of the Kestelman–Borwein–Ditor Theorem, 4.2.1, restated below, as two corollaries of Theorem 10.2.2 by applying it first to the usual and then to the density topology on the reals, $\mathbb{R}$.

For our first application we take $X = \mathbb{R}$ with the density topology, a Baire space. Let $z_n \to 0$ be a null sequence. Put

$$h_n(x) := x - z_n, \text{ so that } h_n^{-1}(x) = x + z_n.$$

The topology is translation-invariant, and so each h_n is a homeomorphism. To verify the category convergence of the sequence h_n, consider U non-empty and $\mathcal{D}$-open; then consider any measurable non-null $V \subseteq U$. To verify (cc) in relation to V, it now suffices to prove the following result, which is of independent interest (cf. Littlewood's First Principle, §1.1).

Theorem 10.3.1 (Verification Theorem for $\mathcal{D}$) *Let V be measurable and non-null. For any null sequence $\{z_n\} \to 0$ and each $k \in \omega$,*

$$H_k := \bigcap\nolimits_{n \geq k} V \backslash (V + z_n) \text{ is null, so meagre in the } \mathcal{D}\text{-topology.}$$

Proof Suppose otherwise. Then for some k, we have $|H_k| > 0$. Write H for H_k. Since $H \subseteq V$, it follows, for $n \geq k$, that $\emptyset = H \cap h_n^{-1}(V) = H \cap (V + z_n)$, and so a fortiori $\emptyset = H \cap (H + z_n)$.

Let u be a density point of H. Thus, for some interval $I_\delta(u) := (u - \delta/2, u + \delta/2)$, we have

$$|H \cap I_\delta(u)| > \frac{3}{4}\delta.$$

Let $E = H \cap I_\delta(u)$. For any z_n, we have $|(E + z_n) \cap (I_\delta(u) + z_n)| = |E| > \frac{3}{4}\delta$. For $0 < z_n < \delta/4$, we have $|(E + z_n) \backslash I_\delta(u)| \leq |(u + \delta/2, u + 3\delta/4)| = \delta/4$. Put $F = (E + z_n) \cap I_\delta(u)$; then $|F| > \delta/2$.

But $\delta \geq |E \cup F| = |E| + |F| - |E \cap F| \geq \frac{3}{4}\delta + \frac{1}{2}\delta - |E \cap F|$. So

$$|H \cap (H + z_n)| \geq |E \cap F| \geq \frac{1}{4}\delta,$$

contradicting $\emptyset = H \cap (H + z_n)$. This completes the proof. $\square$

A similar but simpler proof establishes the following result, which implies (cc) for the Euclidean topology on $\mathbb{R}$; here for given open U we may take any open interval $V \subseteq U$.

Theorem 10.3.1′ (Verification Theorem for $\mathcal{E}$) *Let V be an open interval in $\mathbb{R}$. For any null sequence $\{z_n\} \to 0$ and each $k \in \omega$,*

$$H_k := \bigcap\nolimits_{n \geq k} V \backslash (V + z_n) \text{ is empty.}$$

We now re-state and re-prove Theorem 4.2.1. As with the Category Embedding Theorem, the set T here may be assumed to be non-meagre/non-null, since otherwise there is nothing to prove.

Theorem 10.3.3 (Theorem KBD – Kestelman–Borwein–Ditor Theorem) *Let $\{z_n\} \to 0$ be a null sequence of reals. If T is Baire/Lebesgue measurable, then, for generically all $t \in T$, there is an infinite set $\mathbb{M}_t$ such that*

$$\{t + z_m : m \in \mathbb{M}_t\} \subseteq T.$$

Proof Theorem 10.2.2 may be applied to $h_n(x) := x + z_n$ in view of Theorem 10.3.1 or 10.3.1′, respectively, in the category/measure cases. $\square$

10.4 Kestelman–Borwein–Ditor Theorem: Second Proof

Our second proof in this chapter (and third proof within the book, with Theorem 4.2.1) depends on the interplay of the z-topology (see Remark 4b in §8.5) and the density topology $\mathcal{D}$ (see Chapters 7 and 8). We include this third proof here because (**) and Lemma 10.4.1 below are needed for the results of §10.5. We recall some salient features. The set of density points of T is denoted by $\phi_N(T)$, where this notation refers to the σ-ideal N of Lebesgue null sets and the fact that ϕ is a *lower-density operator* (Chapter 8). (See CieL1990 for a discussion of density topologies generated by σ-ideals.) By the Lebesgue Density Theorem, 7.1.1, almost all points of a measurable set are density points and so $\phi_N(\phi_N(T)) = \phi_N(T)$. See Sze2011 for a study of the exceptional points $E_N(T)$.

If u is a density point of both S and T it follows from the definition that u is a density point of $S \cap T$. It is this fact that justifies the introduction of the $\mathcal{D}$-topology, the density topology, on $\mathbb{R}$. We thus have

$$\mathcal{D} = \{T \in \mathcal{L} : T \subseteq \phi_N(T)\},$$

introduced in GofW1961 (see also HauP1952) and studied also in GofNN1961 (cf. CieLO1994, and, for a textbook treatment, Kec1995).

For $z = \{z_n\} \to 0$ any null sequence, the z-topology is that defined by reference to the base operator

$$b^z(T) := \bigcap_{k \in \omega} \bigcup_{n \geq k} (T - z_n). \qquad (**)$$

We will write

$$\mathbf{z}(T) := T \cap b^z(T).$$

Lemma 10.4.1 (Fundamental Lemma on Genericity) *Let $\mathbf{z} = \{z_n\} \to 0$ be any null sequence. Suppose that $T \in T_N$, i.e. T is density-open (measurable and every point of T is a density point of T). Then u is a density point of $b^z(T)$ for any $u \in T$; in symbols:*

$$u \in \phi_N(\mathbf{z}(T)) = \phi_N(T \cap b^z(T)).$$

Proof Let $u \in T$. Write $I_\delta(u) := (u - \delta, u + \delta)$. Then u is a density point of T (since $T \in \mathcal{D}$) and hence, for any $\varepsilon > 0$, there is $\delta < \varepsilon$ such that

$$\frac{|T \cap I_\delta(u)|}{2\delta} \geq (1 - \varepsilon),$$

i.e. nearly all of $I_\delta(u)$ is in T. Let $\eta = 2\delta\varepsilon$. For $n > N$, $|z_n| < \eta$ and so putting $T_n = T \cap (T - z_n)$,

$$|T_n \cap I_\delta(u)| \geq (1 - \varepsilon)2\delta - 2\eta,$$

i.e. nearly all of $I_\delta(u)$ is in $T - z_n$. Similarly, for $k > N$,

$$\left| \bigcup_{n \geq k} T_n \cap I_\delta(u) \right| \geq (1 - \varepsilon)2\delta - 2\eta.$$

Hence we have (cf. ErdKR1963)

$$|\mathbf{z}(T) \cap I_\delta(u)| = \left| \bigcap_{k \in \omega} \bigcup_{n \geq k} T_n \cap I_\delta(u) \right| \geq (1 - 2\varepsilon)2\delta,$$

i.e. nearly all of $I_\delta(u)$ is covered by $\mathbf{z}(T)$. Thus u is a density point of $\mathbf{z}(T)$ and of course $\mathbf{z}(T) \subset T$. □

Corollary 10.4.2 *For $u \in T$ arbitrarily close to u, there is a point $t \in \mathbf{z}(T)$, i.e. a point t such that $t \in T$ and $\{t + z_m : m \in \mathbb{M}_t\} \subseteq T$, for some infinite $\mathbb{M}_t$.*

We turn to the proof of the Kestelman–Borwein–Ditor Theorem.

Proof of the Kestelman–Borwein–Ditor Theorem In the measure case, denote the set of density points of S by T and apply the Fundamental Lemma on Genericity, 10.4.1.

In the Baire case, if $S = I \backslash M \cup M'$, where I is an interval and M, M' are meagre, take $T = I \backslash M$. We will show that for any $\{u_n\} \to u \in T$, there are $v \in T$ and an infinite $\mathbb{M}_v$ such that

$$\{v + u_m : m \in \mathbb{M}_v\} \subseteq S.$$

Select $\delta > 0$ so that $J = (u - \delta, u + \delta) \subseteq I$. We wish to pick $v, v_n \notin M$, with the aim of later putting $v_n := v + u_n$. This means we require in particular that $v + u_n \notin M$. So pick v in J to avoid the meagre set

$$M \cup \bigcup_{n \in \omega} M - u_n.$$

Now, for n large enough, $u_n \in J$. By choice, $v, v_n \notin M$. Hence, for large enough n, we have $v, v_n \in T$, as required. □

10.5 Near-Closure and No Trumps

Here we work at first in $\mathbb{R}$. Recall that a set A is z-closed if and only if every z-translator into A (i.e. $t \in \mathbb{R}$ such that $t + z_m \in A$ infinitely often) is in A and that the z-topology is translation invariant (see Remark in §8.5). This notion enables us to strengthen Theorem 10.3.3 to say that any non-meagre Baire/non-null measurable set A is almost z-closed for any null sequence z, i.e. every translator into A is in A, modulo a meagre/null set, depending on whether the z-topology is regarded as refining $\mathcal{E}$ or $\mathcal{D}$. Thus A is 'nearly closed' under the z-topology.

Theorem 10.5.1 (Near-Closure Theorem) *For any null sequence z, and A a (non-meagre) Baire or (non-null) measurable set, A is almost z-closed, i.e. modulo a meagre/null set every z translator into A is in A.*

Proof By Theorem 10.3.3 $T \backslash b^z(T)$ is null (since $t \in b^z(T)$ if and only if $t + z_m \in T$ for a subsequence z_m). Recalling the definition (**) of the base operator, the Fundamental Lemma, 10.4.1, implies that $T \subseteq \Phi_N(T \cap b^z(T)) \subseteq \Phi_N(b^z(T))$. But $\Phi_N(b^z(T))$ differs from $b^z(T)$ by a null set (by the Density Theorem, 7.1.1), hence $b^z(T) \backslash T$ is also null. □

It suffices to consider the density topology case.

The theorem has two important corollaries. Our first result involves boundedly shift-compact sets.

Theorem 10.5.2 (Shift-Compactness Theorem) *For any non-negligible Baire/measurable set T and any bounded sequence $\langle u_n \rangle$, for quasi all $t \in T$ there is an infinite $\mathbb{M}_t$ such that $\{t + (u_m - u_0) : m \in \mathbb{M}_t\} \subseteq T$ for some limit u_0 of the sequence.*

Proof Suppose that $\langle u_n \rangle$ is bounded. Passing to a subsequence, we may assume that $\langle u_n \rangle$ converges, say to u_0. Then $z_n := u_n - u_0 \to 0$ and so, for quasi all $t \in T$, there is a subsequence $\langle z_m \rangle_{m \in \mathbb{M}}$ such that $t + z_m \in T$, i.e. $(t - u_0) + u_m \in T$ for $m \in \mathbb{M}_t$. Thus, a shift of a subsequence of $\langle u_n \rangle$ lies in T. □

Corollary 10.5.3 *The restriction to $\mathcal{L}$ of $b^z(.)$ is an outer density.*

Further generality may be achieved by assuming somewhat less than almost z-closure for any z. We have in mind a sense of 'largeness', of the following type: for a countable sequence of sets in the class $\mathcal{L}$ (or $\mathcal{B}a$), if their union is 'large' so is one of them. The following definition replaces null sequences $z = \langle z_n \rangle$ with convergent or bounded sequences $u = \langle u_n \rangle$, and the translator t may be arbitrary.

The origin of the NT terminology below (Jensen's $\diamond$ and Ostaszewski's $\clubsuit$) was clarified in §9.4 from where we recall:

Definition (No Trumps) For a sequence $\{T_k : k \in \omega\}$ of subsets of $\mathbb{R}^d$, the property $\mathbf{NT}(\{T_k : k \in \omega\})$ asserts that:

for any bounded/convergent sequence $u = \langle u_n \rangle$ in $\mathbb{R}^d$ there are $t \in \mathbb{R}^d$, an index $k \in \omega$ and an infinite $\mathbb{M} \subseteq \omega$ such that

$$\{t + u_n : n \in \mathbb{M}\} \subseteq T_k. \qquad \text{(NT)}$$

In words: for every bounded/convergent sequence $u = \langle u_n \rangle$ in $\mathbb{R}^d$, some T_k contains a translate of a subsequence of $\langle u_n \rangle$.

A localized version may be obtained by considering only convergent sequences.

Definition (Local No Trumps) For a sequence $\{T_k : k \in \omega\}$ of subsets of $\mathbb{R}^d$, the property $\mathbf{NT}_L(\{T_k : k \in \omega\})$ asserts that:

for any convergent sequence $\langle u_n \rangle$ in $\mathbb{R}^d$ with limit u_0, arbitrarily close to u_0 there are $t \in \mathbb{R}^d$, an index $k \in \omega$, and an infinite $\mathbb{M} \subseteq \omega$ with

$$\{t + u_n : n \in \mathbb{M}\} \subseteq T_k.$$

In words: for every bounded/convergent sequence $\langle u_n \rangle$ in $\mathbb{R}^d$, some T_k contains a translate of a subsequence of $\langle u_n \rangle$.

Lemma 10.5.4 *In words: for every convergent sequence $\langle u_n \rangle$ in $\mathbb{R}^d$ with limit u_0, some T_k contains a translate of a subsequence of $\langle u_n \rangle$ arbitrarily close to u_0.*

Proof As in Theorem 10.5.2 (the Shift-Compactness Theorem), if $u_n \to u_0$, then $z_n := u_n - u_0 \to 0$. As $T - u_0$ is z-closed for any z, then $s + z_m \in T - u_0$ for $m \in \mathbb{M}$ for some infinite $\mathbb{M}$ and some $s \in T - u_0$. Hence $s + z_m \in T - u_0$, i.e. $s + u_m \in T$ for $m \in \mathbb{M}$. $\qquad \square$

Remark We also have $\lim(s + u_m) = s + u_0 \in T$.

We will need the following result, implicit in CsiE1964.

Theorem 10.5.5 (Strong No Trumps Theorem; CsiE1964) *If T is a non-meagre Baire/non-null measurable set in any interval which it meets and $T = \bigcup_{k \in \omega} T_k$ with each T_k measurable/Baire, then $\mathbf{NT}_L(\{T_k : k \in \omega\})$ holds. Indeed, for every convergent sequence $\{u_n\} \to u_0 \in T$, any neighbourhood of the limit u_0 contains a point s for which there exist $K = K(s) \in \omega$ and an infinite set $\mathbb{M} = \mathbb{M}(s) \subseteq \omega$ such that*

$$s + u_m \in T_K \text{ for } m \in \mathbb{M}.$$

Proof Suppose u_n converges to u_0. Consider any interval $I = (u_0 - \eta, u_0 + \eta)$ with $\eta > 0$. As T meets I, for some $K \in \omega$, the set $T_K \cap I$ is measurable and non-null (resp. Baire non-meagre). Let $z_n := u_n - u$. Then $z_n \to 0$ and so, by the Kestelman–Borwein–Ditor Theorem, for almost all (resp. for quasi all) $t \in T_K \cap I$, there is an infinite set $\mathbb{M}_t$ such that

$$\{t + z_m : m \in \mathbb{M}_t\} \subseteq T_K \cap I.$$

For any such t put $s = t - u$. Then writing $\mathbb{M} = \mathbb{M}(s)$ for $\mathbb{M}_t$, we have

$$\{s + u_m : m \in \mathbb{M}_t\} \subseteq T_K. \qquad \square$$

11

Effros' Theorem and the Cornerstone Theorems of Functional Analysis

11.1 Introduction

We give a short proof of a classic theorem of Effros (Eff1965) stated in a form which holds also beyond its original separable context, namely the general metrizable context. All that is required for the general context is a single modification to the continuity assumptions of group action (see Definition (1) below). Effros' Theorem asserts that a continuous transitive action by a Polish group G on a non-meagre space X is open: the point-evaluation $g \mapsto gx$ is an open map from G to X. Viewed, despite its original separability, as a group-action counterpart to one of the cornerstones of functional analysis, the Open Mapping Theorem (OMT) (that a surjective continuous linear map between Fréchet spaces is open – cf. Rud1973), it has come to be called the *Open Mapping Principle* (OMP) – see Anc1987, §1.

The paradigm for OMP was an earlier result, due to Glimm (Gli1961), which was restricted to locally compact groups and was directed at resolving a conjecture of Mackey concerned with representation of a C^*-algebra A on (an infinite-dimensional) Hilbert space H. A key notion was that of a quotient space A^i/G, for A^i the irreducible representations and G the group of unitary operators on H, being countably separated (separation of points by a countable family of Borel sets): briefly, *smooth*. Group action takes (g, x) to the g-conjugate $g^{-1}xg$ of x, for $g, x \in H$, i.e. similarity, and smoothness is an equivalent of openness. We note that the underlying (unitary) equivalence of representations, viewed as a set in $A^i \times A^i$, is closed (so Borel). Glimm observed that the openness property excludes the group of those sequences of 0s and 1s which have all but a finite number of their terms equal to 0 when this group acts on the space of all sequences of 0s and 1s by coordinate-wise addition (modulo 2). In another context this is the Vitali equivalence (difference modulo rationals), for which see Kha2004. This is the nub of the *Glimm–Effros dichotomy* in the realm of

Borel equivalences: either an equivalence is smooth or it embeds in itself the Vitali equivalence; see Harrington–Kechris–Louveau (HarrKL1990).

An altogether different area received a huge boost from the Effros result of 1965 some 10 years after its publication: the theory of homogeneous continua (the origins of which hark back to Montgomery; Mon1950). Ungar (Ung1975) in 1975 discovered that if (X, d) is a homogeneous compact metric space, then the group $H(X)$ of self-homeomorphisms of X equipped with the compact-open topology acts *microtransitively* on X, meaning that for each $\delta > 0$ there is an 'Effros number' $e(\delta) > 0$ such that for any $x, y \in X$ with $d(x, y) < e(\delta)$, there is $h \in H(X)$ with $h(x) = y$ and with $d(h(z), z) < \delta$ for all $z \in X$ (a 'δ-push'). A slew of papers followed in the next decade, with early contributions from F.B. Jones (see Jon1975) and Hagopian, J.T. Rogers, Jr, W. Lewis, Phelps, Kennedy. See again Anc1987, §1 and CharM1966.

Not quite a decade later still, van Mill (Mil2004) offered both a generalization of the OMP, extending it to (separable) groups that are analytic (i.e. continuous images of a Polish space), and a clever counterexample to a variety of conjectures (Mil2008). By van Mill's theorem of Mil2004, separable metrizable groups that are analytic and have continuous action are microtransitive (as above) on any non-meagre separable metrizable space. A co-analytic example of a continuous group action which is not microtransitive exists under the Axiom of Constructibility ($V = L$, see Chapter 16), but under the Axiom of Determinacy (AD; see Chapters 14 and again 16) no such example can exist. Under AC there always exist such examples. See Med2022.

To include a 'non-separable' context requires in place of 'global' countability a more 'local' notion: a sequential property related to the Steinhaus-type Sum–Set Theorem (that 0 is an interior point of $A - A$, for non-meagre A with BP, the Baire property – (Pic1939; Pic1942); see Chapter 15), because of the following argument (which goes back to Pettis, Pet1950).

Consider $L: E \to F$, a linear, continuous surjection between Fréchet spaces, and U a neighbourhood of the origin. Choose A an *open* neighbourhood of the origin with $A - A \subseteq U$; as $L(A)$ is non-meagre (since $\{nL(A) : n \in \mathbb{N}\}$ covers F) and has the Baire property (see Proposition 11.4.3 in §11.4), $L(A) - L(A)$ is a neighbourhood of the origin by the Sum–Set Theorem. But of course

$$L(U) \supseteq L(A) - L(A),$$

so $L(U)$ is a neighbourhood of the origin. So L is an open mapping.

The sequential approach followed here is based on Ost2015b. Throughout this chapter, without further comment, all spaces considered will be metrizable. We recall the Birkhoff–Kakutani Theorem (Chapter 6; cf. HewR1979, II.8.6, pp. 70, 83), that a metrizable group G with neutral element e_G has a right-invariant metric d_{R}^G. Passage to $\|g\| := d_{\mathrm{R}}^G(g, e_G)$ yields a (group) norm (invariant under inversion, satisfying the triangle inequality), which justifies calling these *normed groups*; any Fréchet space qua additive group, equipped with an F-norm (KaltPR1984, Ch. 1, §2), is a natural example (cf. the autohomeomorphism group Auth, see p. 176). Below we need the following.

Definitions   1. (Cf. Pet1950) For G a metrizable group, say that the group action $\varphi \colon G \times X \to X$ has the *Nikodym property* or, briefly, is *Nikodym*, if for every non-empty open neighbourhood U of e_G and every $x \in X$, the set $Ux = \varphi_x(U) := \varphi(x, U)$ contains a non-meagre *Baire set.*

2. A^q denotes the *quasi-interior* of A – the largest open set U with $U \backslash A$ meagre (cf. Ost2011, §4).

Thanks to the Nikodym property, introduced above, we are able to state the Effros Theorem in a form that embraces both the separable and the non-separable contexts. We establish the property first in the simpler separable context in §11.2 and then again in the non-separable context in Chapter 13. The main results below are Theorems 11.1.1 and 11.1.2 (for later convenience, also referred to symbolically as Theorems Sh and E resp.), with corollaries in §11.4 including OMT; see below for commentary.

Theorem 11.1.1 (Theorem Sh, Shift-Compactness Theorem)   *For T a Baire non-meagre subset of a metric space X and G a group, Baire under a right-invariant metric, and with separately continuous and transitive Nikodym action on X:*

> *for every convergent sequence x_n with limit x and any Baire non-meagre $A \subseteq G$ with $e_G \in A^q$ and $A^q x \cap T^q \neq \emptyset$, there are $\alpha \in A$ and an integer N such that $\alpha x \in T$ and*
>
> $$\{\alpha(x_n) : n > N\} \subseteq T.$$

In particular, this is so if G is analytic and all point-evaluation maps $\varphi_x \colon g \to g(x)$ are base-σ-discrete.

This theorem has wide-ranging consequences, including Steinhaus' Sum–Set Theorem; see the survey article Ost2013a; Ost2013b and Chapter 14. See also BinO2024.

Theorem 11.1.2 (Theorem E, Effros' Theorem – Baire Version)   *If*

(i) *the normed group G has separately continuous and transitive Nikodym action on X, and*

(ii) *G is Baire under the norm topology and X is non-meagre,*

then for any open neighbourhood U of e_G and any $x \in X$ the set $Ux := \{u(x) :$
$u \in U\}$ is a neighbourhood of x, so that in particular the point-evaluation maps
$g \mapsto g(x)$ are open for each x. That is, the action of G is microtransitive.

In particular, this holds if G is Polish and all point-evaluation maps φ_x are
continuous.

More generally, this holds if G is analytic and Baire, and all point-evaluation
maps φ_x are continuous and base-σ-discrete (for which see Chapter 12). This
last property holds automatically when G is separable.

By Proposition 11.2.3 X, being non-meagre here, is also a Baire space.

The classical counterpart of Theorem E has G a Polish group; van Mill's
version (Mil2004) requires the group G to be analytic (i.e. the continuous image
of some Polish space). The Baire version above improves the version given in
Ost2013b, where the group is almost complete. (The two cited sources taken
together cover the literature.)

A result due to Hoffmann-Jørgensen (Hof1980, Th. 2.3.6, p. 355) asserts that
a Baire, separable, analytic *topological group* is Polish (as a consequence of an
analytic group being metrizable – for which see again Hof1980, Th. 2.3.6), so
the analytic separable case of Theorem E reduces to its classical version.

Unlike the proof of the Effros Theorem attributed to Becker in Kec1994,
Th. 3.1, the one offered here does not employ the Kuratowski–Ulam Theorem
of §9.5 (the Category version of the Fubini Theorem), a result known to fail
beyond the separable context (as shown in Pol1979, cf. MilP1986, but see
FreNR2000 and Chapter 9).

For further commentary (connections between convexity and the Baire property, relation to van Mill's separation property in Mil2009, certain specializa-
tions) see §11.4.

11.2 Action, Microtransitive Action, Shift-Compactness

We recall some group-related notions.

A normed group G *acts continuously* on X if there is a continuous mapping
$\varphi \colon G \times X \to X$ such that $\varphi(e_G, x) = x$ and $\varphi(gh, x) = \varphi(g, \varphi(h, x))$ for $x \in X$
and $g, h \in G$. The action φ is *separately continuous* if $g \colon x \mapsto \varphi(g, x)$ is
continuous for each g and $\varphi_x \colon g \mapsto \varphi(g, x)$ is continuous for each x; in such
circumstances:

(i) the elements $g \in G$ yield autohomeomorphisms of X via $g \colon x \mapsto g(x) :=$
$\varphi(g, x)$ (as g^{-1} is continuous); and

(ii) point evaluation of these homeomorphisms, $\varphi_x(g) = g(x)$, is continuous.

In certain situations joint continuity of action is implied by separate continuity (see Bouz1993 and literature cited in Ost2012).

The action is *transitive* if for any x, y in X there is $g \in G$ such that $g(x) = y$. For later purposes (§11.3), say that the action of G on X is *weakly microtransitive* if, for $x \in X$ and each neighbourhood A of e_G, the set

$$\mathrm{cl}(Ax) = \mathrm{cl}\{ax : a \in A\}$$

is a neighbourhood of x. The action is *microtransitive* ('transitive in the small' – for details see Mil2004) if for $x \in X$ and each neighbourhood A of e_G the set

$$Ax = \{ax : a \in A\}$$

is a neighbourhood of x. This (norm) property implies that Ux is open for U open in G (i.e. that here each φ_x is an open mapping). We refer to Ax as an *x-orbit* (the A-orbit of x). The following group action connects the Open Mapping Theorem to the present context.

Example (Induced Homomorphic Action)   A surjective, continuous homomorphism $\lambda : G \to H$ between normed groups induces a transitive action of G on H via $\varphi^\lambda(g, h) := \lambda(g)h$ (cf. Ost2012, Th. 5.1), specializing to

$$\varphi^L(a, b) := L(a) + b$$

for G, H Fréchet spaces (regarded as normed, additive groups) and $\lambda = L : G \to H$ linear (Anc1987; Mil2004). Of course for Fréchet spaces, by the Open Mapping Theorem itself, φ^L has the Nikodym property.

Definitions   1. Auth(X) denotes the autohomeomorphisms of a metric space (X, d^X); this is a group under composition. $\mathcal{H}(X)$ comprises those $h \in$ Auth(X) of bounded norm:

$$\|h\| := \sup_{x \in X} d^X(h(x), x) < \infty.$$

2. For a normed group G acting on X, say that X has the *crimping property* (property C for short) with respect to G if, for each $x \in X$ and each sequence $\{x_n\} \to x$, there exists in G a sequence $\{g_n\} \to e_G$ with $g_n(x) = x_n$. (This and a variant occur in Ban1932, Ch. III; Th. 4; CharC2001; for the term see BinO2009a.)

For a subgroup $\mathcal{G} \subseteq \mathcal{H}(X)$, say that X has the *crimping property* with respect to $\mathcal{G}$ if X has the crimping property with respect to the natural action $(g, x) \to g(x)$ from $\mathcal{G} \times X \to X$. (This action is continuous relative to the left or right norm-topology on $\mathcal{G}$ – cf. Dug1966, XII.8.3, p. 271.)

3. As a matter of convenience, say that the *Effros property* (or *property E*) holds for the group G acting on X if the action is microtransitive, as above.

4. For a subgroup $\mathcal{G} \subseteq \text{Auth}(X)$ say that X is $\mathcal{G}$-*shift-compact* (or, shift-compact under $\mathcal{G}$) if, for any convergent sequence $x_n \to x_0$, any open subset U in X, and any Baire set T co-meagre in U, there is $g \in \mathcal{G}$ with $g(x_n) \in T \cap U$ along a subsequence (cf. Chapter 10, p. 161). Call the space *shift-compact* if it is $\mathcal{H}(X)$-shift-compact (cf. Mille02012; Ost2013c).

In such a space, any Baire non-meagre set is locally co-meagre (co-meagre on open sets) in view of Proposition 11.2.3.

We shall prove in §11.3.1 equivalence between the Effros and crimping properties.

Theorem 11.2.1 (Theorem EC)   *The Effros property holds for a group G acting on X if and only if X has the crimping property with respect to G.*

We now clarify the role of shift-compactness.

Proposition 11.2.2   *For any subgroup $\mathcal{G} \subseteq \mathcal{H}(X)$, if X is $\mathcal{G}$-shift-compact, then X is a Baire space.*

Proof   We argue as in Mil2004, Prop 3.1(1). Suppose otherwise; then X contains a non-empty meagre open set. By Banach's Category Theorem, the union of all such sets is a largest open meagre set M, and is non-empty. Thus $X \backslash M$ is a co-meagre Baire set. For any $x \in M$ the constant sequence $x_n \equiv x$ is convergent and, since $X \backslash M$ is co-meagre in X, there is $g \in G$ with $g(x) \in X \backslash M$. But, as g is a homeomorphism, $g(M)$ is a non-empty open meagre set, so is contained in M, implying $g(x) \in M$, a contradiction.   □

A similar argument gives the following and clarifies an assumption in Theorem 11.1.2 (cf. Mil2004; Hof1980, Prop. 2.2.3).

Proposition 11.2.3   *If X is non-meagre and G acts transitively on X, then X is a Baire space.*

Proof   As above, refer again to M, the union of all meagre open sets, which, being meagre, has non-empty complement. For x_0 in this complement and any non-empty open U pick $u \in U$ and $g \in G$ such that $g(x_0) = u$. Now as g is continuous, $g^{-1}(U)$ is a neighbourhood of x_0, so is non-meagre, since every neighbourhood of x_0 is non-meagre. But g is a homeomorphism, so $U = g(g^{-1}(U))$ is non-meagre. So X is Baire, as every non-empty open set is non-meagre.   □

11.2.1 Nikodym Actions: Separable Context

The following result is usually a first step in proving the weakly microtransitive variant of the classical Effros Theorem (cf. Anc1987, Lemma 3; Ost2013a, Th. 2). Indeed, one may think of it as giving a form of 'very weak microtransitivity'. We will later see a direct generalization to the non-separable context: cf. Mil2004, Lemma 3.2.

Proposition 11.2.4 *If G is a separable normed group acting transitively on a non-meagre space X with each point-evaluation map $\varphi_x : g \mapsto g(x)$ continuous, then, for each non-empty open U in G and each $x \in X$, the set Ux is non-meagre in X.*

In particular, if G is analytic, then G is a Nikodym action.

Proof We first work in the right norm-topology, i.e. derived from the assumed right-invariant metric $d_R^G(s,t) = \|st^{-1}\|$. Suppose that $u \in U$, and so without loss of generality assume that $U = B_\varepsilon(u) = B_\varepsilon(e_G)u$ (open balls of radius some $\varepsilon > 0$): indeed,

$$B_\varepsilon(e_G)u = \{xu : d_R^G(x,e) < \varepsilon\} = \{z : d_R^G(zu^{-1},e) < \varepsilon\}$$
$$= \{z : d_R^G(z,u) < \varepsilon\} = U.$$

Now put $y := ux$ and $W = B_\varepsilon(e_G)$. Then $Ux = B_\varepsilon(e_G)ux = Wy$. Next work in the left norm-topology, derived from $d_L^G(s,t) = \|s^{-1}t\| = d_R^G(s^{-1},t^{-1})$ (for which $W = B_\varepsilon(e_G)$ is still a neighbourhood of e_G). As each set hW for $h \in G$ is now open (since now the left shift $g \to hg$ is a homeomorphism), the open family $\mathcal{W} = \{gW : g \in G\}$ covers G.

As G is separable, the cover $\mathcal{W}$ has a countable subcover, say $\mathcal{V}$. Thus $X := \bigcup\{Vy : V \in \mathcal{V}\}$, as $X = Gy$, and so Vy is non-meagre for some $V \in \mathcal{V}$, say for $V = \hat{V}$. As $\mathcal{V}$ is a subcover, there is some $\hat{g} \in G$ with $\hat{V} = \hat{g}W$, so $\hat{V}y = \hat{g}Wy$, and so $\hat{g}Wy$ is non-meagre. As $\hat{g}^{-1}$ is a homeomorphism of X, $Wy = Ux$ is also non-meagre in X.

If G is analytic, then as U is open, it is also analytic (since open sets are $\mathcal{F}_\sigma$ and Souslin-$\mathcal{F}$ subsets of analytic sets are analytic, cf. RogJ1980), and hence so is $\varphi_x(U)$. Indeed, since φ_x is continuous, Ux is analytic, so Souslin-$\mathcal{F}$, and so Baire by Nikodym's Theorem.   □

11.3 Effros and Crimping Properties: E, EC and C

11.3.1 Proof that E $\Longleftrightarrow$ C

We first show that if the Effros property holds for the action of a group G on X, then X has the crimping property with respect to G. Indeed, suppose that

$x = \lim x_n$. For each n, take $U = B^G_{1/n}(e_G)$; then $Ux := \{u(x) : u \in U\}$ is an open neighbourhood of x, and so there exists $h_{n,m} \in U$ with $h_{n,m}(x) = x_m$ for all m large enough, say for all $m > m(n)$. Without loss of generality we may assume that $m(1) < m(2) < \cdots$. Put $h_m := e_G$ for $m < m(1)$, and for $m(k) \le m < m(k+1)$ take $h_m := h_{k,m}$. Then $h_m \in B^G_{1/k}(e_G)$, so h_m converges to e_G and $h_m(e_G) = x_m$.

For the converse, suppose that the Effros property fails for G acting on X. Then for some open neighbourhood U of e_G and some $x \in X$, $Ux := \{u(x) : u \in U\}$ is not an open neighbourhood of x. So for each n there is a point $x_n \in B_{1/n}(x) \backslash Ux$. As x_n converges to x, there are homeomorphisms h_n converging to the identity e_G with $h_n(x) = x_n$. As U is an open neighbourhood of e_G and since h_n converges to e_G, there is N such that $h_n \in U$ for $n > N$. In particular, for any $n > N$, $h_n(x) = x_n \in Ux$, a contradiction, and we are done.

11.3.2 Proof of the Shift-Compactness Theorem, Theorem Sh

We view Theorem 11.1.1 as having 'two tasks': to find a 'translator of the sequence' τ, and to locate it in a given Baire non-meagre subset of the group – provided that subset satisfies a consistency condition (a necessary condition).

For clarity we break the tasks into two steps – the first delivering a weaker version of Theorem 11.1.1 in Proposition 11.3.3 . The arguments are based on the following lemma. We note a corollary, observed earlier by van Mill in the case of metric topological groups (Mil2008, Prop. 3.4), which concerns a co-meagre set, but we need its refinement to a localized version for a non-meagre set.

Lemma 11.3.1 (Separation Lemma)   *Let G be a normed group, with separately continuous and transitive Nikodym action on a non-meagre space X. Then, for any point x and any F closed nowhere dense,*

$$W_{x,F} := \{\alpha \in G : \alpha(x) \notin F\}$$

is dense open in G. In particular, G separates points from nowhere dense closed sets.

Proof   The set $W_{x,F}$ is open, being of the form $\varphi_x^{-1}(X \backslash F)$ with φ_x continuous (by assumption). By the Nikodym property, for U any non-empty open set in G, the set Ux is non-meagre, and so $Ux \backslash F$ is non-empty, as F is meagre. But then for some $u \in U$ we have $u(x) \notin F$. $\square$

Corollary 11.3.2   *If G is a normed group, Baire in the norm topology with transitive and separately continuously Nikodym action on a non-meagre space X space, and T is co-meagre in X – then, for countable $D \subseteq X$, the set $\{g : g(D) \subseteq T\}$ is a dense $\mathcal{G}_\delta$.*

In particular, this holds if G is analytic and each point-evaluation map
$\varphi_x \colon g \to g(x)$ *is base-σ-discrete.*

Proof Without loss of generality, the co-meagre set is of the form $T = U \setminus \bigcup_{n \in \omega} F_n$ with each F_n closed and nowhere dense, and U open. Then, by Lemma 11.3.1 (the Separation Lemma) and as G is Baire,

$$\{g \in G : g(D) \subseteq T\} = \bigcap_{n \in \omega} \{g : g(D) \cap F_n = \emptyset\}$$
$$= \bigcap_{d \in D, n \in \omega} \{g : g(d) \notin F_n\}$$

is a dense $\mathcal{G}_\delta$. For the final assertion, see p. 209, point (2).   □

Proposition 11.3.3 *If T is a Baire non-meagre subset of a metric space X and G a normed group, Baire in its norm topology, acting separately continuously and transitively on X, with the Nikodym property – then, for every convergent sequence x_n with limit x_0, there is $\tau \in G$ and an integer N with $\tau x_0 \in T$ and*

$$\{\tau(x_n) : n > N\} \subseteq T.$$

Proof Write $T := M \cup (U \setminus \bigcup_{n \in \omega} F_n)$ with U open, M meagre and each F_n closed and nowhere dense in X. Let $u_0 \in T \cap U$. By transitivity there is $\sigma \in G$ with $\sigma x_0 = u_0$. Put $u_n := \sigma x_n$. Then $u_n \to u_0$. Put

$$C := \bigcap_{m,n \in \omega} \{\alpha \in G : \alpha(u_m) \notin F_n\},$$

a dense $\mathcal{G}_\delta$ in G; then, by the Separation Lemma, 11.3.1, as G is Baire,

$$\{\alpha \in G : \alpha(u_0) \in U\} \cap C$$

is non-empty. For α in this set we have $\alpha(u_0) \in U \setminus \bigcup_{n \in \omega} F_n$. Now $\alpha(u_n) \to \alpha(u_0)$, by continuity of α, and U is open. So for some N we have for $n > N$ that $\alpha(u_n) \in U$. Since $\{\alpha(u_m) : m = 1, 2, \ldots\} \in X \setminus \bigcup_{n \in \omega} F_n$, we have, for $n > N$, that $\alpha(u_n) \in U \setminus \bigcup_{n \in \omega} F_n \subseteq T$.

Finally, put $\tau := \alpha\sigma$. It then follows that $\tau(x_0) = \alpha\sigma(x_0) \in T$ and $\{\tau(x_n) : n > N\} \subseteq T$.   □

Proof of Theorem 11.1.1, Theorem Sh We work in the right norm-topology and use the notation of the preceding proof (of Proposition 11.3.3), so that U here is the quasi-interior of T and $\sigma x_0 = u_0$. As $e_G \in A^q$ and A is a non-meagre Baire set, we may without loss of generality write $A = B_\varepsilon(e_G) \setminus \bigcup_n G_n$, where each G_n is closed nowhere dense with $e_G \notin G_n$ and $B_\varepsilon(e_G)$ is the quasi-interior of A.

As $A^q x_0 \cap T^q$ is non-empty, there is $\alpha_0 \in B_\varepsilon(e_G)$ with $\alpha_0 x_0 \in U$ (but, we want a better α so that $\alpha x_0 \in T$ and $\alpha \in A$). Put $\beta_0 = \alpha_0 \sigma^{-1}$; then

$$
\begin{aligned}
\beta_0 = \alpha_0 \sigma^{-1} &\in B_\varepsilon(e_G)\sigma^{-1} \cap \{\alpha : \alpha(x_0) \in U\}\sigma^{-1} \\
&= B_\varepsilon(e_G)\sigma^{-1} \cap \{\beta : \beta(\sigma x_0) \in U\} \\
&= B_\varepsilon(e_G)\sigma^{-1} \cap \{\beta : \beta(u_0) \in U\},
\end{aligned}
$$

i.e. the open set $\{\beta : \beta(u_0) \in U\} \cap B_\varepsilon(e_G)\sigma^{-1}$ is non-empty. So

$$
\left(C \backslash \bigcup_n G_n \sigma^{-1}\right) \cap \{\beta : \beta(u_0) \in U\} \cap B_\varepsilon(e_G)\sigma^{-1} \neq \emptyset,
$$

since G is a Baire space and each $G_n \sigma^{-1}$ is closed and nowhere dense in G (as the right shift $g \to g\sigma^{-1}$ is a homeomorphism).

So there is β with $\beta(u_0) \in U$ such that $\alpha := \beta\sigma \in B_\varepsilon(e_G) \backslash \bigcup_n G_n = A$. That is, $\alpha x_0 = \beta u_0 \in U$; so $\beta(u_n) \in U$ for large n, for $n > N$ say, as $\alpha x_0 = \lim \alpha x_n = \lim \beta\sigma x_n = \lim \beta u_n$. But $\{\beta(u_m) : m = 1, 2, \ldots\} \in X \backslash \bigcup_n F_n$, as $\beta \in C$; so $\beta(u_n) \in U \backslash \bigcup_n F_n \subseteq T$ for $n > N$.

Finally, $\alpha(x_0) = \beta\sigma(x_0) \in T$ and $\{\alpha(x_n) : n > N\} \subseteq T$. $\qquad \square$

We recall that Theorem Sh refers to Theorem 11.1.1.

Proof that Sh $\Longrightarrow$ E   Assume G acts transitively on X and that X is non-meagre. Let $B := B_\varepsilon(e_G)$ and suppose that for some x the set Bx is not a neighbourhood of x. Then there is $x_n \to x$ with $x_n \notin Bx$ for each n. Take $A := B_{\varepsilon/2}(e_G)$ and note first that A is a symmetric open set ($A^{-1} = A$, since $\|g\| = \|g^{-1}\|$), and second that by the Nikodym property Ax contains a non-meagre, Baire subset T. So by Theorem Sh, as Ax meets T^q, there are $a \in A$ (which, being open, has the Baire property) and a co-finite $\mathbb{M}_a$ such that $ax_m \in Ax$ for $m \in \mathbb{M}_a$. For any such m, choose $b_m \in A$ with $ax_m = b_m x$. Then $x_m = a^{-1}b_m x \in A^2 x \subseteq Bx$, contradicting $x_m \notin Bx$ (note that $a^{-1} \in A$, by symmetry). So Bx is a neighbourhood of x. $\qquad \square$

Remark   As earlier, in the special case that G is (metrizable and) analytic, A is analytic, since open sets are $\mathcal{F}_\sigma$ and Souslin-$\mathcal{F}$ subsets of analytic sets are analytic, cf. RogJ1980, Th. 2.5.3. So by Proposition 11.3.3 Ax is Baire non-meagre, as φ_x is base-σ-discrete, cf. page 209, point (2).

11.4 From Effros to the Open Mapping Theorem

Definition (Anc1987)   Call the map φ_x *countably covered* if there exist self-homeomorphisms h_n^x of X for $n \in \mathbb{N}$ such that for any open neighbourhood U in G the sets $\{h_n^x(\varphi_x(U)) : n \in \mathbb{N}\}$ cover X.

Proposition 11.4.1 (cf. Anc1987)   *For the action $\varphi\colon G \times X \to X$ with X non-meagre, if each map φ_x is countably covered and takes open sets to sets with the Baire property, then the action has the Nikodym property.*

Proof   If φ_x is countably covered, then there exist self-homeomorphisms h_n^x of X for $n \in \mathbb{N}$ such that for any open neighbourhood U in G the sets $\{h_n^x(\varphi_x(U)) : n \in \mathbb{N}\}$ cover X. Then for X non-meagre, there is $n \in \mathbb{N}$ with $h_n^x(\varphi_x(U))$ non-meagre, so $Ux = \varphi_x(U)$ is itself non-meagre, being a homeomorphic copy of $h_n^x(\varphi_x(U))$. As Ux is assumed Baire, the action has the Nikodym property. $\square$

For E separable, an immediate consequence of *continuous* maps taking open sets to analytic sets (which are Baire sets) and of Proposition 11.4.1 is that φ^L is a Nikodym action.

For the general context, one needs *demi-open* continuous maps, which preserve *almost completeness* (absolute G_δ sets modulo meagre sets – see Mich1991 and its antecedent Nol1990), as it is not known which linear maps are base-σ-discrete – a delicate matter to determine, since the former include continuous linear surjections (by Lemma 11.4.2) and preserve almost analyticity as opposed to analyticity.

For present purposes, however, the *monotonicity property* below suffices. We omit the proof of the following observation (for which see the opening step in Rud1973, 2.11 or Conw1990, Ch. 3, §12.3). The open balls below refer to the underlying translation-invariant metric of a Fréchet space.

Lemma 11.4.2   *For a continuous linear map $L\colon X \to Y$ from a Fréchet space X to a normed space Y, for $s < t < r$,*

$$\mathrm{int}(\mathrm{cl}\, L(B_s(0)) \subseteq L(B_t(0)) \subseteq L(B_r(0)).$$

Hence, if $L(B_r(a))$ is convex, either it is meagre or it differs from $\mathrm{int}\, L(B_r(a))$ by a meagre set.

Proposition 11.4.3   *For L a continuous linear surjection from a Fréchet space E to a non-meagre normed space F, the action φ^L has the Nikodym property.*

Proof   We first show that as in Proposition 11.2.4 for $L\colon E \to F$ a continuous linear surjection, $\{\varphi_x^L : x \in F\}$ are countably covered. Indeed, fixing $x \in F$,

$$h_n^x(z) := n(z - x) \qquad (n \in \mathbb{N} \text{ and } z \in F)$$

is, on the one hand, a self-homeomorphism satisfying $h_n^x(\varphi_x(L(V))) = L(nV)$, since $n[(L(v) + x) - x] = nL(v) = L(nv)$; on the other hand, the family

$$\{h_n^x(L(V) + x) : n \geq 1\}$$

covers F, as $\{nV : n \in \mathbb{N}\}$ covers E where V is any open neighbourhood of the origin in E (by the 'absorbing' property, cf. Conw1990, 4.1.13; Rud1973, 1.33). In particular, $nL(B_1(0))$ is non-meagre for some n, and so $L(B_s(0))$ is non-meagre for any s. By Lemma 11.4.2, $L(B_t(0))$, for any $t > s$, contains the non-meagre Baire set cl $L(B_s(0))$. □

Corollary 11.4.4 is now immediate; it is used in Ost2012, Th. 5.1 to prove the 'Semi-Completeness Theorem', an Ellis-Type Theorem (Elli1953, Cor. 2) (cf. Ost2013d) giving a one-sided continuity condition which implies that a right-topological group generated by a right-invariant metric is a topological group (cf. 13.3.3).

Corollary 11.4.4 *If the continuous surjective homomorphism λ between normed groups G and H, with G analytic and H a Baire space, is base-σ-discrete, then λ is open; in particular, for λ bijective, λ^{-1} is continuous.*

Corollary 11.4.5 *For $L: E \to F$ a continuous surjective linear map between Fréchet spaces, the point evaluations φ_b^L for $b \in F$ are open, and so L is an open mapping.*

Proof By surjectivity of L, the action is transitive, and by Proposition 11.4.3 the action φ^L has the Nikodym property. So by Theorem E, 11.1.2, above the point-evaluations maps φ_b^L are open. Hence so also is L. □

12

Continuity and Coincidence Theorems

12.1 Continuity Theorems: Introduction

We begin with an overview of the range of results of concern to us.

Theorem 12.1.1 (Discontinuity-Set Theorem; Kur1966, 33, p. 397; Kur1924)

(i) *For $f: X \to Y$ Baire-measurable, with X, Y metric, the set of discontinuity points is meagre; in particular,*

(ii) *for $f: X \to Y$ Borel-measurable of class 1, with X, Y metric and Y separable, the set of discontinuity points is meagre.*

The following theorem, from somewhat more recent literature, usefully overlaps with the last result and will be proved in the next section.

Theorem 12.1.2 (Banach–Neeb Theorem; Banach (Ban1931; Ban1932), extended by Neeb (Nee1997))

(i) *A Borel-measurable $f: X \to Y$ with X, Y metric and Y separable is Baire-measurable.*

(ii) *A Baire-measurable $f: X \to Y$ with X a Baire space and Y metric is Baire-continuous.*

Remarks   1. In fact Banach shows that a Baire-measurable function is Baire-continuous on each perfect set – see Ban1932, vol. II, p. 206. For the distinction between Baire and Baire-measurable, see the two paragraphs below and §12.2.

2. In (i) if X, Y are completely metrizable, topological groups and f is a homomorphism, Neeb's additional assumption in Nee1997 that Y is arcwise connected becomes unnecessary, as Pestov (Pes1998) remarks in his MathSciNet commentary to Nee1997.

The following 'portmanteau theorem' summarizes what is in the literature. Theorem 12.1.3 concerns the Baire functions (with preimages of open sets having the Baire property). The later Theorem 12.2.2 concerns Baire-measurable functions (obtained from the continuous functions by iterating pointwise limits). See the comment ahead of Theorem 12.1.4.

Theorem 12.1.3 (Baire Continuity Theorem – Baire Version; cf. BinO2010g, 4, Th. 11.8)   *A Baire function* $f \colon X \to Y$ *is Baire-continuous in the following cases:*

(i) *Baire's condition (see, e.g., Hof1980, Th. 2.2.10, p. 346): Y is a second-countable space;*

(ii) *Emeryk–Frankiewicz–Kulpa (EmeFK1979): X is Čech-complete and Y has a base of cardinality not exceeding the continuum;*

(iii) *Pol's condition (Pol1976): f is Borel, X is Borelian-$\mathcal{K}$ and Y is metrizable and of non-measurable cardinality, see §12.5;*

(iv) *Hansell's condition (Hanse1971): f is σ-discrete and Y is metric.*

We will say that the pair (X, Y) *enables Baire continuity* if the spaces X, Y satisfy either of the two conditions (i) or (ii). One might include (iii), albeit the Borel assumption is strong.

Building on EmeFK1979, Fremlin (Fre1987, Section 10) characterizes a space X such that every Baire function $f \colon X \to Y$ is Baire-continuous for all metric Y in the language of 'measurable spaces with negligibles'; reference there is made to disjoint families of negligible sets all of whose subfamilies have a measurable union. (One may term this *completely additive measurable* by analogy with the established phrase *completely additive analytic*, which we will meet in §13.1.3, Remark 5.) For a discussion of discontinuous homomorphisms, especially counterexamples on $C(X)$ with X compact (e.g. employing Stone–Čech compactifications, $X = \beta\mathbb{N}\backslash\mathbb{N}$), see Dal1978, 10, Section 9.

Remarks   Hansell's condition, requiring the function f to be σ-discrete, is implied by f being analytic when X is absolutely analytic (i.e. Souslin-$\mathcal{F}(Y)$ in any complete metric space Y into which it embeds). Frankiewicz and Kunen (FraK1987) study the consistency relative to ZFC of the existence of a Baire function failing to have Baire continuity. See also Fra1982.

Theorem 12.1.4 (Hartman–Mycielski Embedding Theorem; HartM1958)   *Every topological group is a closed subgroup of a group G^* which is arcwise connected and locally arcwise connected.*

In particular, any separable (invariantly) metrizable group G is embeddable as a subgroup of an arcwise connected separable (invariantly) metrizable group.

Proof   Let G^* be the collection of range-finite functions $f : [0, 1) \to G$ which are càdlàg (continue à droite, limite à gauche), piecewise constant (i.e. there is a partition of $[0, 1)$ into a finite number of contiguous half-open intervals $[u, v)$) with pointwise product as the group operation, so that

$$g h(x) = g(x)h(x), \qquad g^{-1}(x) = g(x)^{-1}, \qquad e(x) \equiv 1_G.$$

Endow G^* with a topology in which the neighbourhoods of f take the form

$$B_V(f, r) := \{h : |x : h(x)f(x)^{-1} \in V| < r\}$$

for V open in G with $|.|$ Lebesgue measure on $[0, 1)$. An arc joining two functions $f, g \in G^*$ may be defined by

$$h_t(x) := \begin{cases} f(x), & x \in [0, t), \\ g(x), & x \in [t, 1), \end{cases}$$

so that G^* is both arcwise- and by the same token locally arcwise-connected. Identifying G with the constant functions $f_g(x) \equiv g$ embeds G into G^*.

If G has metric d^G, then d^* defined below metrizes G^*:

$$d^*(f, g) := \int_0^1 d^G(f(x), g(x)) \, \mathrm{d}x \qquad (f, g \in G^*),$$

and agrees with d^G on the constant functions. If d^G is right/left invariant, then so is d^*. If G is separable, so is G^*.   □

Remark   *Arcwise connectedness* occurs in Dixmier's Theorem on the structure theory of locally compact abelian groups and embedding of infinitely divisible probability measures on groups; see, e.g., Hey1977, pp. 8, 220. For background see Kur1968, Ch. 6.

Theorem 12.1.5 (Banach–Mehdi Theorem)   (Cf. Ban1932, 1.3.4, p. 40, albeit for 'Baire-measurable' functions, Meh1964.) *An additive Baire function between complete normed vector spaces is continuous, and so linear, provided the image space is separable.*

Proof   Suppose k is a Baire function, in the sense that inverse images under k of open sets are sets with the Baire property.

By the Baire Continuity Theorem, 12.1.3, k is continuous on some co-meagre set D. Suppose further that k is additive. If $x_n \to x_0$ and $\mathbb{M} \subseteq \mathbb{N}$, then, since

D is shift-compact, for some t and infinitely many $m \in \mathbb{M}$, say $m \in \mathbb{M}' \subseteq \mathbb{M}$, it follows that $t + (x_m - x_0) \in D$. So, by continuity on D,

$$k(t) = \lim_{m \in \mathbb{M}'} k(t + (x_m - x_0)) = \lim_{m \in \mathbb{M}'} k(t) + k(x_0) - k(x_m),$$

so that $k(x_m) \to k(x_0)$ for $m \in \mathbb{M}'$. Thus k is continuous (by the 'subsequence theorem', 1.3.1). From additivity one has $k(rx) = rk(x)$, for r rational, and so from continuity for all real r. That is, k is linear. □

The proof above is essentially due to Banach (Ban1932), although the concept of shift-compactness had then not been recognized.

The Souslin criterion and the next theorem together have as an immediate corollary the classical Souslin-graph Theorem (RogJ1980, §2.10). In this connection recall (see the corollary of Hof1980, Th. 2.3.6, p. 355) that a normed topological group which is Baire and analytic is Polish. Our proof, which is for normed groups, is inspired by the topological vector space proof in RogJ1980, §2.10, of the Souslin-graph theorem; their proof may be construed as having two steps: one establishing their Souslin criterion, the other the Baire homomorphism theorem. They state without proof the topological group analogue. For a non-separable analogue, see Chapter 13.

Theorem 12.1.6 (Baire Homomorphism Theorem; cf. RogJ1980, §2.10) *Let X and Y be normed groups with X non-meagre and analytic (e.g. topologically complete) and Y separable. If $f : X \to Y$ is a Baire homomorphism, then f is continuous. In particular, if f is a homomorphism with a Souslin-$\mathcal{F}(X \times Y)$ graph and Y is in addition a $\mathcal{K}$-analytic space, then f is continuous.*

Proof For $f : X \to Y$ the given homomorphism, it is enough to prove continuity at e_X, i.e. that for any $\varepsilon > 0$ there is $\delta > 0$ such that $B_\delta(e_X) \subseteq f^{-1}[B_\varepsilon(e_X)]$. So let $\varepsilon > 0$. We work with the right norm-topology.

Being $\mathcal{K}$-analytic, Y is Lindelöf (cf. RogJ1980, Th. 2.7.1, p. 36) and metric, so separable; so choose a countable dense set $\{y_n\}$ in $f(X)$ and select $a_n \in f^{-1}(y_n)$. Put $T := f^{-1}[B_{\varepsilon/4}(e_Y)]$. Since f is a homomorphism, $f(Ta_n) = f(T)f(a_n) = B_{\varepsilon/4}(e_Y)y_n$. Note also that $f(T^{-1}) = f(T)^{-1}$, so

$$TT^{-1} = f^{-1}[B_{\varepsilon/4}(e_Y)]f^{-1}[B_{\varepsilon/4}(e_Y)^{-1}] = f^{-1}[B_{\varepsilon/4}(e_Y)^2]$$
$$\subseteq f^{-1}[B_{\varepsilon/2}(e_Y)],$$

by the triangle inequality.

Now

$$f(X) \subseteq \bigcup_n B_{\varepsilon/4}(e_Y)y_n,$$

so

$$X = f^{-1}(Y) = \bigcup_n T a_n.$$

But X is non-meagre, so for some n the set $T a_n$ is non-meagre, and so too is T (as right shifts are homeomorphisms). By assumption f is Baire. Thus T is Baire and non-meagre. By the Squared Pettis Theorem, 6.3.2, $(TT^{-1})^2$ contains a ball $B_\delta(e_X)$. Thus we have

$$B_\delta(e_X) \subseteq (TT^{-1})^2 \subseteq f^{-1}[B_{\varepsilon/4}(e_Y)^4] \subseteq f^{-1}[B_\varepsilon(e_Y)]. \qquad \square$$

Theorem 12.1.7 (Souslin-Graph Theorem; Schw1966, cf. Pet1974; RogJ1980, p. 50)   *Let X and Y be normed groups with Y a $\mathcal{K}$-analytic space and X non-meagre. If $f : X \to Y$ is a homomorphism with Souslin-$\mathcal{F}(X \times Y)$ graph, then f is continuous.*

Proof   This follows from the preceding result and the Banach–Mehdi Theorem, 12.1.5. $\qquad \square$

12.2 Banach–Neeb Theorem

We begin with a simple result.

Lemma 12.2.1   *For X a Baire space and meagre $Y \subseteq X$, $X \setminus Y$ is a Baire subspace. Moreover, each subset that is meagre in $X \setminus Y$ is also meagre in X.*

Proof   Let $F \subseteq X \setminus Y$ be meagre in X. Choose nowhere dense subsets $F_n \subseteq X \setminus Y$ with $F = \bigcup_n F_n$. Let cl_X, resp. $\mathrm{cl}_{X \setminus Y}$, denote the closure of a set in X, resp. $X \setminus Y$. Then the fact that F_n is nowhere dense in $X \setminus Y$ implies that $\mathrm{cl}_{X \setminus Y} F_n = (X \setminus Y) \cap \mathrm{cl}_X F_n$ has empty interior. We conclude that, for each open subset $U \subseteq X$ with $U \subseteq \mathrm{cl}_X F_n$, we have $U \cap (X \setminus Y) = \emptyset$, i.e. $U \subseteq Y$. Since Y is of first category in X, the assumption that X is Baire implies that U is empty. This shows that $\mathrm{cl}_X F_n$ has empty interior, i.e. F_n is also nowhere dense in X. So F is meagre in X. $\qquad \square$

The following result should be compared with Baire's Continuity Theorem, 12.1.3, where Y is second countable and f is a Baire function in the sense that inverse images under f have the Baire property.

Theorem 12.2.2 (Baire Continuity Theorem – Baire-Measurable Version)   *If X is a Baire space, Y a metric space and $f : X \to Y$ is Baire measurable, then there exists a meagre $M \subseteq X$ such that $f \mid X \setminus M$ is continuous.*

Proof (Nee1997, cf. Ban1932.) Since the set of Baire-measurable functions is the smallest class of all functions containing the continuous functions closed under pointwise limits, it is enough to show that the class of functions satisfying the condition of the theorem is closed under pointwise limits, since the class trivially contains the continuous functions.

Suppose that the restriction of $f_n \colon X \to Y$ to $X \backslash M_n$ is continuous, where M_n is meagre in X. Then $M := \bigcup_{n \in \mathbb{N}} M_n$ is meagre in X, and all functions f_n are continuous on $X_1 := X \backslash M$. Suppose that $f = \lim_{n \to \infty} f_n$ holds pointwise on X. Denote the metric on Y by d^Y. Since the functions f_n are continuous on X_1, the sets

$$A_{n,\varepsilon} := \{x \in X_1 : (\forall m \geq n) d^Y(f_n(x), f_m(x)) \leq \varepsilon\}$$

are closed in X_1 and $f = \lim_{n \to \infty} f_n$ implies that $X_1 = \bigcup_{n \in \mathbb{N}} A_{n,\varepsilon}$. We put $B_\varepsilon := \bigcup_{n \in \mathbb{N}} A_{n,\varepsilon}^0$, where A^0 denotes the interior of A in X_1. Then B_ε is open and we claim that B_ε is dense in X_1. In fact, let $U \subseteq X_1$ be open. By Lemma 3.1.5 X_1 is a Baire space and so, since U is open, U is also a Baire space (Chapter 2). Hence $U = \bigcup_{n \in \mathbb{N}} (U \cap A_{n,\varepsilon})$ implies that at least one of the sets $U \cap A_{n,\varepsilon}$ is somewhere dense in U. But these sets are closed subsets of U, so there exists an $n \in \mathbb{N}$ for which $U \cap A_{n,\varepsilon}$ has interior points in U and therefore also in X_1, i.e. $A_{n,\varepsilon}^0 \cap U \neq \emptyset$. Now $B_\varepsilon \cap U \neq \emptyset$ entails that B_ε is dense in X_1. This means that $X_1 \backslash B_\varepsilon$ is closed and has no interior points, i.e. $X_1 \backslash B_\varepsilon$ is nowhere dense. This proves that

$$J := \bigcup_{\varepsilon > 0} X_1 \backslash B_\varepsilon = \bigcup_{n \in \mathbb{N}} X_1 \backslash B_{1/n}$$

is meagre in X_1.

Fix $x \in X_1 \backslash J$ and let $\varepsilon > 0$. Then $x \in B_\varepsilon$ and we can find $m \in \mathbb{N}$ with $x \in A_{m,\varepsilon}^0$. Then

$$d^Y(f(y), f_m(y)) = \lim_{n \to \infty} d^Y(f_n(y), f_m(y)) \leq \varepsilon$$

for all $y \in A_{m,\varepsilon}$ implies

$$d^Y(f(x), f(y)) \leq 2\varepsilon + d^Y(f_n(x), f_m(y)),$$

hence that f is continuous at x because $x \in A_{m,\varepsilon}^0$. This implies that f is continuous on $X_1 \backslash J = X \backslash (M \cup J)$ and so completes the proof as J is meagre in X (by the Lemma 3.1.5).

This proves that the class of all functions satisfying the assumptions of the theorem is closed under pointwise limits and contains the continuous functions, hence also contains the Baire-measurable functions. $\square$

Theorem 12.2.3 (Banach Continuous Homomorphism Theorem) *If G is a metrizable topological group which is a Baire space and H is a metrizable* *topological group, then every Baire-measurable homomorphism* $f: G \to H$ *is continuous.*

Proof (Cf. Ban1932, p. 23, Th. 4; cf. Ban1920) First, Theorem 12.2.2 shows that there exists a meagre subset $M \subseteq G$ such that f is continuous on $G \backslash M$. Let $x_n \to 1_G$ in G. Then the set $x_n . M \subseteq G$ is meagre for each $n \in \mathbb{N}$. Hence the same holds for

$$M \cup \bigcup_{n \in \mathbb{N}} x_n . M,$$

which, since G is non-meagre, implies that this set must be different from G. Let x be in the complement of this set. Then $x \notin M$ and $x_n^{-1} x \notin M$ for all $n \in \mathbb{N}$. Hence the continuity of f on the complement of M implies that $f(x_n)^{-1} f(x) = f(x_n^{-1} x) \to f(x)$, which in turn implies that $f(x_n) \to 1_G$. Since G was assumed metrizable, i.e. has a countable local base in 1_G, we see that f is continuous at 1_G, and so, being a homomorphism, f is continuous. $\square$

Remark In the proof above, it would suffice to prove that $\lim f(x_m)_{m \in \mathbb{M}'} = 1_H$ for some infinite subset of any infinite $\mathbb{M} \subseteq \mathbb{N}$. As $G \backslash M$ is shift-compact, there is $t \in G \backslash M$ and an infinite set $\mathbb{M}' \subseteq \mathbb{M}$ with $t x_m \in G \backslash M'$ for $m \in \mathbb{M}'$. Since $t x_m \to t$, by continuity on $G \backslash M$,

$$f(t) = \lim_{m \in \mathbb{M}'} f(t x_m) = f(t) \lim_{m \in \mathbb{M}'} f(x_m).$$

So $\lim_{m \in \mathbb{M}'} f(x_m) = 1_H$.

We proceed to weaken the assumption that f is Baire measurable.

Lemma 12.2.4 (Neeb's Lemma; Nee1997) *Let X, Y be metric spaces, with Y arcwise connected and separable, and $f: X \to Y$ a Borel function. Then f is Baire measurable.*

Proof First we show that f is the limit of a sequence $(f_n)_{n \in \mathbb{N}}$ of measurable functions with at most countably many values. Let $\varepsilon > 0$ and $(Y_n)_{n \in \mathbb{N}}$ be a basis for the topology consisting of sets whose diameter does not exceed ε. We put $Z_1 := Y_1$ and $Z_n := Y_n \backslash (Y_1 \cup \cdots \cup Y_{n-1})$ for $n > 1$. Deleting any empty Z_n, we may assume that the Z_n are all non-empty. The sets Z_n are Borel, and so the sets $X_n := f^{-1}(Z_n)$ are Borel subsets of X. Choosing $z_n \in Z_n$ we define a new function $f_\varepsilon: X \to Z$ by taking $f_\varepsilon(x) := z_n$ for $x \in X_n$. Then $d^Y(f(x), f_\varepsilon(x)) \le \varepsilon$ for all $x \in X_n$, and $f_\varepsilon(X)$ is countable. Hence f is a uniform limit of functions with at most countably many values. So, without loss of generality, we may now assume that $f(X)$ is countable. We write $f(X) = \{y_n : n \in \mathbb{N}\}$ and, using the arcwise connectedness of Y, find a continuous function $\gamma: \mathbb{R} \to Y$ with $\gamma(n) = y_n$. Next define a real-valued function $h: X \to \mathbb{R}$ by taking $h(x) := n$ whenever $f(x) = y_n$ and n is minimal with respect to this property.

Then h is Borel measurable and $\gamma \circ h = f$.

The set of all functions $u: X \to \mathbb{R}$ for which $\gamma \circ u: X \to Y$ is Baire measurable contains the continuous functions and is closed under pointwise limits. This implies that for each Baire function $u: X \to \mathbb{R}$ the function $\gamma \circ u: X \to Y$ is Baire. Since h is a limit of finite linear combinations of characteristic functions, to show that h is a Baire function, it suffices to see that indicator functions χ_B of Borel $B \subseteq X$ are Baire. In fact, the set of all subsets $B \subseteq X$ for which χ_B is Baire contains all open subsets, because for an open subset B, we have

$$\chi_B(x) = \lim_{n \to \infty} \min\{1, n \cdot \operatorname{dist}(x, X \backslash B)\},$$

where $\operatorname{dist}(x, C) := \inf\{d(x, y) : y \in C\}$. Also $\chi_{X \backslash B} = 1 - \chi_B$ and

$$\chi_{\cap B_n} = \lim_{n \to \infty} \prod_{k=1}^{n} \chi_{B_n}.$$

So, since the Baire-measurable functions $X \to \mathbb{R}$ form a σ-algebra, $\{B \subseteq X : \chi_B \text{ Baire measurable}\}$ is a σ-algebra containing all the Borel sets. Thus characteristic functions of Borel sets are Baire measurable. This proves that h is Baire measurable, and hence that f is Baire measurable. □

Theorem 12.2.5 (Banach–Neeb Theorem; Ban1932; Nee1997) *Every Borel-measurable group homomorphism $f: G \to H$ from a completely metrizable separable topological group into a separable metrizable group is continuous.*

Proof Since G is completely metrizable it is a Baire space. By the Hartman–Mycielski Theorem, 12.1.4, any separable metrizable group embeds as a topological subgroup into an arcwise connected separable metrizable group, so without loss of generality we may assume that H is arcwise connected. Thus by Lemma 12.2.4, f is Baire measurable. Hence Banach's Continuous Homomorphism Theorem implies that f is continuous. □

For connections between Borel functions and Baire functions, see Fos1993. We note in passing the next result (HewR1979, Th. 22.18).

Theorem 12.2.6 *Let G be a locally compact group with λ a left Haar measure and H a topological group which is σ-compact or separable, and suppose $f: G \to H$ is a group homomorphism for which there exists a λ-measurable subset $A \subseteq G$ with $0 < \lambda(A) < \infty$ such that for each open subset $U \subseteq H$ the set $f^{-1}(U) \cap A$ is λ-measurable. Then f is continuous.*

12.3 Levi Coincidence Theorem

This section is inspired by Sandro Levi's article (Lev1983) titled 'On Baire Cosmic Spaces', where he derives an Open Mapping Theorem (a result of the Direct Baire Property given below in Theorem 12.3.1) and a useful corollary on a comparison of topologies: *if one refines the other, then they must coincide on a subspace*. The treatment in this section assumes the spaces to be separable. To demonstrate their usefulness, recall an important result of Ellis asserting that if a metric group is endowed with a topology under which the group is locally compact and for which inversion is continuous while multiplication is separately continuous, then in fact multiplication is jointly continuous. In brief, *a group with a semi-topological structure has a topological group structure*.

We will deduce in §12.4 a result similar to Ellis' for groups with a metric which is right-invariant: a right-topological group generated by a right-invariant metric (i.e. a normed group in the terminology of Chapter 6) is a topological group. Unlike Ellis we do not assume that the group is abelian, nor that it is locally compact and instead a form of analyticity suffices.

In Chapter 13, we develop non-separable generalizations of Levi's results and with them also non-separable versions of our Ellis-Type Theorem, 12.4.1, with a 'one-sided' continuity condition implying that a right-topological group generated by a right-invariant metric (i.e. a normed group) is a topological group. Again, unlike Ellis, we do not assume that the group is abelian, nor that it is locally compact; the non-separable context requires some preservation of σ-discreteness as a side-condition (see the Remarks after Corollary 13.3.3).

Given that the application in mind is metrizable, references to non-separable descriptive theory remain, for transparency, almost exclusively in the metric realm, though we do comment on the regular Hausdorff context in the Remark following Theorem 13.3.8.

Levi's work draws together two notions: BP – the Baire *set* property (i.e. that a set is open modulo a meagre set, so 'almost open'), and BS – the Baire *space* property (i.e. that Baire's Theorem holds in the space). Below we keep the distinction clear by using the terms 'Baire property' and 'Baire space'. The connection between BP and BS is not altogether surprising, and the two are 'almost' the same in a precise sense, at least in the context of normed groups (cf. Ost2013b, where this closeness is fully exploited). For, in an almost-complete space, the terms 'Baire set', 'set with the Baire property' and 'Baire space' are almost-synonyms in the sense that, for B non-meagre, B has the Baire property if and only if B is a Baire space if and only if B is almost-complete (Ost2013b, Th. 7.4).

We recall from the Chapter 3 (where we introduce analyticity) that a subspace S of a metric space X has a *Souslin-$\mathcal{F}(X)$ representation* if there is a 'determining' system $\langle F(i \mid n) \rangle := \langle F(i \mid n) : i \in \mathbb{N}^{\mathbb{N}} \rangle$ of sets in $\mathcal{F}(X)$ (the closed sets) with

$$S = \bigcup_{i \in I} \bigcap_{n \in \mathbb{N}} F(i \mid n), \text{ where } I = \mathbb{N}^{\mathbb{N}}$$

and $i \mid n$ denotes $(i_1, \ldots, i_n)$. We will say that a topological space is *classically analytic* if it is the continuous image of a Polish space (Levi terms these 'Souslin') and not necessarily metrizable, in distinction to an (absolutely) *analytic* space, i.e. one that here *is* metrizable and is embeddable as a Souslin-$\mathcal{F}$ set in its own metric completion; in particular, in a complete metric space, $\mathcal{G}_\delta$-subsets (being $\mathcal{F}_{\sigma\delta}$) are analytic. We call a Hausdorff space *almost analytic* if it is analytic modulo a meagre set. Similarly, a space X' is absolutely $\mathcal{G}_\delta$, or an *absolute-$\mathcal{G}_\delta$*, if X' is a $\mathcal{G}_\delta$ in all spaces X containing X' as a subspace. (This is equivalent to complete metrizability in the narrowed realm of metrizable spaces (Eng1989, Th. 4.3.24), and to topological/Čech-completeness in the narrowed realm of completely regular spaces, Eng1989, §3.9.) So a metrizable absolute-$\mathcal{G}_\delta$ is analytic; we use this fact in Lemma 12.4.3.

Levi's results follow from the following routine observation.

Theorem 12.3.1 (Direct Baire Property; Lev1983) *Let X be a classically analytic space and Y a Hausdorff space. Every continuous map $f : X \to Y$ has the direct Baire property: the image of any open set in X has the Baire property in Y.*

Proof Suffice to note that an open set in a metric space X is $\mathcal{F}_\sigma$ and so, being Souslin-$\mathcal{F}$ in an analytic space, is itself analytic. The result is then immediate from Nikodym's Corollary, 3.1.3. □

The nub of the theorem is that, with X as above, continuity preserves various analyticity properties such as that open, and likewise closed, sets are taken to analytic sets, in brief: a continuous map is *open-analytic* and *closed-analytic* in the terminology of Hanse1974, and so preserves the Baire property. (See Remark 2 in §13.1.3 for a reprise of this theme.) Levi deduces the following characterization of Baire spaces in the category of classically analytic spaces.

Theorem 12.3.2 (Levi Open Mapping Theorem; Lev1983) *Let Y be a regular classically analytic space. Then Y is a Baire space if and only if $Y = f(X)$ for some continuous map f on some complete separable metric space X with the property that, for some dense metrizable absolute-$\mathcal{G}_\delta$ subspace $Y' \subseteq Y$ and $X' = f^{-1}(Y')$, the restriction map $f \mid X' : X' \to Y'$ is open.*

The notation in the proof below may seem inefficient; however, our purpose is to make its later non-separable variant more intelligible, as there all the sets appearing here need to be partitioned into σ-discrete parts. This is necessary because use of Urysohn's metrization theorem here needs to be replaced in the non-separable case by Bing's Characterization Theorem. Recall that Bing's Theorem (Eng1989, Th. 4.4.8) asserts that a regular space is metrizable if and only if it has a σ-discrete base. The latter property requires the apparatus of σ-discrete families as mentioned above. See the Remark immediately after the proof.

Proof One direction is clear from the density statement. For the converse, let $\mathcal{A} = \{A_n : n = 1, 2, \ldots\}$ be an open base for X. Then $E := f(A)$ for $A \in \mathcal{A}$ is analytic (see also Remark 5 in §13.1.3), so has the Baire property (by Nikodym's Corollary, 3.1.3, for analytic sets). Put $\mathcal{E} = \{f(A) : A \in \mathcal{A}\} = \{f(A_n) : n = 1, 2, \ldots\}$. For each $E \in \mathcal{E}$ pick an open set U_E and meagre sets N_E and M_E such that

$$E = (U_E \backslash N_E) \cup M_E,$$

with M_E disjoint from M_E and with $N_E \subseteq U_E$. Put

$$M := \bigcup \{M_E : E \in \mathcal{E}\} \text{ and } N := \bigcup \{N_E : E \in \mathcal{E}\},$$

both being meagre, as $\mathcal{E}$ is countable. Now put $Y' := Y \backslash (M \cup N)$ and

$$W = \bigcup \{U_E : E \in \mathcal{E}\},$$

which is open in Y. Then, for $A \in \mathcal{A}$ with $E = f(A)$,

$$f(A) \cap Y' = E \cap Y' = U_E \cap Y',$$

so that $f(A)$ is open relative to Y'.

For $G \subseteq X$ open, since $\mathcal{A}$ is a base, we may write

$$G := \bigcup_n \{A_n : A_n \subseteq G\}.$$

Then

$$f(G) := \bigcup_n \{E_n : E_n = f(A_n) \,\&\, A_n \subseteq G\}.$$

So, for $X' := f^{-1}(Y')$,

$$f(G \cap X') = Y' \cap \bigcup \{U_{f(A)} : A \subseteq G \,\&\, A \in \mathcal{A}\},$$

which is open in Y'.

Now Y' is second countable: the family $\bigcup \{Y' \cap U_E : E \in \mathcal{E}\}$ is a countable base for Y'. As Y is regular (Eng1989, Th. 5.1.5), the subspace Y' is regular (Eng1989, Th. 2.1.6). Being regular and second countable, Y' is metrizable by

Urysohn's Theorem. Finally, by replacing the meagre sets M, N by larger sets that are unions of closed nowhere dense sets, we obtain in place of Y' a smaller, metrizable, dense $\mathcal{G}_\delta$-subspace. □

Remark To generalize one must take $\mathcal{A} = \bigcup_n \mathcal{A}_n$ with each $\mathcal{A}_n = \{A_{tn} : t \in T_n\}$ discrete (which in the separable case reduces to the singletons $\mathcal{A}_n = \{A_n\}$); then $\mathcal{E}_n = \{f(A) : A \in \mathcal{A}_n\}$ replaces $\mathcal{E}$ and will have a σ-discrete base $\mathcal{B}_n$ comprising sets with the Baire property. In place of the sets E above one works with sets B in $\mathcal{B} = \bigcup_n \mathcal{B}_n$.

The result may be regarded as implying an 'inner regularity' property (compare the capacitability property) of a classically analytic space Y: if Y is a Baire space, then Y contains a dense absolute-$\mathcal{G}_\delta$ subspace, so a Baire space. Compare the result due to Roy Davies (Dav1952) that an analytic set of infinite Hausdorff measure contains closed sets of any desired finite measure. The existence of a dense completely metrizable subspace – making Y *almost complete* in the sense of Frolík (Fro1960, though the term is due to Michael, Mich1991) – is a result that implicitly goes back to Kuratowski (Kur1966, IV.2, p. 88, because a classically analytic set has the Baire property in the restricted sense – Cor. 1, p. 482). Generalizations of the latter result, including the existence of a restriction map that is a homeomorphism between a $\mathcal{G}_\delta$-subset and a dense set, are given by Michael (Mich1986); but there the continuous map f requires stronger additional properties such as openness on X (unless X is separable), which Levi's result delivers.

Theorem 12.3.2 has a natural extension characterizing a Baire space (in the same way) when it is almost analytic. Indeed, with Y' as above, the space Y is *almost complete* and so almost analytic. On the other hand, if Y is a Baire space and almost analytic, then by supressing a meagre $\mathcal{F}_\sigma$ and passing to an absolutely $\mathcal{G}_\delta$-subspace, we may assume that Y is a Baire space which is analytic, so has the open mapping representation of the theorem, and in particular is almost complete (for more background, see Ost2013b: cf. Cor. 1.8).

Since an analytic space is a continuous image, Theorem 12.3.2 may be viewed as an 'almost preservation' result for complete metrizability under continuity in the spirit of the classical theorem of Hausdorff (resp. Vainstein) on the preservation of complete metrizability by open (resp. closed) continuous mappings – see the Remarks in §13.3.3 for the most recent improvements and the literature of preservation. We note that Michael (Mich1991, Prop. 6.5) shows that almost completeness is preserved by demi-open maps (i.e. continuous maps under which inverse images of dense open sets are dense). Theorem 12.3.2 has an interesting corollary on the comparison of refinement topologies. For a discussion of refinements, see Ost2013b, §7.1 (for examples of completely metrizable and of analytic refinements, see Kec1995, Th. 13.6, Th. 25.18, Th. 25.19).

Theorem 12.3.3 (Levi Coincidence Theorem; Lev1983)   *For $\mathcal{T}, \mathcal{T}'$ two topologies on a set Y with $(Y, \mathcal{T}')$ classically analytic (e.g. Polish) and $\mathcal{T}'$ refining $\mathcal{T}$ (i.e. $\mathcal{T} \subseteq \mathcal{T}'$), if $(Y, \mathcal{T}')$ is a regular Baire space, then there is a $\mathcal{T}$-dense $\mathcal{G}(\mathcal{T})_\delta$-set on which $\mathcal{T}$ and $\mathcal{T}'$ coincide.*

Proof   As $\mathcal{T}'$ refines $\mathcal{T}$, to prove the theorem one may pass to any dense $\mathcal{G}_\delta$ subset of $(Y, \mathcal{T}')$. We claim that we may pass to such a subset that is also Polish. Indeed, we may pick a Polish X and $g : X \to (Y, \mathcal{T}')$ continuous with $g(X) = Y$. Now the embedding $j : (Y, \mathcal{T}') \to (Y, \mathcal{T})$ with $j(y) = y$ is continuous. Taking for f the composition jg, which is continuous and gives $f(X) = Y$, apply Levi's Open Mapping Theorem to obtain a $\mathcal{G}_\delta$ set $X' \subseteq X$ such that $f \mid X'$ is open and $f(X')$ is a dense $\mathcal{G}_\delta$ in $(Y, \mathcal{T}')$. By Hausdorff's Theorem that an open continuous image of a Polish space is open (cf. for example Anc1987, or, for a more recent account, HoliP2010), $f(X')$ is Polish.

So we now assume that $(Y, \mathcal{T}')$ is Polish and again apply Levi's Open Mapping Theorem, 12.3.2, this time taking X to be $(Y, \mathcal{T}')$, Y to be $(Y, \mathcal{T})$ and f to be j. Then, for some $\mathcal{G}_\delta$ subset Z of X, the map $j \mid Z$ is open. Writing $\mathcal{T}_Z$ and $\mathcal{T}_Z'$ for the subspace topologies induced on Z by $\mathcal{T}$ and $\mathcal{T}'$ respectively, j takes the sets of $\mathcal{T}_Z'$ to the sets $\mathcal{T}_Z$. But $\mathcal{T}_Z'$ refines $\mathcal{T}_Z$, so the two topologies coincide (and are again Polish by Hausdorff's Theorem).   □

12.4 Semi-Polish Theorem

Below we are concerned with the join of the two metric topologies (their coarsest joint refinement), which is generated by the symmetrized metric

$$d_S^X := \max\{d_R^X, d_L^X\},$$

where d_R^X and d_L^X are respectively a right- and a left-invariant metric on X. See Itz1972 for connections with uniform spaces.

When X is Polish/analytic under the topology of d_S^X we shall say that X is *semi-Polish/semi-analytic* under the topology of d_R^X. The term was suggested by Anatole Beck.

Notation   We use the subscripts R, L, S as in $x_n \to_R x$ etc. to indicate convergence in the corresponding metrics d_R^X, d_L^X, d_S^X.

Notice that if $d_S^X(x, y) < r$, then $d_R^X(x, y) < r$, and so $B_S(x; r) \subseteq B_R(x; r)$. Thus any R-open set is S-open. In other terms, if $x_n \to_S x$, then $x_n \to_R x$, and so R-closed sets are S-closed (for, if an R-closed set F were not S-closed, then there would be a sequence x_n in F with $x_n \to_S x$ and $x \notin F$, contradicting $x_n \to_R x$). Passing to complements shows that an R-open set is S-open. In summary: the topology generated by d_S^X is finer (has more sets) than that generated by d_R^X.

Its general significance comes from the theorem that, for (T, d^T) any complete metric space, the group of bounded self-homeomorphisms of T is complete under the symmetrization of the supremum metric (cf. §6.1; for details see Ost2013c; Dug1966, Th. XIV.2.6, p. 296).

Theorem 12.4.1 *For a group X equipped with a right-invariant metric d_R^X: if the space X is non-meagre and semi-Polish (more generally, semi-analytic) under the topology of d_R^X, then it is a Polish topological group (i.e. under the d_R^X topology, X is completely metrizable and a topological group).*

We recall that every metrizable topological group has an equivalent right-invariant metric, by the Birkhoff–Kakutani Theorem, 6.1.4, and a Polish group is non-meagre (by Baire's Theorem), so this theorem covers all Polish groups.

The theorem also generalizes a result due to Hoffman-Jørgensen that a Baire analytic *topological* group is Polish, because an analytic group is separable and metrizable (see Hof1980, Th. 2.3.6, p. 355). The theorem addresses the question: when does one-sided continuity of multiplication imply its joint continuity and further its *admissibility,* i.e. endowment of a topological group structure? As noted above, that question was considered in the *abelian* context by Ellis in Elli1957 (see in particular his Th. 2, where the topology is locally compact – cf. §13.1), but otherwise the existing literature, which goes back to Montgomery (Mon1936) and also Ellis (Elli1953) via Namioka (Nam1974), considers some form of weak bilateral continuity, usually separate continuity, supported by additional topological features, including some form of completeness. See Bouziad's two papers (Bouz1993; Bouz1996) for the state-of-the-art results, deducing automatic joint continuity from separate continuity (and for a review of the historic literature), and the later paper of Solecki and Srivastava (SolS1997), where separate continuity is weakened; cf. CaoDP2010; CaoM2004; Chr1981. For the broader context of automatic continuity, see TopH1980 (e.g. p. 338), and for the interaction of topology and algebra, Dales (Dal2000).

Namioka's Theorem, cited above, giving conditions under which a separately continuous function is jointly continuous on a dense $\mathcal{G}_\delta$, is reminiscent of the coincidence themes of Chapter 12. For some developments of his theme, see HansT1992.

By contrast to these bilateral conditions, Theorem 12.4.1 assumes only a particular form of one-sided continuity, supported by additional topological properties. An advantage of this approach is to replace the use of local compactness (or even subcompactness, for which see Bouz1996) by the much weaker notion of shift-compactness in groups in the form of Theorem 6.2.3, the Analytic Shift Theorem.

We will need two lemmas. The first is merely a sharpening appropriate for groups equipped with a right-invariant metric of an old result of Levi. For completeness we give the (direct) proof; our main work begins in earnest in Lemma 12.4.3.

Lemma 12.4.2 (cf. Lev1983, Th. 2 and Cor. 4) *For a group X equipped with a right-invariant metric d_R^X, if (X, d_S^X) is Polish, i.e. separable and topologically complete, or more generally analytic, and (X, d_R^X) non-meagre, then there is a subset Y of X which is a dense absolute-$\mathcal{G}_\delta$ in (X, d_R^X), and on which the d_S^X and d_R^X topologies agree.*

Proof We may apply Levi's Coincidence Theorem, taking T' to be the topology generated by the metric d_S^X and T by the metric d_R^X, to obtain a dense $\mathcal{G}_\delta(X, d_R^X)$, and on which the d_S^X and d_R^X topologies agree. Working in Y, we have $y_n \to_R y$ if and only if $y_n \to_S y$ if and only if $y_n \to_L y$.   □

Observe that above, since $X \backslash Y$ is meagre under d_R^X, the space (X, d_R^X) is almost complete (see Chapter 13). We use almost completeness to extract much more.

Lemma 12.4.3 *If, in the setting of Lemma 12.4.2, the three topologies generated by d_R^X, d_L^X, d_S^X agree on a dense absolutely $\mathcal{G}_\delta$ set Y of (X, d_R^X), then for any $\tau \in Y$ the conjugacy $\gamma_\tau(x) := \tau x \tau^{-1}$ is continuous.*

Proof We work in (X, d_R^X). Let $\tau \in Y$. We first establish the continuity on X at e of the conjugacy $x \mapsto \tau^{-1} x \tau$ (by shifting into Y). Let $z_n \to e$ be any null sequence in X. Fix $\varepsilon > 0$; then $T := Y \cap B_\varepsilon^L(\tau)$ is analytic, since T is d_R^Y-open in Y, and is non-meagre, as X is Baire. By the Analytic Shift Theorem, 6.2.3, there is $t \in T$ and t_n in T with t_n converging to t (in d_R^X, so also in d_L^X) and an infinite $\mathbb{M}_t$ such that $\{tt_m^{-1} z_m t_m : m \in \mathbb{M}_t\} \subseteq T$. Since the three topologies agree on Y and as the subsequence $tt_m^{-1} z_m t_m$ lies in Y and converges to t in Y under d_R^Y (see Remark on p. 97), it also converges to t under d_L^Y. Using the identity

$$d_L^X(tt_m^{-1} z_m t_m, t) = d_L^X(t_m^{-1} z_m t_m, e) = d_L^X(z_m t_m, t_m),$$

we note that

$$\|t^{-1} z_m t\| = d_L^X(t, z_m t) \leq d_L^X(t, t_m) + d_L^X(t_m, z_m t_m) + d_L^X(z_m t_m, z_m t)$$
$$\leq d_L^X(t, t_m) + d_L^X(tt_m^{-1} z_m t_m, t) + d_L^X(t_m, t) \to 0,$$

as $m \to \infty$ through $\mathbb{M}_t$. So $d_L^X(t, z_m t) < \varepsilon$ for large enough $m \in \mathbb{M}_t$. Then, as $d_L^X(\tau, t) < \varepsilon$, for any such m one has

$$\|\tau^{-1} z_m \tau\| = d_L^X(z_m \tau, \tau) \le d_L^X(z_m \tau, z_m t) + d_L^X(z_m t, t) + d_L^X(t, \tau)$$
$$\le d_L^X(\tau, t) + d_L^X(t, z_m t) + d_L^X(t, \tau) \le 3\varepsilon.$$

Thus for any $\varepsilon > 0$ and any k there is $m = m(k, \varepsilon) > k$ with $\|\tau^{-1} z_m \tau\| \le 3\varepsilon$. Inductively, taking successively $\varepsilon = 1/n$ and $k(n) := m(k(n-1), \varepsilon)$, one has $\|\tau^{-1} z_{k(n)} \tau\| \to 0$. By the weak continuity criterion, Lemma 6.1.10 (cf. BinO2010g, Lemma 3.5, p. 37), $\gamma(x) := \tau^{-1} x \tau$ is continuous. Since (X, d_R^X) is analytic and metric, each open set U is analytic, so $\gamma_\tau^{-1}(U) = \gamma(U)$ is analytic, so has the Baire property, by Nikodym's Theorem, 3.1.3. So $\gamma_\tau(x) = \tau x \tau^{-1} = \gamma^{-1}(x)$ is a Baire homomorphism, and so is continuous – by the Baire Homomorphism Theorem, 12.1.6.   □

Proof of Theorem 12.4.1   Under d_R^X, the set $Z_\Gamma := \{x : \gamma_x \text{ is continuous}\}$ is a closed subsemigroup of X (BinO2010g, Prop. 3.43). By Lemmas 12.4.2 and 12.4.3, $X = \operatorname{cl}_R Y \subseteq Z_\Gamma$, i.e. γ_x is continuous for all x, and so (X, d_R^X) is a topological group. So $x_n \to_R x$ if and only if $x_n^{-1} \to_R x^{-1}$ if and only if $x_n \to_L x$ if and only if $x_n \to_S x$. So, being homeomorphic to (X, d_S^X), the space (X, d_R^X) has the structure of a Polish topological group.   □

12.5 * Pol's Continuity Theorem

This section is devoted to a sketch proof of Theorem 2.1.6 (recalled below), namely Pol's version of Baire's Continuity Theorem 2.1.4. Pol's version is concerned with a stronger Baire property. Recall from §2.1 that $A \subseteq X$ has the *restricted* Baire property in X if $A \cap Z$ is Baire in any subspace $Z \subseteq X$: below we will say A is *hereditarily Baire*. Likewise, a map is restricted Baire (or hereditarily Baire) if preimages of open sets are restricted Baire. Below compact is taken to imply Hausdorff. We start with a combinatorial result; for *measurable cardinals*, see §16.3 for a discussion and literature. In brief: these are cardinals κ which support $\{0, 1\}$-valued measure on $\wp(\kappa)$ that vanish on singletons and are not just σ-additive but also κ-additive, i.e. the measure is additive over any disjoint family of cardinality less than κ. By contrast, a cardinal κ is said to be *non-measurable* if the only σ-additive $\{0, 1\}$-valued measure on $\wp(\kappa)$ vanishing on singletons is trivial, i.e. identically zero (see, e.g., KurM1968, Ch. IX.3 and Bog2007, 1.12(x), p. 79). This is not quite the negation of the notion of measurable cardinal. However, the least cardinal κ that is not non-measurable supports a non-trivial 2-valued measure that is

κ -additive (see, e.g., Jec2002, Lemma 10.2; Dra1974, Ch. 6 Th. 1.4), so that κ is a measurable cardinal.[1]

Proposition 12.5.1 (Pol1976, Th. 1.)   *Let $\mathcal{E}$ be a disjoint family of meagre subsets of a compact space X with non-meagre union. If the family $\mathcal{E}$ has non-measurable cardinality, then there is a subfamily $\mathcal{E}'$ whose union fails to be hereditarily Baire.*

We will give a sketch proof below after first proving a lemma. For the purposes of the lemma, observe that if S is a non-meagre Baire set, then modulo a meagre set it has the form of a $\mathcal{G}_\delta$-set H with

$$\emptyset \neq H \subseteq \text{int}(\text{cl}(H)).$$

Indeed, there are meagre sets M, M' with

$$S = (G \backslash M) \cup M',$$

and without loss of generality M may be taken to be a countable union of closed nowhere dense sets (increasing M' by a meagre set, if needed). Here G is non-empty (otherwise S is meagre). Now take $H = G \backslash M$, then $G = \text{int}(\text{cl}(H))$, as claimed. The following lemma enables an inductive binary-tree construction in X of a Cantor set by selecting descending $\mathcal{G}_\delta$-subsets with disjoint closures (so forming a binary tree under inclusion).

Lemma 12.5.2 (Pol1976, Lemma 1)   *Suppose a $\mathcal{G}_\delta$-set $H \subseteq X$ with $\emptyset \neq H \subseteq \text{int}(\text{cl}(H))$ is covered by an open family $\mathcal{U}$ and also by a disjoint family $\mathcal{A}$, of non-measurable cardinality, comprising meagre sets. Then there is a decomposition $\mathcal{A} = \mathcal{A}_0 \cup \mathcal{A}_1$ and two open sets V_i, refining $\mathcal{U}$ and with disjoint closures, such that each of $V_i \cap H \cap \bigcup \mathcal{A}_i$ is non-meagre. In particular, if both unions are Baire, each covers a non-meagre $\mathcal{G}_\delta$-set H_i with $\emptyset \neq H_i \subseteq \text{int}(\text{cl}(H_i))$.*

Proof   Key here is the family $\mathcal{I} := \{\mathcal{B} \subseteq \mathcal{A} : H \cap \bigcup \mathcal{B} \text{ is meagre}\}$, which is a σ-ideal in $\wp(\mathcal{A})$, containing all the singleton subsets of $\mathcal{A}$, and is proper, since H is non-meagre. So, as the cardinality of $\mathcal{A}$ is non-measurable, there is a decomposition $\mathcal{A} = \mathcal{A}_0 \cup \mathcal{A}_1$ with each $\mathcal{A}_i \notin \mathcal{I}$. (Otherwise, for each $\mathcal{B} \subseteq \mathcal{A}$, the decomposition $\mathcal{B} \cup \mathcal{A} \backslash \mathcal{B}$ yields that either $\mathcal{B} \in \mathcal{I}$ or $\mathcal{A} \backslash \mathcal{B} \in \mathcal{I}$. Then $\mathcal{F} = \{A \subseteq X : A \notin \mathcal{I}\}$ is an ultrafilter see, e.g., Jec2002, Ch. 10, p. 126. So the function μ defined on $\mathcal{A}$ by $\mu(A) = 1$ if $A \in \mathcal{F}$ and $\mu(A) = 0$ if $A \in \mathcal{I}$, is a 2-valued measure on $\mathcal{A}$, as $\mathcal{I}$ is a σ-ideal.) By the Banach Category

---

[1] A σ-additive prime ideal on a non-measurable cardinal κ is κ-additive; see KurM1969, Ch. IX.3, Th.3, Bog2007, 1.12(x), p. 79. Usage of the term *measurable cardinal* in the early literature followed Ulam's original approach in Ula1930.

Theorem (§5.2.2), we may choose for $i = 0, 1$ a (heavy) point $x_i \in H \cap \bigcup \mathcal{A}_i$, i.e. one such that $H \cap \bigcup \mathcal{A}_i$ is non-meagre in any neighbourhood of x_i. Finally, by compactness, we may choose closed neighbourhoods V_i separating the two points.   □

Sketch Proof of Proposition 12.5.1 Suppose the Proposition fails for some family $\mathcal{E}$. Then $\mathcal{E}$ covers a non-meagre Baire set; moreover, all its subfamilies have Baire union. Thus Lemma 12.5.2 may be used in an induction on n to construct a binary-tree system of non-meagre G_δ-sets $H_{(i_1,...,i_n)} = \bigcap_m G^m_{(i_1,...,i_n)}$ for $(i_1,...,i_n) \in \{0,1\}^n$ and corresponding subfamilies $\mathcal{E}_{(i_1,...,i_n)}$ covering $H_{(i_1,...,i_n)}$. Take for $i \in \{0,1\}^{\mathbb{N}}$

$$\mathcal{E}(i) = \bigcap_n \mathcal{E}_{(i_1,...,i_n)}.$$

Also, for $i \in \{0,1\}^{\mathbb{N}}$, put

$$C(i) := \bigcap_n \mathrm{cl} H(i|n) \subseteq \mathcal{E}(i) \text{ with } K := \bigcup \{C(i) : i \in \{0,1\}^{\mathbb{N}}\}.$$

The set K is compact in X and the map $f \colon K \to \{0,1\}^{\mathbb{N}}$ defined by $f^{-1}\{i\} = C(i)$ is continuous, by construction. For $A \subseteq \{0,1\}^{\mathbb{N}}$,

$$f^{-1}(A) = K \cap \bigcup \mathcal{E}_A, \text{ with } \mathcal{E}_A := \bigcup_{i \in A} \mathcal{E}(i),$$

is hereditarily Baire, since each subfamily $\mathcal{E}_A$ has Baire union.

So arbitrary subsets of K are Baire. From here we derive a contradiction, namely that all subsets of C are Baire.

Indeed, as K is compact, f is closed (maps closed sets to closed sets) Kel1955. Choose a minimal closed subset $M \subseteq K$ with $f(M) = C$. Then the restriction map $g := f|M$ is irreducible and closed, hence maps Baire sets to Baire sets (cf. Grue1998; Sem1971, Ex. 25.2.3, p. 447). By hypothesis, $f^{-1}(A) \cap M$ is Baire for any $A \subseteq C$. Hence $A = g(f^{-1}(A) \cap M)$ is Baire.   □

Proposition 12.5.1 is used below to deduce that a certain family has meagre union.

Theorem 12.5.3 (Pol's Theorem; Pol1976, Th. 2.) *Let $f \colon X \to Y$ be a mapping from a compact space X to a metrizable space Y of non-measurable cardinality. Then f is hereditarily Baire if and only if for each subspace $Z \subseteq X$ the restriction $g := f|Z\backslash M$ is continuous for some $M \subseteq Z$ meagre in Z.*

Proof The condition is clearly sufficent. For its necessity, it is enough to consider only closed subspaces $Z \subseteq X$. So without loss of generality we may take $Z = X$. Now suppose $f \colon X \to Y$ has the Baire property hereditarily and Y is metrizable of non-measurable cardinality. In Y select a base $\mathcal{B} = \bigcup_n \mathcal{B}_n$

with each $\mathcal{B}_n$ discrete and let $\mathcal{U}_n = \{f^{-1}(V) : V \in \mathcal{B}_n\}$. For $U \in \mathcal{U}_n$, as f has the Baire property, in X there are open sets G_U and disjoint meagre sets M_U, M'_U with $M_U \subseteq G_U$, $M'_U \subseteq X \backslash G_U$ such that

$$U = (G_U \backslash M_U) \cup M'_U.$$

For distinct $U, W \in U_n$, since $G_U \backslash M_U$ and $G_W \backslash M_W$ are disjoint, so also are G_U and G_W (a non-empty open common part would be of second category, so not covered by $M_U \cup M_W$). Put

$$M_n = \bigcup\{M_U : U \in \mathcal{U}_n\}, \qquad M'_n = \bigcup\{M'_U : U \in \mathcal{U}_n\}\backslash M_n.$$

By the Banach Category Theorem (§5.2.2), M_n is meagre since

$$M_n \cap G_U = M_U$$

for $U \in \mathcal{U}_n$. For $\mathcal{U} \subseteq \mathcal{U}_n$,

$$\bigcup_{U \in \mathcal{U}} M'_U \backslash M_n = \bigcup_{U \in \mathcal{U}} M'_U \backslash \bigcup_{U \in \mathcal{U}} G_U = \bigcup_{U \in \mathcal{U}} U \backslash \bigcup_{U \in \mathcal{U}} G_U.$$

So, given the assumptions on f, this set is hereditarily Baire. By Proposition 12.5.1, the family of meagre sets

$$\mathcal{E}_n := \{M'_U \backslash M_n; U \in \mathcal{U}_n\}$$

has meagre union, which is equal to M'_n. Then $M := \bigcup_n M_n \cup M'_n$ is meagre.

Consider the restriction $g := f|(X\backslash M)$. To see that g is continuous, for any n and any $V \in \mathcal{B}_n$, take $U = f^{-1}(V) \in \mathcal{U}_n$ and note that

$$g^{-1}(V) = f^{-1}(V) \cap (X\backslash M) = U\backslash M = G_U \backslash M$$

is open in $X\backslash M$.   □

13

* Non-separable Variants

13.1 Non-separable Analyticity and the Baire Property

This section enables some standard (separable) analyticity and category arguments to be lifted from Chapter 12 to the *non-separable* context. The requisite concepts and their definitions rely on various forms of countability typified by σ-discrete families (abbreviated occasionally to σ-d): these are delayed till after the statements of theorems and given in §13.1.2.

13.1.1 Classical Souslin Representation

We recall that a subspace S of a metric space X has a *Souslin-$\mathcal{H}$ representation* if there is a *determining system* $\langle H(i \mid n) \rangle := \langle H(i \mid n) : i \in \mathbb{N}^{\mathbb{N}} \rangle$ of sets in $\mathcal{H}$ with (RogJD1980; Hanse1973a)

$$S = \bigcup_{i \in I} \bigcap_{n \in \mathbb{N}} H(i \mid n), \quad (I := \mathbb{N}^{\mathbb{N}}, \ i \mid n := (i_1, \ldots, i_n)).$$

A topological space S is an (absolutely) *analytic* space if it is embeddable as a Souslin-$\mathcal{F}$ set in its own metric completion S^* (with $\mathcal{F}$ the closed sets); in particular, in a complete metric space $\mathcal{G}_\delta$-subsets (being $\mathcal{F}_{\sigma\delta}$) are analytic. For more recent generalizations, see, e.g., NamP1969. According to Nikodym's Theorem, 3.1.3, if $\mathcal{H}$ above comprises Baire sets, then also S is Baire (RogJ1980, §2.9 or Kec1995, Th. 29.14), i.e. the Baire property is preserved by the Souslin operation.

Central to the needs of a non-separable context are three results: Hansell's Characterization Theorem, 13.1.1, yielding representation of analytic sets in the form $\bigcup_{j \in \kappa^{\mathbb{N}}} H(j)$ with H upper semi-continuous and compact-valued (defined on the completely metrizable, countable product of discrete spaces of cardinality κ), Nikodym's Theorem, 3.1.3, recalled and proved below as 13.1.2, implying their Baire property, and the conclusion that *analytic base-σ-discrete* maps (A-σ-d maps, for short, to be defined below) are the only ones that matter (cf. 13.1.8).

We content ourselves mostly with a metric context, though a wider one is feasible (consult Hanse1992). A Hausdorff space S is $\mathcal{K}$-*analytic* if $S = \bigcup_{i \in I} K(i)$ for some upper semi-continuous map K from $I = \mathbb{N}^{\mathbb{N}}$ to $\mathcal{K}(S)$, the compact subsets of S. In a *separable* metric space, an *absolutely analytic* subset is $\mathcal{K}$-analytic (RogJ1980, Cor. 2.4.3 plus Th. 2.5.3). In a *non-separable* complete metric space X, it is not possible to represent a Souslin-$\mathcal{F}(X)$ subset S of X as a $\mathcal{K}$-analytic set relative to $I = \mathbb{N}^{\mathbb{N}}$. Various generalizations of countability now enter the picture, as we now recall, referring also to two survey papers: Sto1980 and Hanse1992.

13.1.2 Extended Souslin Representation: Consequences

Denoting by wt(X) the *weight* of the space X (i.e. the smallest cardinality of a base for the topology), and replacing $I = \mathbb{N}^{\mathbb{N}}$ by $J = \kappa^{\mathbb{N}}$ for $\kappa = $ wt(X), with basic open sets $J(j \mid n) := \{j' \in J : j' \mid n = j \mid n\}$ (as in Chapter 3), consider sets S represented by the following notion of a Souslin representation, broader than that above, namely by the *extended κ-Souslin operation* (briefly: the *extended Souslin operation*):

$$S = \bigcup_{j \in J} H(j), \quad \text{where } H(j) := \bigcap_{n \in \mathbb{N}} H(j \mid n),$$

applied to a *determining system* $\langle H(j \mid n) \rangle := \langle H(j \mid n) : j \in \kappa^{\mathbb{N}} \rangle$ of sets from a family $\mathcal{H}$ subject to the requirement that:

(i) $\{H(j \mid n) : j \mid n \in \kappa^n\}$ is σ-discrete for each n.

We will usually also require that the determining system is *shrinking*, meaning:

(ii) diam$_X H(j \min n) < 2^{-n}$, so that $H(j)$ is empty or single-valued, and so compact.

For $\mathcal{H} = \mathcal{F}$ the corresponding extended Souslin-$\mathcal{F}$ sets reduce to the κ-*Souslin* sets of Hanse1992. (This slightly refines Hansell's terminology, and abandons Stone's term 'κ-restricted Souslin' of Sto1980.)

With X above complete (e.g. $X = S^*$, the completion of S) and for $\mathcal{H} = \mathcal{F}(X)$, the mapping $H \colon J \to \mathcal{K}(X)$ evidently yields a natural upper semi-continuous representation of S. We refer to it below, in relation to the Analytic Cantor Theorem, 13.1.5, cf. 3.1.6, in the separable context of Chapter 3, and also in Proposition 13.1.8. There the fact that $C := \{j : H(j) \neq \emptyset\}$ is closed in $\kappa^{\mathbb{N}}$ yields *a natural representation* of S as the image of C under a map h

defined by $H(j) = \{h(j)\}$. The map h is continuous and, as will be defined below, index-σ-discrete with countable fibres (preimages of single points), by (i) above, as noted in opening remarks of §13.1.2.

Theorem 13.1.1 (Hansell's Characterization Theorem) *In a metric space X, the Souslin-$\mathcal{F}(X)$ subsets of X are precisely the sets S represented by a shrinking determining system of closed sets through the extended κ-Souslin representation above with $\kappa = \mathrm{wt}(X)$.*

For other equivalent representations, including a weakening of σ-discreteness in X above to σ-d relative to its union, as well as to σ-d decompositions, see Hanse1972; Hanse1973b; Hanse1973a. Thus, working relative to J, the corresponding extended Souslin sets exhibit properties similar to the $\mathcal{K}$-analytic sets relative to I. In particular of interest here is Nikodym's Corollary, Theorem 3.1.3, which we recall and prove.

Theorem 13.1.2 (Nikodym's Corollary) *In a metric space S, analytic sets have the Baire property.*

Proof Since $S \cap \mathcal{F}(S^*) = \mathcal{F}(S)$, the theorem follows immediately from the definition of analytic sets as Souslin-$\mathcal{F}(S^*)$ and from Nikodym's theorem, 3.1.3, asserting that the Baire property is preserved by the usual Souslin operation, with the consequence that Souslin-$\mathcal{F}$ sets have the Baire property (since a closed set differs from its interior by a nowhere dense set). □

Using Hansell's Characterization Theorem and again Nikodym's Theorem, one also has the equally thematic result.

Theorem 13.1.3 *In a metric space, sets with a shrinking extended Souslin-$\mathcal{F}$ representation have the Baire property.*

Actually, this is a direct consequence of the following result, apparently first noted in Ost2012.

Theorem 13.1.4 *In a topological space, the extended Souslin operation applied to a determining system of sets with the Baire property yields a set with the Baire property.*

Proof We follow the classical 'separable' proof given for the usual Souslin operation as given in RogJ1980, Th. 2.9.2, pp. 43–44, checking that it continues to hold mutatis mutandis for the choice $\mathcal{B}a := \mathcal{B}a(X)$ of the family of sets with the Baire property and $\mathcal{M}$ of the meagre subsets of the metric space X. In particular, we must interpret $\mathbb{N}^{\mathbb{N}}$ there as $\kappa^{\mathbb{N}}$ throughout, with $\kappa^{(\mathbb{N})}$ denoting finite sequences with terms in κ.

By Banach's Category Theorem, $\mathcal{M}$ is closed under σ-discrete unions, and hence so is $\mathcal{B}a$ (open sets being closed under arbitrary unions).

Assume the extended Souslin operation above is applied to a determining system of sets $\langle B(\sigma \mid n) \rangle$ in $\mathcal{B}a$, giving rise to a set

$$A = \bigcup\nolimits_{j \in J} B(j), \quad \text{where} \quad B(j) := \bigcap\nolimits_{n \in \mathbb{N}} H(j \mid n),$$

where $\{B(\sigma \mid n) : \sigma \mid n \in \kappa^n\}$ is a σ-discrete family for each n.

For $\sigma \mid n \in \kappa^n$, put

$$A(\sigma \mid n) := \bigcup\nolimits_{j \in J} \{B(j) : j \in J, \, j \mid n = \sigma \mid n\}.$$

Choose $C(\sigma \mid n) \in \mathcal{B}a$ with $A(\sigma \mid n) \subseteq C(\sigma \mid n)$ that is a (Baire) hull (cf. §3.4) for $A(\sigma \mid n)$ with the 'approximation property' that, for $B' \in \mathcal{B}a$, if

$$A(\sigma \mid n) \subseteq B' \subseteq C(\sigma \mid n),$$

then $C(\sigma \mid n) \backslash B' \in \mathcal{M}$. Write

$$D(\sigma \mid n) := \bigcap\nolimits_{k \leq n} [B(\sigma \mid k) \cap C(\sigma \mid k)].$$

As the system $\langle D(\sigma \mid n) \rangle$ is a refinement of the $\langle B(\sigma \mid n) \rangle$ system, $\{D(\sigma \mid n) : \sigma \mid n \in \kappa^n\}$ is also σ-discrete for each n, and so the union $\bigcup \{D(\sigma \mid n, t) : t \in \kappa\}$ is in $\mathcal{B}a$. (Note that the sets $D(\sigma \mid n)$ are defined as finite intersections of sets in $\mathcal{B}a$.)

For $\sigma \mid n \in \kappa^n$,

$$A(\sigma \mid n) = \bigcup\nolimits_{t \in \kappa} A(\sigma \mid n, t) \subseteq \bigcup\nolimits_{t \in \kappa} D(\sigma \mid n, t) \subseteq D(\sigma \mid n),$$

and each $D(\sigma \mid n)$ is a hull for $A(\sigma \mid n)$ with the same approximation property as $C(\sigma \mid n)$.

Each set $M(\sigma \mid n) := D(\sigma \mid n) \backslash \bigcup \{D(\sigma \mid n, t) : t \in \kappa\}$ is in $\mathcal{M}$, as $\mathcal{M}$ is closed under subset formation, and again the family $\{M(\sigma \mid n) : \sigma \mid n \in \kappa^n\}$ is σ-discrete for each n, as before by refinement: $M(\sigma \mid n) \subseteq D(\sigma \mid n)$. Hence

$$L := \bigcup \{M(\sigma) : \sigma \in \kappa^{(\mathbb{M})}\} = \bigcup\nolimits_{n \in \mathbb{M}} \bigcup \{M(\sigma \mid n) : \sigma \mid n \in \kappa^n\}$$

is in $\mathcal{M}$, again by Banach's Category Theorem. Denoting the empty sequence by $\langle \rangle$, put

$$D = D(\langle \rangle) = C(\langle \rangle).$$

We prove that $D \backslash L \subseteq A$. Let x be any point of D that is not in L. As

$$x \in D(\langle \rangle), \quad \text{but} \quad x \notin D(\langle \rangle) \backslash \bigcup\nolimits_{t \in \kappa} D(t),$$

we may choose $t_1 \in \kappa$ so that

$$x \in D(t_1),$$

but then

$$x \notin D(t_1) \backslash \bigcup_{t \in \kappa} D(t_1, t).$$

Proceeding inductively, we can choose $t_1, t_2, t_3, \ldots$, all in κ, so that

$$x \in D(t_1, t_2, \ldots, t_n) \quad \text{for } n \geq 1.$$

Then, taking $\tau = (t_1, t_2, t_3, \ldots) \in J$, we have

$$x \in D(\tau \mid n) \subseteq B(\tau \mid n), \quad \text{for all } n \geq 1.$$

Hence $x \in A$, so $D \backslash L \subseteq A$ and

$$D \backslash L \subseteq A \in \mathcal{M}.$$

As $\mathcal{M}$ is (subset) hereditary, $D \backslash A \in \mathcal{M}$ and

$$X \backslash (D \backslash A) \in \mathcal{M}.$$

Now $A = A(\langle \rangle) \subseteq C(\langle \rangle) = D$ so that

$$A = D \cap (X \backslash (D \backslash A)) \in \mathcal{M}.$$

Thus $\mathcal{M}$ is closed under the Souslin operation. □

A similar, but simpler, argument with $\mathcal{B}a$ replaced by $\mathcal{M}$, the *Radon measurable* sets, shows these to be stable under the extended Souslin operation (using measure completeness and local determination, for which see Fre2003, 412J, cf. 431A, and measurable hulls, Fre2001, 213L).

Evidently, the standard separable category arguments may also be applied to σ-discrete decompositions of a set, in view again of Banach's Category Theorem.

Finally, since $H : J \to \mathcal{K}(X)$ above is upper semi-continuous (for X complete and $\mathcal{H} = \mathcal{F}$), the following theorem, used in the separable context of Ost2011, §2 and Ost2013c, Th. AC, continues to hold in the non-separable context (by the same proof), which permits us to quote freely some of its consequences as established in Chapter 6.

Theorem 13.1.5 *Let X be a Hausdorff space and $A = K(J)$, with $K : J \to \mathcal{K}(X)$ compact-valued and upper semi-continuous.*

If F_n is a decreasing sequence of (non-empty) closed sets in X such that $F_n \cap K(J(j_1, \ldots, j_n)) \neq \emptyset$, for some $j = (j_1, \ldots) \in J$ and each n, then $K(j) \cap \bigcap_n F_n \neq \emptyset$.

Here, beyond upper semi-continuity, we do not need properties related to the notion of σ-discrete possessed by the mapping H (for which see HanseJR1983).

13.1.3 Supporting Notions

We will need the following definitions (see below for comments). Recall that a Hausdorff space X is *paracompact* (Eng1989, Ch. 5) if every open cover of X has a locally finite open refinement (so that a (regular) Lindelöf space is paracompact (Eng1989, 5.1.2), as in the context of Levi's Open Mapping Theorem). We employ the terminology of the preceding section.

Definitions

(1) (Hanse1974, §3, cf. Hanse1971, §3.1 and Mich1982, Def. 3.3) Call $f : X \to Y$ *base-σ-discrete* (or *co-σ-discrete*) if the image under f of any discrete family in X has a σ-discrete base in Y. We need two refinements that are more useful and arise in practice: call $f : X \to Y$ an *analytic* (resp. *Baire*) *base-σ-discrete map* (henceforth A-σ-d, resp. B-σ-d map) if, in addition, for any discrete family $\mathcal{E}$ of *analytic* sets in X, the family $f(\mathcal{E})$ has a σ-d base consisting of analytic sets (resp. sets with the Baire property) in Y. We explain in §13.1.4 (Theorem 13.1.6 and thereafter) why A-σ-d maps, though not previously isolated, are really the only base-σ-discrete maps needed in practice in analytic space theory.

(2) (Hanse1974, §2) An indexed family $\mathcal{A} := \{A_t : t \in T\}$ is *σ-discretely decomposable* (σ-d decomposable) if there are discrete families $\mathcal{A}_n := \{A_{tn} : t \in T\}$ such that $A_t = \bigcup_n A_{tn}$ for each t. (The open family $\{(-r, r) : r \in \mathbb{R}\}$ on the real line has a σ-d base, but is not σ-d decomposable – see Hanse1973b, §3.)

(3) (Mich1982, Def. 3.3) Call $f : X \to Y$ *index-σ-discrete* if the image under f of any discrete family $\mathcal{E}$ in X is σ-d decomposable in Y. (Note $f(\mathcal{E})$ is regarded as indexed by $\mathcal{E}$, so could be discrete without being index-discrete; this explains the prefix 'index-' in the terminology here.) An index-σ-discrete function is A-σ-d (analytic base-σ-discrete): see Theorem 13.1.6.

Remarks   These concepts are motivated Bing's Theorem (see Chapter 1) that a regular space is metrizable if and only if it has a σ-discrete base. In a separable space discrete sets are at most countable. So all the notions above generalize various aspects of countability; in particular, in a separable metric setting all maps are (Baire) base-σ-discrete. We now comment briefly on their standing. (The paper Hanse1974 is the primary source for these.)

1. In (3) above f has a stronger property than base-σ-discreteness. For a proof see Hanse1974, Prop 3.7(i). Compare Mich1982, Prop 4.3, which shows that f with closed fibres has (3) if and only if it is base-σ-discrete and has fibres that are $\aleph_1$-compact, i.e. separable (in the metric setting). The stronger property is often easier to work with than Baire base-σ-discreteness. In any case, the concepts are

close, since for metric spaces and κ an infinite cardinal: X is a base-σ-discrete continuous image of $\kappa^{\mathbb{N}}$ if and only if X is an index-σ-discrete continuous image of a closed subset of $\kappa^{\mathbb{N}}$, both equivalent to analyticity (Hanse1974, Th. 4.1). See Proposition 13.1.8, where it is shown that analytic metric spaces are the A-σ-d continuous images of $\kappa^{\mathbb{N}}$.

See Hanse1992, Th. 4.2 or Hanse1974, Th. 4.1. In fact the natural continuous index-σ-discrete representation of an analytic set has separable fibres; for a study of fibre conditions, see Hanse1999.

2. Base-σ-discrete *continuous* maps (in particular, Baire base-σ-discrete and index-σ-discrete continuous maps) preserve analyticity (Hanse1974, Cor. 4.2).

3. If X is metric and (absolutely) analytic and $f: X \to Y$ is injective and *closed-analytic* (or *open-analytic*), then f is base-σ-discrete (Hanse1974, Prop. 3.14). Base-σ-discreteness is key to this section just as open-analyticity is key to the *separable* context of the Levi results in Chapter 12.

4. If $\mathcal{B} = \bigcup_{n \in \mathbb{N}} \mathcal{B}_n$, with each $\mathcal{B}_n$ discrete, is a σ-d base for the metrizable space X and each $f(\mathcal{B}_n)$ is σ-d decomposable, then f is index-σ-discrete, and so base-σ-discrete (Hanse1974, Prop. 3.9).

5. A discrete collection $\mathcal{A} = \{A_t : t \in T\}$ comprising analytic sets has the property that any subfamily has analytic union, i.e. is *completely additive analytic*. It turns out that in an analytic space a disjoint (or a point-finite) collection $\mathcal{A}$ is completely additive analytic if and only if it is σ-d decomposable (see KaniP1975, generalizing the disjoint case in Hanse1971, Th. 2; see also FroH1981). By the proof of Theorem 13.1.6 the decompositions can be into analytic sets.

It is Nikodym's Corollary, 13.1.2, that motivates our interest in analyticity as a carrier of the Baire property, especially as continuous images of separable analytic sets are separable, hence Baire.

However, the continuous image of an analytic space is not in general analytic – for an example of failure, see Hanse1974, Ex. 3.12. But analyticity does obtain when, additionally, the continuous map is base-σ-discrete, as defined below (Hanse1974, Cor. 4.2). This technical condition is the standard assumption for preservation of analyticity and holds automatically in the separable realm. Special cases include *closed surjective* maps and *open-to-analytic injective* maps (taking open sets to analytic sets).

We may turn now to a discussion of analytic base-σ-discrete maps.

13.1.4 A-σ-Discrete Maps

Their definition requires in addition to base-σ-discreteness that, *for any discrete family* $\mathcal{E}$ of analytic sets in X, the family $f(\mathcal{E})$ has a σ-discrete base consisting of sets with the Baire property. The remaining results in this section are gleaned from a close reading of the main results in Hanse1974 regarding base-σ-discrete maps, i.e. Hansell's sequence of results 3.6–3.10 and his Th. 4.1, all of which derive the required base-σ-discrete property by arguments that combine σ-discrete decompositions with discrete collections of singletons. We shall see below that all these results may be refined to the A-σ-d context.

Theorem 13.1.6   *An index-σ-discrete function is Analytic base-σ-discrete* (A-σ-d).

Proof   Suppose that an indexed family $\mathcal{A} := \{A_t : t \in T\}$ has a σ-discrete decomposition using discrete families $\mathcal{A}_n := \{A_{tn} : t \in T\}$, with $A_t = \bigcup_n A_{tn}$ for each t, and that all the sets A_t have the Baire property. A particular case occurs when these sets are analytic – by Nikodym's Corollary, 13.1.2. Putting $\tilde{A}_{tn} := A_t \cap \bar{A}_{tn}$ and $\tilde{\mathcal{A}}_n := \{\tilde{A}_{tn} : t \in T\}$, which is discrete, we obtain a B-σ-d decomposition for $\mathcal{A}$ (or an A-σ-d one in the case of an analytic σ-d decomposition), and so a fortiori a B-σ-d base for $\mathcal{A}$ (or an A-σ-d one).

Thus, if $\mathcal{E}$ is a discrete family of analytic sets and f is index-σ-discrete, then $\mathcal{A} := f(\mathcal{E})$ comprises analytic sets (see Remark 2 in §13.1.3); $\mathcal{A}$ has, for its σ-d base, the family $\bigcup_n \tilde{\mathcal{A}}_n$ of analytic sets, so with the Baire property, where $\tilde{\mathcal{A}}_n$ are as just given above.   □

The A-σ-d maps are closed under composition (cf. Hanse1974, Prop. 3.4); also by the construction in Theorem 13.1.6, the conclusions of Hanse1974, Prop. 3.7, Cors. 3.8 and 3.9 yield A-σ-discrete bases for $f(\mathcal{E})$, when $\mathcal{E}$ comprises analytic sets. In similar vein are the next two refinements of results due to Hansell – together verifying the adequacy of A-σ-d maps for analytic-sets theory.

Theorem 13.1.7 (Hanse1974, Prop. 3.10)   *A closed surjective map onto a metrizable space is Analytic base-σ-discrete.*

Proof   Since singletons are analytic, a base that is a discrete family of singletons is an A-σ-d base. This, combined with the construction in Theorem 13.1.6, refines the argument for Hansell's cited Prop. 3.10 proving that a closed surjective map onto a metrizable space is an A-σ-d map.   □

Proposition 13.1.8   *Analytic metric spaces are the A-σ-d continuous images of $\kappa^{\mathbb{N}}$.*

Proof By Remark 2 in §13.1.3 again, there is only one direction to consider. So let S be analytic; we refine Hansell's argument. Observe first that the argument in his Prop. 3.5(ii) proves more: if Y is σ-discrete, then any map into Y is A-σ-d (as in Theorem 13.1.6). Next, let us again use the notation H for analytic sets established above (in a complete context, with $\mathcal{H} = \mathcal{F}$). We will work in the closed subspace $C \subseteq \kappa^{\mathbb{N}}$ comprising those j with $H(j) \neq \emptyset$, and define, as above, the (continuous) map h on C via $H(j) = \{h(j)\}$. Observe that h takes, for each n, the discrete family of basic open sets $J(j \mid n)$, relativized to C, to the σ-d family of analytic sets $h(J(j \mid n) \cap C)$, and so h is an A-σ-d map (by Hanse1974, Cor. 3.9, reported in Remark 4 in §13.1.3). Stone's canonical retraction r of $\kappa^{\mathbb{N}}$ onto any closed subspace as applied to the closed subspace C has σ-discrete range on $\kappa^{\mathbb{N}} \backslash C$ and is the identity homeomorphism on C – for details see Eng1969 (where r is also shown to be a closed map). So, in view of the preceding two observations, r is an A-σ-d map. Hence $h \circ r$ is A-σ-d, since composition of A-σ-d maps is A-σ-d and provides the required characterization.   □

13.2 Nikodym Actions: Non-separable Case

This short section is devoted to providing a non-separable analogue for the Effros Theorem of Chapter 11, now that we have the tools to generalize Proposition 11.2.4 to the non-separable context. The wording of its statement is nearly verbatim: we omit separability and add base-σ-discreteness to the assumption that point-evaluation maps are continuous; but its proof is more involved.

Proposition 13.2.1 ('Weak Microtransitivity') *Suppose G is a normed group, acting transitively on a non-meagre space X with each point-evaluation map $\varphi_x \colon g \mapsto g(x)$ continuous and base-σ-discrete. Then, for each non-empty open U in G and each $x \in X$, the set Ux is non-meagre in X.*

In particular, if G is analytic, then G is a Nikodym action.

Proof We first work in the right-sided topology, i.e. derived from the assumed right-invariant metric $d_{\mathrm{R}}^G(s,t) = \|st^{-1}\|$. Suppose that $u \in U$, and so without loss of generality assume that $U = B_\varepsilon(u) = B_\varepsilon(e_G)u$ (open balls of radius some $\varepsilon > 0$): indeed,

$$B_\varepsilon(e_G)u = \{xu : d_{\mathrm{R}}^G(x,e) < \varepsilon\} = \{z : d_{\mathrm{R}}^G(zu^{-1},e) < \varepsilon\}$$
$$= \{z : d_{\mathrm{R}}^G(z,u) < \varepsilon\} = U.$$

Now put $y := ux$ and $W = B_\varepsilon(e_G)$. Then $Ux = B_\varepsilon(e_G)ux = Wy$. Next work in the left norm-topology, derived from $d_{\mathrm{L}}^G(s,t) = \|s^{-1}t\| = d_{\mathrm{R}}^G(s^{-1},t^{-1})$ (for which $W = B_\varepsilon(e_G)$ is still a neighbourhood of e_G). As each set hW for $h \in G$ is now open (since now the left shift $g \to hg$ is a homeomorphism), the open family $\mathcal{W} = \{gW : g \in G\}$ covers G. As G is metrizable (and so has a σ-discrete base), the cover $\mathcal{W}$ has a σ-discrete refinement, say $\mathcal{V} = \bigcup_{n \in \mathbb{N}} \mathcal{V}_n$, with each $\mathcal{V}_n$ discrete. Put $X_n := \bigcup\{Vy : V \in \mathcal{V}_n\}$; then $X = \bigcup_{n \in \mathbb{N}} X_n$, as $X = Gy$, and so X_n is non-meagre for some n, say $n = N$. Since φ_y is base-σ-discrete, $\{Vy : V \in \mathcal{V}_N\}$ has a σ-discrete base, say $\mathcal{B} = \bigcup_{m \in \mathbb{N}} \mathcal{B}_m$, with each $\mathcal{B}_m$ discrete. Then, as $\mathcal{B}$ is a base for $\{Vy : V \in \mathcal{V}_N\}$,

$$X_N = \bigcup_{m \in \mathbb{N}} \left(\bigcup\{B \in \mathcal{B}_m : (\exists V \in \mathcal{V}_N)B \subseteq Vy\} \right).$$

So for some m, say $m = M$,

$$\bigcup\{B \in \mathcal{B}_m : (\exists V \in \mathcal{V}_N)B \subseteq Vy\}$$

is non-meagre. But as $\mathcal{B}_M$ is discrete, by Banach's Category Theorem (cf. Proposition 11.2.2), there are $\hat{B} \in \mathcal{B}_M$ and $\hat{V} \in \mathcal{V}_N$ with $\hat{B} \subseteq \hat{V}y$ such that $\hat{B}$ is non-meagre. As $\mathcal{V}$ refines $\mathcal{W}$, there is some $\hat{g} \in G$ with $\hat{V} \subseteq \hat{g}W$, so $\hat{B} \subseteq \hat{V}y \subseteq \hat{g}Wy$, and so $\hat{g}Wy$ is non-meagre. As $\hat{g}^{-1}$ is a homeomorphism of X, $Wy = Ux$ is also non-meagre in X.

If G is analytic, then as U is open, it is also analytic (since open sets are $\mathcal{F}_\sigma$ and Souslin-$\mathcal{F}$ subsets of analytic sets are analytic, cf. RogJ1980), and hence so is $\varphi_x(U)$. Indeed, since φ_x is continuous and base-σ-discrete, Ux is analytic (Hansell's Theorem, §13.1), so Souslin-$\mathcal{F}$, and so Baire by Nikodym's Corollary, 13.1.2.   □

13.3 Coincidence Theorems: Non-separable Case

We begin with Theorem 13.3.1, a non-separable variant of the Levi Open Mapping Theorem of Chapter 12, a *Generalized Levi Open Mapping Theorem* applicable to the broader category of (absolutely) analytic spaces – be they separable or non-separable metric spaces. We recall that $\mathcal{K}$ *-analytic sets* were defined in §13.1.

Theorem 13.3.1 *Let Y be an analytic space (more generally, paracompact and $\mathcal{K}$-analytic). Then Y is a Baire space if and only if $Y = f(X)$ for some continuous, index-σ-discrete map f on a completely metrizable space X with the property that there exists a dense completely metrizable $\mathcal{G}_\delta$ subset Y' of Y such that, with $X' := f^{-1}(Y')$, the restriction $f|X' : X' \to Y'$ is an open mapping, with X' again completely metrizable.*

This is proved below (in §13.3.1). For related results on restriction maps of other special maps, see Mich1991, §7. (Compare also Hanse1992, Ths. 6.4 and 6.25.) As immediate corollaries one has:

Theorem 13.3.2 (Generalized Levi Coincidence Theorem)   *If $\mathcal{T}, \mathcal{T}'$ are two topologies on a set X with $(X, \mathcal{T})$ a regular Baire space, and $\mathcal{T}'$ an absolutely analytic (e.g. completely metrizable) refinement of $\mathcal{T}$ such that every $\mathcal{T}'$-index-discrete collection is σ-discretely decomposable under $\mathcal{T}$, then there is a $\mathcal{T}$-dense $\mathcal{G}(\mathcal{T})_\delta$-set on which $\mathcal{T}$ and $\mathcal{T}'$ agree.*

Proof   For f take the identity map from $(X, \mathcal{T}')$ to $(X, \mathcal{T})$, which is continuous and index-σ-discrete.   □

Corollary 13.3.3 (Almost Completeness Theorem)   *A space X is almost analytic and a Baire space if and only if X is almost complete.*

Proof   If X contains a co-meagre analytic subspace A, then, by Theorem 13.3.1, A, being a Baire space, contains a dense completely metrizable subspace X' which is co-meagre in X. So X is almost complete.   □

For a sharper characterization in the case of normed groups, see Ost2013c, Th. 1. Theorem 13.3.2 will enable us to prove (in §13.3.3) the automatic continuity result of the *Semi-Completeness Theorem*, 13.3.9, for right-topological groups with a right-invariant metric d_R (the normed groups of Chapter 6). We write $d_L(x, y) := d_R(x^{-1}, y^{-1})$, which is left-invariant, and $d_S := \max\{d_R, d_L\}$ for the symmetrized ('ambidextrous') metric. The basic open sets under d_S take the form $B_\varepsilon^R(x) \cap B_\varepsilon^L(x)$, i.e. an intersection of balls of ε-radius under d_R and d_L centred at x, giving the join (coarsest common refinement) of d_R and d_L. Following Ost2013d, for P a topological property it is convenient to say that the metric space (X, d_R) is topologically *symmetrized-P*, or just *semi-P*, if (X, d_S) has property P. In particular (X, d_R) is *semi-complete* if (X, d_S) is topologically complete. As d_R and d_L are isometric under inversion, (X, d_R) is semi-complete if and only if (X, d_L) is.

The following result will be needed later in conjunction with Theorem 13.3.7.

Lemma 13.3.4   *For a normed group X, if the continuous (identity) embedding map, $j : (X, d_S) \to (X, d_R)$, is index-$\sigma$-discrete (resp. base-$\sigma$-discrete), then so also is the inversion mapping from (X, d_R) to (X, d_R); i.e. $i : x \to x^{-1}$.*

Proof   Suppose $\mathcal{V}$ is a family of sets that is discrete in (X, d_R). Then $\mathcal{V}^{-1} := \{V^{-1} : V \in \mathcal{V}\}$ is a family of sets that is discrete in (X, d_L). As the d_S topology refines the d_L topology, $\mathcal{V}^{-1}$ is discrete in (X, d_S). Assuming that j is index-σ-discrete (resp. base-σ-discrete), the family $j(\mathcal{V}^{-1}) = \mathcal{V}^{-1}$ is σ-d decomposable (has a σ-d base) in (X, d_R). So inversion maps $\mathcal{V}$ to a family $\mathcal{V}^{-1}$ that is σ-d decomposable (has σ-d base) in (X, d_R).   □

13.3.1 Generalized Levi Theorem

The generalized Levi characterization of Baire spaces in terms of open mappings in Theorem 13.3.1 is a consequence of the following result, which we also apply in §13.3.3.

Lemma 13.3.5 (Generalized Levi Lemma) *If $f : X \to Y$ is surjective, continuous and Baire base-σ-discrete (in particular, index-σ-discrete) from X metric and analytic to Y a paracompact space, then there is a dense metrizable $\mathcal{G}_\delta$-subspace $Y' \subseteq Y$ such that, for $X' := f^{-1}(Y')$, the restriction map $f \mid X' : X' \to Y'$ is open.*

Proof Let $\mathcal{A} = \bigcup_n \mathcal{A}_n = \{A_t : t \in T\}$ be an open base for X with $\mathcal{A}_n = \{A_{tn} : t \in T_n\}$ discrete. Then $E_t := f(A_t)$ is analytic (see Remarks in §13.1.2), so has the Baire property by Nikodym's Corollary, 13.1.2. Let $\mathcal{E}_n = \{f(A) : A \in \mathcal{A}_n\}$ and let $\mathcal{B}_n$ be a σ-d base for $\mathcal{E}_n$ comprising sets with the Baire property. Put $\mathcal{B}_n = \bigcup_m \mathcal{B}_{nm}$ with each $\mathcal{B}_{nm}$ discrete. Thus for each $t \in T$ and $E_t \in \mathcal{E}_n$ one has

$$E_t = \bigcup_m \bigcup \{B : B \subseteq E_t \text{ and } B \in \mathcal{B}_{nm}\}.$$

Put $\mathcal{B} = \bigcup_{nm} \mathcal{B}_{nm}$. For each $B \in \mathcal{B}$ pick an open set U_B and meagre sets M'_B and M_B such that

$$B = (U_B \backslash M'_B) \cup M_B$$

with M_B disjoint from M'_B and with $M'_B \subseteq U_B$. As $\{B : B \in \mathcal{B}_{nm}\}$ is discrete, the set

$$M := \bigcup_{mn} \bigcup \{M_B : B \in \mathcal{B}_{nm}\}$$

is meagre. By paracompactness of Y (cf. Eng1989, Th. 5.1.18), since $\{U_B \backslash M'_B : B \in \mathcal{B}_{nm}\}$ is discrete, for $B \in \mathcal{B}_{nm}$ we may select open sets W_B with $U_B \backslash M'_B \subseteq W_B$ with $\{W_B : B \in \mathcal{B}_{nm}\}$ discrete. Without loss of generality $W_B \subseteq U_B$ (otherwise replace W_B by $W_B \cap U_B$). So $U_B \backslash M'_B \subseteq W_B \subseteq U_B$ and hence $U_B \backslash M'_B = W_B \backslash M'_B$. Then

$$M' := \bigcup_{n,m} \bigcup \{W_B \cap M'_B : B \in \mathcal{B}_{nm}\}$$

is also meagre. Now put $Y' := Y \backslash (M \cup M')$ and $W := \bigcup_{m,n} \bigcup \{W_B : B \in \mathcal{B}_{nm}\}$, which is open. Then, for $B \in \mathcal{B}_{nm}$, one has

$$B \cap Y' = W_B \cap Y',$$

so that B is open relative to Y' and also $\{Y' \cap W_B : B \in \mathcal{B}_{nm}\}$ is open and discrete in Y'. Now, for $t \in T$, since $E_t \in \mathcal{E}_n$ for some n and $\mathcal{B}_n$ is a base for $\mathcal{E}_n$,

$$E_t \cap Y' = \bigcup_{n,m} \bigcup \{B \cap Y' : B \subseteq E_t \ \& \ B \in \mathcal{B}_{nm}\}$$
$$= \bigcup_{n,m} \bigcup \{W_B \cap Y' : B \subseteq E_t \ \& \ B \in \mathcal{B}_{nm}\}$$

is open in Y'.

For $G \subseteq X$ open, since $\mathcal{A}$ is a (topological) base, we may write

$$G := \bigcup_n \mathcal{A}_n^G \text{ with } \mathcal{A}_n^G := \{A : A \subseteq G \ \& \ A \in \mathcal{A}_n\}.$$

Then

$$f(G) := \bigcup_n f(\mathcal{A}_n^G)$$

with

$$f(\mathcal{A}_n^G) := \{E : E = f(A) \ \& \ A \subseteq G \ \& \ A \in \mathcal{A}_n\}.$$

So, for $X' := f^{-1}(Y')$,

$$f(G \cap X') = Y' \cap \bigcup_{n,m} \mathcal{W}_{nm}^G,$$

with

$$\mathcal{W}_{nm}^G := \{W_B : B \subseteq f(A) \ \& \ B \in \mathcal{B}_{nm} \ \& \ A \subseteq G \ \& \ A \in \mathcal{A}_n\},$$

which is open in Y'.

Since f is continuous and $\mathcal{W}_{nm} := \{W_B : B \in \mathcal{B}_{nm}\}$ is discrete for each m, n, this also shows that the family $\bigcup_{n,m}\{Y' \cap W_B : B \in \mathcal{B}_{nm}\}$ is a σ-d base for Y'. Being paracompact, Y is regular (Eng1989, Th. 5.1.5), so the subspace Y' is regular (Eng1989, Th. 2.1.6), and so Y' is metrizable by Bing's Characterization Theorem (see Eng1989, Th. 4.4.8, cf. Remarks in §13.1.2). Finally, by replacing the meagre sets M, M' by larger sets that are unions of closed nowhere dense sets, we obtain, in place of Y', a smaller, metrizable, dense G_δ-subspace. $\qquad \square$

Remarks 1. For Lemma 13.3.5 we replaced each system $\{U_B : B \in \mathcal{B}_{nm}\}$ to obtain discrete systems $\{W_B : B \in \mathcal{B}_{nm}\}$ to reduce the sets M'_B to the sets $M'_B \cap W_B$ and only then did we take unions. This circumvents the suggested approach (in a parenthetical remark) to the proof of Th. 6.4(c) in Hanse1992 (by way of representing a Baire set as $E = (G_E \backslash P_E) \cup Q_E$ with $P_E \subseteq E$ (sic) for G_E open and P_E, Q_E meagre).

2. Given an arbitrary base $\mathcal{B}$ one may replace each $B \in \mathcal{B}$ with a Baire hull (§3.4) B^+ such that $B \subseteq B^+ \subseteq \bar{B}$. Then $\mathcal{B}^+ = \bigcup_n \mathcal{B}_n^+$ is σ-d and $E_t \subseteq E_t^+ :=$

$\bigcup_n \bigcup \{B^+ : B \subseteq E_t$ and $B \in \mathcal{B}_n\}$, with $E_t^+ \backslash E_t$ meagre. However, it is not clear that the union of these meagre sets is meagre.

Proof of Theorem 13.3.1 Let X be analytic of weight κ. Then for some closed subset P of $\kappa^{\mathbb{N}}$ there is a continuous index-σ-discrete map $f : P \to X$. Form P' and X' analogously to X' and Y' in the preceding lemma. As X is a Baire space and X' is co-meagre, without loss of generality X' is a dense $\mathcal{G}_\delta$ and is metrizable. Also P' is a $\mathcal{G}_\delta$ subspace of the complete space $\kappa^{\mathbb{N}}$, hence is also topologically complete. So P' has the desired properties. As X' is metrizable, the result now follows from Hausdorff's Theorem that the image under an open continuous mapping of the completely metrizable space P' onto a metrizable space X' is also completely metrizable (for a proof see, e.g., Anc1987, or, for a more recent account, e.g., HoliP2010).

For the converse, as X' is metrizable, the result again follows from Hausdorff's Theorem. Thus X' is completely metrizable. But its complement in X is meagre. So X is a Baire space – in fact an almost complete space. □

13.3.2 Automatic Continuity of Homomorphisms

In the proof of the Semi-Completeness Theorem, 13.3.9, we will need to know that the inverse of a continuous bijective homomorphism is continuous. In the separable case this follows by noting that the graph of the homomorphism is closed and, as a consequence of Theorem 12.1.7 (the Souslin-Graph Theorem), the inverse is a Baire homomorphism (meaning that preimages of open sets have the Baire property) and hence continuous. However, in the non-separable case the paradigm falls foul of the technical requirement for σ-discreteness. We will employ a modified approach based on the following.

Theorem 13.3.6 (Open Homomorphism Theorem) *For normed groups X, Y with X analytic and Y a Baire space, let $f : X \to Y$ be a surjective, continuous homomorphism, which is base-σ-discrete. Then f is open. So if also $X = Y$ and f is bijective, then f^{-1} is continuous.*

Proof Suppose that B and A are arbitrary open balls around e_X with $B^4 \subseteq A$. Let D be a dense set in X (e.g. X itself). Now $\mathcal{U} = \{Bd : d \in D\}$ is an open cover of X. (Indeed, if $x \in X$ and $d \in D \cap Bx$, then $x \in Bd$, by symmetry.) Let $\mathcal{V}$ be a σ-d open refinement of $\mathcal{U}$, and say $\mathcal{V} = \bigcup_n \mathcal{V}_n$, with each $\mathcal{V}_n$ discrete. Then, as $X = \bigcup_n (\bigcup \{V : V \in \mathcal{V}_n\})$, we have $Y = \bigcup_n (\bigcup \{f(V) : V \in \mathcal{V}_n\})$. As f is base-σ-discrete each $\mathcal{W}_n := f(\mathcal{V}_n) = \{f(V) : V \in \mathcal{V}_n\}$ has a σ-d base $\mathcal{B}_n$; write $\mathcal{B}_n := \bigcup_m \mathcal{B}_{nm}$ with each $\mathcal{B}_{nm}$ discrete. So for each $V \in \mathcal{V}_n$ one has $f(V) := \bigcup_m \{B \in \mathcal{B}_{nm} : B \subseteq f(V)\}$, and so

$$Y = \bigcup\nolimits_{nm} \{B \in \mathcal{B}_{nm} : B \subseteq f(V) \text{ for some } V \in \mathcal{V}_n\}.$$

As Y is non-meagre, there are $n, m \in \mathbb{N}$, $V \in \mathcal{V}_n$ and $B \in \mathcal{B}_{nm}$ such that $B \subseteq f(V)$ and B is non-meagre; for otherwise, since $\mathcal{B}_{nm}$ is discrete, by Banach's Category Theorem, $\{B \in \mathcal{B}_{nm} : B \subseteq f(V) \text{ for some } V \in \mathcal{V}_n\}$ is meagre, implying the contradiction that also Y is meagre. Pick such m, n and B, V such that $B \subseteq f(V) \subseteq f(Bd)$. Now $V \subseteq Bd$ for some $d \in D$, as $\mathcal{V}_n$ refines $\mathcal{U}$, and so $B \subseteq f(V) \subseteq f(Bd) = f(B)f(d)$ is non-meagre. So $f(B)$ is non-meagre and analytic (as B is analytic). By the Squared Pettis Theorem, 6.3.1, $(f(B)(f(B))^{-1})^2 = f(B)f(B)^{-1}f(B)f(B)^{-1} = f(BB^{-1}BB^{-1})$ is a neighbourhood of e_Y contained in $f(A)$. $\qquad\qquad\square$

The following corollary will be used together with Lemma 13.3.4, where not every conjugacy will be guaranteed to be continuous.

Theorem 13.3.7 (Continuous Inverse Theorem) *If, under d_R, the normed group X is an analytic Baire space and the inversion map $i : x \to x^{-1}$ preserves σ-discreteness (takes discrete families to σ-discrete families), then the inverse of any continuous conjugacy, $\gamma_\tau(x) := \tau x \tau^{-1}$, is also continuous.*

Proof If $\{V_t : t \in T\}$ is σ-d, then, for any $\tau \in X$, so is $\{V_t \tau^{-1} : t \in T\}$, as right shifts are homeomorphisms. Applying our assumption about the inversion map, for any τ the family $\{\tau V_t^{-1} : t \in T\}$ is σ-d, hence $\{\tau V_t^{-1} \tau^{-1} : t \in T\}$ is σ-discrete, and so again $\{\tau V_t \tau^{-1} : t \in T\}$ is σ-d. This means γ_τ is index-σ-discrete. By Theorem 13.3.6, if γ_τ is continuous, then so is its inverse. $\qquad\square$

With some minor amendments and from somewhat different hypotheses, the same proof as in the Open Homomorphism Theorem, 13.3.6, demonstrates the following generalization of a separable result (given in BinO2010g, Th. 11.11), but unfortunately without any prospect for achieving the Baire property (see the Remark at the end of this subsection). Here again the assumed discreteness preservation is fulfilled in the realm of separable spaces. We give the proof for the sake of comparison and because of its affinity with a result due to Noll (Nol1992, Th. 1) concerning topological groups (not necessarily metrizable), in which the map f has the property that $f^{-1}(U)$ is analytic for each open $\mathcal{F}_\sigma$-set U. In our metric setting, when preimages under the homomorphism f of open sets are analytic, f is Baire by Nikodym's Theorem and, since $f^{-1}(\mathcal{A})$ is a disjoint. Also, referring to Remark 4 on page 209, completely additive analytic for $\mathcal{A}$ discrete, the σ-d decomposability condition given below is satisfied by Hansell's result (Hanse1971, Th. 2) cited in the Remarks of §13.1.2. Noll shows the σ-d decomposability condition below is satisfied when X is a topologically complete topological group (using the FroH1981 generalization of Hansell's result and of KaniP1975 – cf. Remark 5 on page 209).

Theorem 13.3.8 (Baire Homomorphism Theorem) *For normed groups X, Y with X analytic, a surjective Baire homomorphism $f: X \to Y$ is continuous provided $f^{-1}(\mathcal{A})$ is σ-discretely decomposable for each σ-discrete family $\mathcal{A}$ in Y.*

Proof We proceed as above but now in Y. For $\varepsilon > 0$, with $B = B_{\varepsilon/4}(e_Y)$ open and D any dense set in Y choose a_d with $f(a_d) = d \in D$. Put $T := f^{-1}(B)$, which has the Baire property (as f is Baire). As Y is metrizable, the open cover $\{Bd : d \in D\}$ has a σ-d refinement $\mathcal{A} = \bigcup_n \mathcal{A}_n$ with $\mathcal{A}_n := \{A_{tn} : t \in T_n\}$ discrete. For each n, by assumption, we may write $\{f^{-1}(A_{tn}) : t \in T_n\} = \bigcup_{nm}\{B_{tnm} : t \in T_n\}$ with $\{B_{tnm} : t \in T_n\}$ discrete in X for each m and n. Now

$$X = f^{-1}(Y) = \bigcup_n \{f^{-1}(A_{tn}) : t \in T_n\} = \bigcup_{nm}\{B_{tnm} : t \in T_n\}.$$

As X is a Baire space, there are n, m such that $\bigcup\{B_{tnm} : t \in T_n\}$ is non-meagre. Again by Banach's Category Theorem, and since $\{B_{tnm} : t \in T_n\}$ is discrete, there is t with B_{tnm} non-meagre. But $B_{tnm} \subseteq f^{-1}(A_{tn}) \subseteq f^{-1}(Bd) = Ta_d$ for some $d \in D$, as $\mathcal{A}$ refines $\{Bd : d \in D\}$. Thus Ta_d and so T is non-meagre, as the right shift $\rho_{a_d}(x) = xa_d$ is a homeomorphism. But T has the Baire property and X is analytic, so T contains a non-meagre analytic subset. By the Squared Pettis Theorem, 6.3.2, $(TT^{-1})^2$ contains a ball $B_\delta(e_X)$. Then

$$B_\delta(e_X) \subseteq f^{-1}[(B_{\varepsilon/4})^4] = f^{-1}[B_\varepsilon(e_Y)],$$

proving continuity at e_X. $\qquad\qquad\qquad\qquad\qquad\qquad\qquad$ □

Remark In the separable case, by demanding that the graph Γ of a homomorphism be Souslin-$\mathcal{F}(X \times Y)$ one achieves the Baire property of sets $f^{-1}(U)$, for U open in Y, by projection parallel to the Y-axis of $\Gamma \cap (X \times U)$, provided that Y is a $\mathcal{K}$-analytic space. For Y absolutely analytic, one has an extended Souslin representation, and hence a representation of Y as an upper semi-continuous image of some product space $\kappa^{\mathbb{N}}$. But the proof of the projection theorem in RogJ1980, Ths. 2.6.5 and 2.6.6, now yields that the projection of a Souslin-$\mathcal{F}(X \times Y)$ set has only a Souslin-$\mathcal{F}(X)$ representation relative to $\kappa^{\mathbb{N}}$, without guaranteeing the σ-discreteness requirement. In the non-separable context the Baire property can be generated by a projection theorem, provided one has both that the graph is absolutely analytic and that the relevant projection, namely $(x, f(x)) \to x$, is base-σ-discrete (cf. Hanse1992, Th. 4.6; Hanse1974, §6; Hanse1971, §3.5).

These general results have their roots in much simpler contexts. See, for instance, BinO2009e; BinO2015.

13.3.3 Semi-Completeness Theorem: From Normed to Topological Groups

In this section the generalized Levi result in Theorem 13.3.2 is the key ingredient; we will use it and results of earlier sections to prove the following. Here d_S is as in §12.4.

Theorem 13.3.9 (Semi-Completeness Theorem; Ost2013d) *For a normed group X, if (X, d_R) is semi-complete and a Baire space, and the continuous embedding map $j: (X, d_S) \to (X, d_R)$ is Baire base-σ-discrete (e.g. index-σ-discrete), in particular if X is separable, then the right and left uniformities of d_R and d_L coincide and so (X, d_R) is a topologically complete topological group.*

Remarks   1. Theorem 13.3.9 generalizes a classical result for *abelian* locally compact groups due to Ellis (Elli1953).

2. Remarks in §13.1.2 noted that the 'index-σ-discreteness condition' imposed in the theorem is a natural one from the perspective of the non-separable theory of analytic sets, and Lemma 13.3.4 interprets this in terms of inversion, cf. Remark 4.

3. For separable spaces, where discrete families are countable and so the embedding j above automatically preserves σ-discreteness, the result here was proved in Ost2013d (to which we refer for the literature) in the form that a semi-Polish, normed group X, Baire in the right norm-topology, is a topologically complete topological group. Rephrased in the language of uniformities generated by the norm (Kel1955, Ch. 6, Pb. O), this says that a normed group, Polish in the ambidextrous uniformity and Baire in either of the right or left uniformities, has coincident right and left uniformities, and so is a topological group. Key to its proof is that a continuous image of a complete separable metric space is a classically analytic space. So the 'index-σ-discreteness' condition is exactly the condition that secures preservation of analyticity. In the non-separable context continuity is not enough to preserve analyticity, and an additional property is needed, involving σ-d as above: see Hanse1999, Example 4.2 for a non-analytic metric space that is a one-to-one continuous image of $\kappa^{\mathbb{N}}$ for some uncountable cardinal κ (so a continuous image of a countable product of discrete, hence absolutely analytic, spaces of cardinal κ).

4. More recent work by Holický and Pol (HoliP2010), in response to Ostrovsky's insights and based on Mich1986, especially §6 (which itself goes back to GhoM1985), connects preservation of (topological) completeness under continuous maps between metric spaces to the classical notion of *resolvable* sets. (The latter notion provides the natural generalization to Ostrovsky's setting;

see also Holi2010 for non-metrizable spaces.) They find that a map f preserves completeness if it 'resolves countable discrete sets', i.e. for every countable metrically discrete set C and open neighbourhood V of C there is L with $C \subseteq L \subseteq V$ such that $f(L)$ is a *resolvable* set.

Consider the implications for a group X with right-invariant metric d_R (see above), when, for f, one takes j, the identity embedding $j: (X, d_S) \to (X, d_R)$ and when $C = \{c_n : n \in \mathbb{N}\}$ is a d_S-discrete set (so that C and C^{-1} are d_R-discrete). To obtain the desired resolvability for j, it is necessary and sufficient, for each C as above and each assignment $r: \mathbb{N} \to \mathbb{R}_+$ with $r_n \to 0$, that there exist d_R-resolvable sets $L_n \subseteq B^R_{r_n}(c_n) \cap B^L_{r_n}(c_n)$. Since $B^L_r(c) = \{x : d_R(c^{-1}, x^{-1}) < r\} = \{x : d_R(c^{-1}, y) < r$ and $y = x^{-1}\} = \{y^{-1} : d_R(c^{-1}, y) < r\}$, this is yet another condition relating inversion to the d_R-topology, via the sets $B^R_r(c^{-1})^{-1}$.

We turn now to the proof of Theorem 13.3.9. The proof layout (preparatory lemmas followed by proof) and strategy are the same as in Chapter 12, but, as some of the details differ, it is convenient to repeat the short common part (most of the proof of Lemma 13.3.11).

Lemma 13.3.10 *For a normed group X, if (X, d_S) is topologically complete and the continuous embedding map $j: (X, d_S) \to (X, d_R)$ is Baire base-σ-discrete (e.g. A-σ-d, in particular index-σ-discrete), then there is a dense absolute-$\mathcal{G}_\delta$ subset Y' in (X, d_R) such that the restriction map $j: (Y', d_S) \to (Y', d_R)$ is open, and so all the three topological spaces, (Y', d_S), (Y', d_R) and (Y', d_S), are homeomorphic (and topologically complete).*

Proof Take $\mathfrak{X} = (X, d_S)$ which is metric and analytic and $\mathfrak{Y} = (X, d_R)$ which is paracompact; then $j: \mathfrak{X} \to \mathfrak{Y}$ is continuous, surjective and Baire base-σ-discrete. By Lemma 13.3.5 there is a dense $\mathcal{G}_\delta$-subspace Y' of (X, d_R) such that $j \mid Y'$ is an open mapping from (Y', d_S) to (Y', d_R). Now $j^{-1}(Y')$ is a $\mathcal{G}_\delta$-subspace of (X, d_S), so it follows that (Y', d_S) is topologically complete, so an absolute-$\mathcal{G}_\delta$. So $j \mid Y'$ embeds Y' as a subset of (X, d_S) homeomorphically into Y' as an absolute-$\mathcal{G}_\delta$ subset of (X, d_R). Since $x_n \to_S x$ if and only if $x_n \to_R x$ and $x_n \to_L x$, all three topologies on Y' agree. $\square$

Lemma 13.3.11 *If in the setting of Lemma 13.3.10 the three topologies d_R, d_L, d_S agree on a dense absolutely-$\mathcal{G}_\delta$ set Y of (X, d_R), then for any $\tau \in Y$ the conjugacy $\gamma_\tau(x) := \tau x \tau^{-1}$ is continuous.*

Proof We work in (X, d_R) which is thus analytic (as a base-σ-discrete continuous image – Remark 2 in §13.1.3). Let $\tau \in Y$. We will first show that the conjugacy $x \to \tau^{-1} x \tau$ is continuous in X at e, and then deduce that its inverse,

$x \to \tau x \tau^{-1}$, is continuous. So let $z_n \to e$ be any null sequence in X. Fix $\varepsilon > 0$; then $T := Y \cap B_\varepsilon^L(\tau)$ is analytic and non-meagre, since X is a Baire space (and $Y \cap B_\varepsilon^L(\tau)$ is d_R-open in Y with Y an absolute $\mathcal{G}_\delta$). By the Analytic Shift Theorem, 6.2.3, there are $t \in T$ and t_n in T with t_n converging to t (in d_R) and an infinite $\mathbb{M}_t$ such that $\{t t_m^{-1} z_m t_m : m \in \mathbb{M}_t\} \subseteq T$. Since the three topologies agree on Y and the subsequence $t t_m^{-1} z_m t_m$ of points of Y converges to t in Y under d_R (see the Remark after Theorem 6.2.3), the same is true under d_L. Using the identity $d_L(t t_m^{-1} z_m t_m, t) = d_L(t_m^{-1} z_m t_m, e) = d_L(z_m t_m, t_m)$, we note that

$$\|t^{-1} z_m t\| = d_L(t, z_m t) \le d_L(t, t_m) + d_L(t_m, z_m t_m) + d_L(z_m t_m, z_m t)$$
$$\le d_L(t, t_m) + d_L(t t_m^{-1} z_m t_m, t) + d_L(t_m, t) \to 0,$$

as $m \to \infty$ through $\mathbb{M}_t$. So $d_L(t, z_m t) < \varepsilon$ for large enough $m \in \mathbb{M}_t$. Then for such m, as $d_L(\tau, t) < \varepsilon$,

$$\|\tau^{-1} z_m \tau\| = d_L(z_m \tau, \tau) \le d_L(z_m \tau, z_m t) + d_L(z_m t, t) + d_L(t, \tau)$$
$$\le d_L(\tau, t) + d_L(t, z_m t) + d_L(t, \tau) \le 3\varepsilon.$$

Thus there are arbitrarily large m with $\|\tau^{-1} z_m \tau\| \le 3\varepsilon$. Inductively, taking successively $\varepsilon = 1/n$ and $k(n) > k(n-1)$ to be such that $\|\tau^{-1} z_{k(n)} \tau\| \le 3/n$, one has $\|\tau^{-1} z_{k(n)} \tau\| \to 0$. By the weak continuity criterion Lemma 6.1.10 (cf. BinO2010g, Lemma 3.5, p. 37), $\gamma(x) := \tau^{-1} x \tau$ is continuous. Hence, by Lemma 13.3.4 and Theorem 13.3.7, $\gamma^{-1}(x)$ is also continuous. $\qquad \square$

Proof of the Semi-Completeness Theorem, 13.3.9 Under d_R, the set $Z_\Gamma := \{x : \gamma_x \text{ is continuous}\}$ is a *closed* subsemi-group of X (BinO2010g, Prop. 3.43, but using the Open Homomorphism Theorem, 13.3.6, in place of the Souslin-Graph Theorem, 12.1.7). So, as Y is dense, $X = \text{cl}_R Y \subseteq Z_\Gamma$, i.e. γ_x is continuous for all x, and so (X, d_R) is a topological group, by the Equivalence Theorem, 6.1.9. So $x_n \to_R x$ if and only if $x_n^{-1} \to_R x^{-1}$ if and only if $x_n \to_L x$ if and only if $x_n \to_S x$. So (X, d_R) is homeomorphic to (X, d_S). Hence the topological group (X, d_R) is topologically complete, being homeomorphic to (X, d_S). $\qquad \square$

14

Contrasts between Category and Measure

14.1 Classical Results

We have seen above, and can also see in Oxtoby's book (Oxt1980), that there are extensive analogies and similarities between measure and category. For instance, one has (Oxt1980, Ch. 19) the *Sierpiński–Erdős duality principle*: that under the Continuum Hypothesis (CH) any valid statement involving only null sets (or their complements) may be translated into its analogue involving only meagre sets (or their complements), and conversely. Similarly (Oxt1980, Ch. 17), in ergodic theory, duality extends to some but not all forms of the Poincaré recurrence theorem, and, in probability theory (Oxt1980, Ch. 21), duality extends as far as the zero–one law (RaoR1974) but not to the strong law of large numbers. One may loosely summarize this as expressing a duality between *category theory and qualitative measure theory*.

A number of distinguished authors have commented on how later developments forced a reassessment. Some quotes, in chronological order:

Bartoszyński and Judah in 1995 The Preface to their book *Set theory: On the structure of the real line* (BartoJ1995) begins 'This book reflects the current progress in an important segment of descriptive set theory. Its main focus is measure and category in set theory, most notably asymmetry results.'

Bagaria and Woodin in 1997 (à propos of 1984 work by Shelah) 'Thus, an interesting asymmetry between measure and category was uncovered. Up to these results, measure and category were regarded as symmetrical properties . . .' (BagW1997, 1380).

Fremlin in 2008 on Cichoń's diagram (Fre2008, 522 Notes and comments) (Referring forward to §14.2.) 'For many years it appeared that "measure" and "category" on the real line, or at least the structures $(\mathbb{R}, \mathcal{B}, \mathcal{N})$

and $(\mathbb{R}, \mathcal{B}, \mathcal{M})$ where $\mathcal{B}$ is the Borel σ-algebra of $\mathbb{R}$, were in a symmetric duality. It was perfectly well understood that the algebras $\mathcal{A} = \mathcal{B}/\mathcal{B} \cap \mathcal{N}$ and $C = \mathcal{B}/\mathcal{B} \cap \mathcal{M} \dots$ are very different, but $\dots$ encouraged us to suppose that anything provable in ZFC relating measure to category ought to respect the symmetry. It therefore came as a surprise to most of us when Bartoszyński and Raisonnier & Stern showed that $\mathrm{add}(\mathcal{N}) \leq \mathrm{add}(\mathcal{M})$ in all models of set theory.'

Of course, probability theory abounds in (*quantitative*) statements that have no category (*qualitative*) analogue. In the classical limit theorems, for example, one has the law of large numbers mentioned above, which involves the mean μ, the central limit theorem, which involves the mean and the variance σ^2, and similarly for the law of the iterated logarithm, etc. This illustrates a sense in which measure (and probability) theory is *stronger* than category theory, breaking the symmetry between the two that has been such a striking feature of the results above.

We shall deal with *random series*, in particular random trigonometric series, areas much influenced by Jean-Pierre Kahane (1926–2017) (see also Kahane's survey Kah1997). Kahane was very struck by Körner's description (Korn1995) of the Baire category theorem as 'a profound triviality', and by Kaufman's use of Baire category methods in Fourier analysis (Kau1967; Kau1995). Of Baire's theorem, he says 'Its proof is trivial, and its use by Kaufman and Körner is profound' (Kah2001, §10 p. 70).

We turn now to classical instances of this asymmetry, in situations where both category and measure theories can be used, but say different (sometimes very different) things.

Liouville Numbers   A striking classical example is that of the *Liouville numbers* (Lio1851; for textbook accounts, see Bug2012; Per1960, §46; Oxt1980, Ch. 2): irrationals for which for each positive integer n there are integers $q > 1$ and p with

$$|z - p/q| < 1/q^n.$$

Every Liouville number is transcendental. The set of Liouville numbers is co-meagre (large in category) but Lebesgue-null (small in measure). Indeed, it has s-dimensional Hausdorff measure zero for every $s > 0$.

Banach–Tarski ('Paradoxical') Decompositions   The results here are particularly dramatic. The Banach–Tarski decomposition (BanT1924) is justly famous. We refer to Tomkowicz and Wagon (TomW2016) for a textbook exposition. It concerns the (apparently 'paradoxical') possibility of decomposing a Lebesgue-measurable set (a Euclidean ball, say) into finitely many pieces, and reassembling these pieces into a measurable set of (say) double the measure. Of course, the constituent pieces are necessarily non-measurable, or additivity of measure would be violated. But it was shown by Dougherty and Foreman (DouF1994) that *they may have the Baire property.*

Restriction and Continuity The condition for a function to have the Baire property is that it be continuous off some meagre set (see, e.g., Kur1966, §28, II; Ch. 12). The corresponding result for measurability is *Lusin's Theorem*: that the function be continuous off some sets of *arbitrarily small positive measure* (Oxt1980, Th. 8.2). This condition cannot be relaxed to continuity off some null set, as Oxtoby shows by example just after the proof of Lusin's Theorem.

Random Series We follow Kahane's book (Kah1985) and papers (Kah1997; Kah2000; Kah2001).

Consider an infinite sequence of independent *Rademacher* random variables $\{\epsilon_n\}$ – tosses of a fair coin, taking values ± 1 with probability $\frac{1}{2}$ each. For a sequence $\sum_0^\infty c_n$ of reals c_n, we may ask for the probability that the corresponding Rademacher series $\sum \epsilon_n c_n$ – which we write more suggestively as $\sum \pm c_n$ – converges. By the zero–one law (see, e.g., Kah1985, p. 7), convergence is a *tail event* (invariant under deletion of finitely many terms), and so has probability 0 or 1: convergence is almost sure (a.s.), or divergence is a.s. (This is true generally; specializing to the Rademacher case reduces the measure theory from quantitative to qualitative, where category and measure can be 'fairly compared'.) As above, duality extends as far as the zero–one law; the series either converges quasi-surely (q.s.) or diverges q.s. But in stark contrast, the conditions on $c = (c_n)$ are far apart:

$$\sum \pm c_n \text{ converges a.s. [diverges a.s.]} \Leftrightarrow c \in [\notin]\ell_2 : \sum c_n^2 < \infty \ [= \infty].$$

But

$$\sum \pm c_n \text{ converges q.s. [diverges q.s.]} \Leftrightarrow c \in [\notin]\ell_1 : \sum |c_n| < \infty \ [= \infty].$$

For the first, the sufficiency of $c \in \ell_2$ for a.s. convergence is due to Rademacher (Rad1922), the necessity to Khintchin and Kolmogorov (KhiK1925). The second, which is simpler, is due to Kahane (Kah2001, §2). To quote Kahane: divergence is quasi-sure as soon as it may happen.

Normal Numbers In the decimal expansion of a real number $x \in [0, 1]$,

$$x = 0.x_1 x_2 \cdots x_n \cdots = \sum_1^\infty x_n/10^n,$$

x is said to be *normal* (to base 10) if all 10 digits occur with equal asymptotic frequency, $1/10$. We probabilize $[0, 1]$ by imposing Lebesgue measure on it.

By the law of large numbers, x is then normal a.s. Similarly for the expansion to base $d = 2, 3, \ldots$. Intersecting over all d, the real number x is normal to all bases simultaneously, a.s. (since for the measures of the exceptional sets, $\sum_1^\infty 0 = 0$). This deals with single digits, but the argument extends to pairs of digits (including one shift), triples (including two shifts), etc. Thus all k-tuples occur with the same asymptotic frequency d^{-k}, even allowing for shifts. We summarize this by saying that x is *strongly normal a.s.* This is *Borel's normal number theorem* (Bore1909). Apart from its intrinsic importance, the result is of historic interest as the first example of a strong (measure-theoretic) law of large numbers in probability theory and the beginning of the metrical theory of numbers (for which see, e.g., Harm1988).

The category analogue of Borel's Theorem fails in a strong sense (or otherwise put, is completely different). It was shown by Šalát (Sal1966) that the set of normal numbers is *meagre*. One can go much further: Olsen (Ols2004) defines and studies 'extremely non-normal numbers', and shows that such behaviour is generic from a category viewpoint: the set of such numbers is residual – while being Lebesgue-null. Hyde et al. (HydLO2010) show that this stark contrast even persists despite the smoothing effect of repeated Cesàro averaging.

Aside on Normal Numbers   Borel's Theorem shows that normal numbers are *generic*, at least in measure (though not, by above, in category). So normal numbers *exist*, in great profusion. So it seems odd – it is certainly very striking – that *not a single example is known explicitly* – though some partial results are known. For instance, Champernowne's number (Cha1933)

$$0.123456789101112 \cdots 99100101 \cdots$$

is normal to all bases d that are powers of 10. If we were to seek for a specific example to test, the prime candidate would surely be π. Its decimal expansion is known to trillions of places (62.8 currently), and passes all tests of equidistribution with flying colours. This of course gives no information whatsoever on the question of its normality. But decimal expansions and base 10, however useful in ordinary life, have no claim to be natural; nor even do binary expansions and base 2, however useful in computer science. There is only one natural way to expand π, as a *continued fraction*. This was done by Brouncker in 1655; see Khrushchev (Khr2008, §2):

$$\frac{4}{\pi} = 1 + \frac{1^2}{2+} \frac{3^2}{2+} \frac{5^2}{2+} \cdots = 1 + \mathbf{K}_{n=1}^\infty \left(\frac{(2n-1)^2}{2} \right),$$

to use both the usual notations for a continued fraction ('**K** for Kettenbruch'). (Brouncker's continued fraction for $4/\pi$ is closely related to Wallis's product for π, in his *Arithmetica Infinitorum* of 1656.) The next candidate might be e, but here again, the only natural way to expand is by a continued fraction, due to Euler in 1737 (and Cotes, in 1714; see Brez1991, 4.1):

$$e = 2 + \frac{1}{1+} \frac{1}{2+} \frac{1}{1+} \frac{1}{1+} \frac{1}{4+} \frac{1}{1+} \frac{1}{1+} \frac{1}{6+} \cdots .$$

Topological and Hausdorff Dimension   For topological dimension, we refer to Hurewicz and Wallman (HurW1941) and Edgar (Edg1990, Ch. 3). Here a theory is developed which gives a natural extension to that of Euclidean dimension. Here dimension takes values in $\mathbb{N} \cup \{-1, 0, \infty\}$; one needs to distinguish between small and large inductive dimension; 0-dimensional spaces (having a topological basis of clopen (closed and open) sets) play a major role.

For the theory of *Hausdorff dimension*, which is much more detailed and quantitative, we refer to Rogers (Rog1970) and Falconer (Fal1985). Here the dimension takes non-negative *real* values; for fixed dimension, a fine-detailed comparison is possible between sets according to the size of their *Hausdorff measure functions*. This gives a valuable and quantitative way to measure and compare the size of small sets: sets such as the sample paths, zero-sets, self-intersections, etc., of stochastic processes, such as Brownian motion (in various dimensions), Lévy processes, etc. It suffices for us here to refer to the extensive work by John Hawkes (1944–2001); for details and references, see his obituary by S.J. Taylor (Tay2004).

For results in the context of normal numbers, see GolSW2000; AlbGI2017; Sal1966.

Random Dirichlet Series   We use the standard notation for Dirichlet series, $\sum_1^\infty a_n n^{-s}$, where $s = \sigma + it \in \mathbb{C}$. Convergence is in a half-plane $\sigma > \sigma_c$, to a function $A(s)$ analytic there. Absolute convergence is in a half-plane $\sigma > \sigma_a$, where $\sigma_a \geq \sigma_c$. One has $A(\sigma + it) = O(|t|)$ for $|t| \to \infty$. The *Lindelöf* (or *order*) *function* is

$$\mu(\sigma) := \inf\{\alpha : A(\sigma + it) = O(|t|^\alpha)\} \quad (|t| \to \infty).$$

Then

$$\mu(\sigma_a) = 0, \qquad \mu(\sigma_c) \leq 1.$$

The *Bohr abscissa* σ_b is the infimum of σ with $\mu(\sigma)$ finite; then

$$\sigma_b \leq \sigma_c \leq \sigma_a.$$

We turn next to how cancellation effects in Dirichlet series may enable analytic continuation. We illustrate with a classic example, the *alternating zeta function* (or *Dirichlet eta function*). Write the Riemann zeta function

$$\zeta(s) := \sum_1^\infty 1/n^s = \sum_{n \text{ odd}} + \sum_{n \text{ even}} = \sum_o + \sum_e,$$

say, and similarly for the alternating zeta function

$$\sum_1^\infty (-)^{n-1}/n^s = \sum_o - \sum_e.$$

Subtract:

$$\zeta(s) - \sum_1^\infty (-)^{n-1}/n^s = 2 \sum_e = 2 \sum_1^\infty 1/(2n)^s = 2^{1-s}\zeta(s):$$

$$\zeta(s) = \frac{\sum_1^\infty (-)^{n-1}/n^s}{1 - 2^{1-s}};$$

Hardy (Har1922) and Titchmarsh (Tit1986, (2.2.1), p. 512). This gives (that ζ has a simple pole at 1 of residue 1 and) the analytic continuation of ζ from $\mathrm{Re}(s) > 1$ to $\mathrm{Re}(s) > 0$.

Kahane (Kah2000, §5) gives two examples with random rather than alternating signs. For 'the Riemann zeta function with random signs', $\sum_1^\infty \pm n^{-s}$, one has

$$\sigma_a = 1, \quad \mu(1) = 0; \quad \sigma_c = \frac{1}{2}, \quad \mu\left(\frac{1}{2}\right) = 0 \text{ a.s.}, \quad but \quad \sigma_c = \sigma_a = 1 \text{ q.s.}$$

For

$$\sum_1^\infty \pm ((2n-1)^{-s} - (2n)^{-s}),$$

one has again $\sigma_a = 1$, $\mu(1) = 0$,

$$\sigma_b = -\frac{1}{2}, \quad \sigma_c = 0, \quad \mu(\sigma) = \left(\frac{1}{2} - \sigma\right)_+ \text{ on } \left[-\frac{1}{2}, \infty\right] \text{ a.s.},$$

but

$$\sigma_b = \sigma_c = 0, \quad \mu(\sigma) = (1 - \sigma)_+ \quad \text{on } [0, \infty] \text{ q.s.}$$

Random Taylor Series; Random Fourier Series; Nowhere Differentiable Functions These matters are studied at length by Kahane (Kah1985; Kah1997). Kahane's interest in category aspects is developed in Kah2000; Kah2001. We refer there for a wealth of results, comparing and contrasting the category and measure cases. See also KahQ1997; Que1980; Korn1995.

For random Taylor series, results include those on the sense in which the circle of convergence of the series being a *natural boundary* – i.e. analytic continuation across any part of it is impossible – is generic.

Erdős–Sierpiński maps   We refer to Oxt1980, Ch. 19 for background, under the Continuum Hypothesis (CH), on the Erdős–Sierpiński Duality Theorem. This duality may be implemented by applying an *Erdős–Sierpiński map*. Bartoszyński (Barto2000) notes an asymmetry between category and measure here, implying in particular that there do not exist additive Erdős–Sierpiński maps.

Filters   Bartoszyński et al. (BartoGJ1993) show that it is consistent with ZFC that all filters with the Baire property are Lebesgue-measurable, but provable in ZFC that there is a Lebesgue-measurable filter without the Baire property.

Genericity (as typicality)   The category property 'quasi everywhere' and its measure analogue 'almost everywhere' provide different (and as above sometimes contrasting) ways in which behaviour or a property can be *generic*. Aspects of genericity have been developed in a number of papers by Grosse-Erdmann; see, e.g., Gros1987; Gros1999.

The Fubini and Kuratowski–Ulam Theorems   Fubini's Theorem may be regarded as having a qualitative variant, the Fubini Null Theorem (§9.5, Douw1989, BinO2016b) stated in the vocabulary of null sets and asserting that a measurable planar set with the Baire property is null if and only almost all its vertical segments are null. This has a category analogue in the Kuratowski–Ulam Theorem (§9.5, Oxt1980, Ch. 15). The latter asserts that a Baire planar set is meagre if and only if all but a meagre set of verticals have meagre vertical sections. So much for similarity. However, despite both being qualitative statements, there is a hidden contrast: the category result is known to fail beyond the separable context (as shown in Pol1979, cf. MilP1986, but see FreNR2000), though not so the Fubini Theorem (see BinO2020c, Th. FN and Chapter 13).

Lipschitz Functions and Rademacher's Theorem   Rademacher's Theorem of 1919 states that every Lipschitz function $f \colon \mathbb{R}^m \to \mathbb{R}^n$ is a.e. differentiable (Rad1919; Fed1969; AlbM2016). The question arises as to a category analogue. Loewen and Wang (LoeW2000, Th. 4) construct a set in a suitably metrized space of Lipschitz functions which is co-meagre, but each member of which is differentiable on an at most meagre set. Thus differentiability of Lipschitz functions is generic in (Lebesgue) measure, yet non-differentiability can be generic in category.

Sets of Uniqueness for Trigonometric Series   While sets of uniqueness for trigonometric series provide an example of category–measure duality (§9.1), they also provide a contrast. The category case (KecL1987, Problem C, p. 4) is much harder to prove (KecL1987, Ch. VIII).

Natural Boundaries of Random Series   Steinhaus in his classic paper (Stei1930) showed that if a power series with radius of convergence $R \in (0, \infty)$ is made random by inserting independent random factors each uniformly distributed on [0, 1] into the coefficients, then the resulting random series has the circle of convergence as its natural boundary (no analytic continuation is possible across it – see, e.g., Simo2015), almost surely. The dual Baire statement, with a meagre rather than a null exceptional set, is also true but again was shown much later, by Breuer and Simon (BreuS2011; Simo2015, p. 241, Problem 9).

For background on such random Taylor series, see, e.g., Kah1985, Ch. 4 (and for more on the Gaussian case, Kah1985, Ch. 13).

We close this section with two insightful comments by Kahane on category–measure contrasts:

Kah2000, p. 169: 'If we measure the thinness of a set through the behaviour of positive measures carried by the set and their Fourier coefficients, the tendency of Baire's methods is to provide sets as thin as possible, and the tendency of probability methods is to provide measures whose Fourier coefficients tend to zero as fast as possible.'

Kah2001, p. 61: 'We see from these examples that the Baire approach emphasizes divergence, singularities and large values, while the probability approach favors convergence, smoothing and regularizing effects.'

14.2 Modern Results: Forcing

We turn now from these 'classical' matters, in analysis and probability – which, though of ongoing interest, have their roots in the work of Borel, Lebesgue, Baire or earlier – to 'modern' aspects, stemming from set theory and logic, with particular reference to forcing, following Cohen's breakthrough in 1964 and 1966. We will not attempt a detailed account, for which we refer the interested reader to our long survey (BinO2019a) and its extensive bibliography.

We recall the view of Judah (formerly Ihoda) and Spinas (JudSp1997):

'Finally Cohen's method of forcing led to a much deeper understanding and shed light on a deep asymmetry between measure and category.'

We begin with Solovay's paper (Solo1970). Solovay assumed that the existence of an inaccessible cardinal (see §16.3) is consistent with ZF. He showed that then ZF is consistent with the conjunction of (1)–(5):

(1) DC, the Axiom of Dependent Choice;
(2) LM: all sets of reals are Lebesgue-measurable;
(3) PB: all sets of reals have the Baire Property;
(4) Every uncountable set of reals has a perfect subset;
(5) If $A := \{A_x : x \in \mathbb{R}\}$ is a class of non-empty sets of reals, then:

(a) there exists a Borel function $h_1 : \mathbb{R} \to \mathbb{R}$ such that $\{x : h_1(x) \notin A_x\}$ is Lebesgue-null; and

(b) there exists a Borel function $h_2 : \mathbb{R} \to \mathbb{R}$ such that $\{x : h_2(x) \notin A_x\}$ is meagre.

Cohen and forcing changed the situation regarding category and measure with the work of Shelah (Shel1984). See also the companion paper by Raisonnier Rai1984; the two are published contiguously, and were reviewed jointly (MR86g:03082a,b) by F. R. Drake. Shelah's task was to show that the inaccessible cardinal used in Solovay's paper (Solo1970) cannot be removed. As Hrbacek remarks in his review (MR800191) of Raisonnier and Stern (RaiS1985):

'the classical results of descriptive set theory concerning measure and category are completely symmetric. Thus it was a surprise when S. Shelah showed that:

(a) Lebesgue measurability of *all* sets of reals implies consistency of inaccessible cardinals,
    but

(b) Models of ZF where *all* sets of reals have the Baire Property can be constructed under ZF alone.'

Thus Solovay's inaccessible cardinal is *necessary for measure but not for category*.

Write $\Sigma_n^1(\mathcal{L})$ as shorthand for 'every Σ_n^1 set is Lebesgue-measurable', and similarly $\Sigma_n^1(\mathcal{B})$ for 'every Σ_n^1 set has the Baire property'. With this notation, Bartoszyński (Barto1984) showed that

$$\Sigma_2^1(\mathcal{L}) \Rightarrow \Sigma_2^1(\mathcal{B}),$$

while Raisonnier and Stern (RaiS1985) proved that the reverse implication fails. This shows again a striking asymmetry between category and measure, and another sense in which *measure is stronger than category*, as expected: measure theory has a quantitative side while category theory does not. One cannot replace Σ_2^1 by Δ_2^1 here: Kanovei and Lyubetskii (KanoL2003, Th. 5.4).

Ihoda and Shelah (IhoS1989) give a study of Δ_2^1 sets of reals (recall that these are the sets particularly suited for the study of regular variation, as noted in Chapter 3). They show (Th. 3.5) that if ZFC is consistent, then:

(i) there exists a model V in which all Δ_2^1 sets are Lebesgue-measurable, but that not all have the Baire property, and not all are Ramsey;

(ii) similar statements hold with Lebesgue-measurable, Baire property and Ramsey cyclically permuted.

There is a fourth property, 'K_σ-regularity', for which see IhoS1989.

They (now as Judah and Shelah, JudS1993a) follow this with a corresponding study of Δ_3^1 sets, with similar conclusions. This in turn is taken further by Fischer, Friedman and Khomskii (FisFK2014), who assume ZFC or ZFC plus the assumption of an inaccessible cardinal. They give some partial results for Δ_4^1.

The interesting recent monograph of Friedman and Schrittesser (FriS2020) is on measurability without the Baire property in the projective hierarchy. They write in the Abstract:

'We prove that it is consistent (relative to a Mahlo cardinal) that all projective sets of reals are Lebesgue measurable, but there is a Δ_3^1 set without the Baire property. The complexity of the set which provides a counterexample to the Baire property is optimal.'

Similar questions were earlier considered by Bagaria and Woodin (BagW1997). They show that if ϕ is a Σ_4^1 sentence consistent with the measurability of all Σ_2^1 sets, then it is still consistent with the measurability of all Σ_2^1 sets *and* the existence of a Δ_2^1 set without the Baire property. This asymmetry is *maximal* (with respect to Σ_4^1) in the sense that there exists a Π_4^1 sentence which implies in ZFC that all projective sets are measurable and have the Baire property.

14.3 Category Duality Revisited

14.3.1 Practical Axiomatic Alternatives

While ZF is common ground in mathematics, the Axiom of Choice(AC) is not, and alternatives to it are widely used, in which for example all sets are Lebesgue-measurable (usually abbreviated to LM) and all sets have the Baire property, sometimes abbreviated to PB (as distinct from BP to indicate individual 'possession of the Baire property'). One such is DC above. As Solovay (Solo1970, p. 25), points out, this axiom is sufficient for the establishment of Lebesgue measure, i.e. including its translation invariance and countable additivity ('... positive results ... of measure theory...'), and may be assumed together with LM. Another is the *Axiom of Determinacy* (AD) introduced by Mycielski and Steinhaus (MycS1962) (see Klei1977 in relation to DC); this implies LM, for which see MycSw1964, and PB, the latter a result, as mentioned in §2.4, due to Banach – see Kec1995, 38.B. Its introduction inspired remarkable and still current developments in set theory concerned with determinacy of 'definable' sets of reals (see ForK2010 and particularly Neem2010) and consequent combinatorial properties (such as the partition relations) of the alephs (see Klei1977); again see §2.4 (games of Banach–Mazur type). Others include the (weaker) *Axiom of Projective Determinacy* (PD) (Kec1995, §38.B) cf. Chapter 16, restricting the operation of AD to the smaller class of projective sets. (The independence and consistency of DC versus AD was established respectively by Solovay, Solo1978 and Kechris, Kec1995 – see also KecS1985; cf. DalW1987; Ost1989.)

14.3.2 LM versus PB

In 1983 Raissonier and Stern (RaiS1983, Th. 2) (cf. Barto1984; Barto2010), inspired by the then-current work of Shelah (circulating in manuscript since 1980) and earlier work of Solovay, showed that *if every Σ_2^1 set is Lebesgue-measurable, then every Σ_2^1 set has BP*, whereas the converse fails – for the latter see Ster1985 – cf. BartoJ1995, §9.3; Paw1985. This shows again that measurability is in fact the stronger notion – see JudS1993b, §1 for a discussion of the consistency of analogues at level 3 and beyond – which is one reason why we regard category rather than measure as *primary*. For, the category version of Berz's Sublinearity Theorem (Berz1975; BinO2019b) implies its measure version: the *stronger hypothesis* weakens the *theorem/conclusion*; see BinO2017; BinO2018a; BinO2020c.

Note that the assumption of Gödel's *Axiom of Constructibility V = L*, viewed as a strengthening of AC, yields Δ_2^1 non-measurable subsets, so that the Fenstad–Normann result on the narrower class of provably Δ_2^1 sets mentioned in §3.3 marks the limit of such results in a purely ZF framework (at level 2).

14.3.3 Consistency and the Role of Large Cardinals

While LM and PB are inconsistent with AC, such axioms can be consistent with DC. Justification with scant exception involves some form of large-cardinal assumption, which in turn, as in §16.3, calibrates relative consistency strengths – see Kan2003; KoeW2010 (cf. Lar2010; KanM1978). Thus Solovay (Solo1970) in 1970 was the first to show the consistency of ZF+DC+LM+PB with that of ZFC+'*there exists an inaccessible cardinal*'. The appearance of the inaccessible in this result is not altogether incongruous, given its emergence in results (from 1930 onwards) due to Banach (Ban1932) (under GCH), Ulam (Ula1930) (under AC), and Tarski (Tar1930), concerning the cardinalities of sets supporting a countably additive/finitely additive [0, 1]-valued/{0, 1}-valued measure (cf. §12.5, §16.3, Bog2007a, 1.12(x); Fre2008). Later, in 1984, Shelah (Shel1984, 5.1) showed in ZF+DC that already the measurability of all Σ_3^1 sets implies that $\aleph_1^V$ is inaccessible in the sense of L (the symbol $\aleph_1^V$ refers to the first uncountable ordinal of V, Cantor's universe – cf. §16.2). As a consequence, Shelah

(Shel1984, 5.1A) showed that ZF+DC+LM is equiconsistent with ZF+'*there exists an inaccessible*', whereas (Shel1984, 7.17) ZF+DC+PB is equiconsistent with just ZFC (i.e. without reference to inaccessible cardinals), so driving another wedge between classical measure–category symmetries (see JudS1993b for further, related 'wedges'). The latter consistency theorem relies on the result (Shel1984, 7.16) that any model of ZFC+CH has a generic (forcing) extension satisfying ZF+'*every set of reals (first-order) defined using a real and an ordinal parameter has BP*'. (Here 'first-order' restricts the range of any quantifiers.) For a topological proof, see Ster1985.

14.3.4 LM versus PB Continued

Raisonnier (Rai1984, Th. 5) (cf. Shel1984, 5.1B) has shown that in ZF+DC one can prove that if there is an uncountable well-ordered set of reals (in particular a subset of cardinality $\aleph_1$), then there is a non-measurable set of reals. (This motivated Judah and Spinas (JudSp1997) to consider generalizations including the consistency of the ω_1-variant of DC.) See also Judah and Rosłanowski (JudR1993) for a model (due to Shelah) in which ZF+DC+LM+¬PB holds, and also Shel1985 where an inaccessible cardinal is used to show consistency of ZF+LM+¬PB+'*there is an uncountable set without a perfect subset*'. For a textbook treatment of much of this material, see again BartoJ1995.

Raisonnier (Rai1984, Th. 3) notes the result, due to Shelah and Stern, that there is a model for ZF+DC+PB+$\aleph_1 = \aleph_1^L$ + '*the ordinally definable subsets of reals are measurable*'. So, in particular by Raisonnier's result, there is a non-measurable set in this model. Shelah's result indicates that the non-measurable set is either Σ_3^1 (light-face symbol: all open sets coded effectively) or Σ_2^1 (bold-face); see the comments at the end of the introduction in Ster1985. Thus here PB+¬LM holds.

14.3.5 Regularity of Reasonably Definable Sets

From the existence of suitably large cardinals flows a most remarkable result due to Shelah and Woodin (ShelW1990) justifying the opening practical remark about BP, which is that every 'reasonably definable' set of reals is Lebesgue-measurable: compare the commentary in BeckK1996 following their Th. 5.3.2. This is a latter-day sweeping generalization of a theorem due to Solovay (cf. Solo1969) that, subject to large-cardinal assumptions, Σ_2^1 *sets are measurable* (and so also *have BP* by Barto1984 and RaiS1983).

14.3.6 Category and Measure: Qualitative versus Quantitative Aspects

Most of the similarities between category and measure (Oxt1980) can now be seen (BinO2018b; BinO2019b; BinO2020c) to flow from *density topology* aspects. As Oxtoby points out (Oxt1980, p. 85), category–measure duality extends as far as qualitative aspects (0−1 laws) but not as far as quantitative aspects (strong law of large numbers, etc.). The differences here can be dramatic. For example, as in §14.1, the requirement on a series for it to converge almost surely when 'random signs' are given to its terms is that it be ℓ_2 (Kah1985); by contrast for convergence off a meagre set, the corresponding convergence criterion is (minimally!) ℓ_1 (Kah2001). On occasion discrepancies can be engineered into realignment by refining the metric – see CalMS2003.

As pointed out earlier, measurability is the stronger notion. Such distinctions give rise to two streams of literature. In one, pathology (strange counterexamples) is pursued; see, e.g., Cie1997. In the other, comparisons are made between the various *cardinal invariants* associated with the σ-ideals of negligible sets; these ask questions, relative to given axioms of set theory, such as: how small may non-negligibles be (the *non* number), how small a family of negligibles has non-negligible union (the *add*itivity number), how small such a family must be to cover the real line (the *cov*ering number), or how small if it is to be cofinal under inclusion (the *cof*inality number). Two further key ingredients are $\mathfrak{b}$, the *bounding* number, and $\mathfrak{d}$, the *dominating* number, corresponding to a smallest unbounded family and a smallest dominating family of functions in ω^ω relative to domination mod-finite. (For the connection between the latter and *maximal almost disjoint* (mad) families of subsets of ω, see, e.g., Douw1984; for the role of mad families in Ramsey properties of ultrafilters, see Mat1979, and for recent developments, Tor2013.) A result on the cardinal invariants, memorable for its symmetries, is summarized in the following *Cichoń diagram*, for which we refer to BartoJ1995, and the somewhat more recent Buk2011.

$$
\begin{array}{ccccccc}
\mathrm{cov}(\mathcal{N}) & \longrightarrow & \mathrm{non}(\mathcal{M}) & \longrightarrow & \mathrm{cof}(\mathcal{M}) & \longrightarrow & \mathrm{cof}(\mathcal{N}) \\
& & \uparrow & & \uparrow & & \\
\uparrow & & \mathfrak{b} & \longrightarrow & \mathfrak{d} & & \uparrow \\
& & \uparrow & & \uparrow & & \\
\mathrm{add}(\mathcal{N}) & \longrightarrow & \mathrm{add}(\mathcal{M}) & \longrightarrow & \mathrm{cov}(\mathcal{M}) & \longrightarrow & \mathrm{non}(\mathcal{N})
\end{array}
$$

Here the arrows $\rightarrow$ indicate $\leq$.

Cardinal invariants can also be studied for the σ-ideal of negligible sets $\mathcal{HN}$, the Haar null sets. See Chapter 16 and Bana2004.

15

Interior-Point Theorems: Steinhaus–Weil Theory

In this long chapter we summarize and develop the work in our 'Steinhaus–Weil quartet' (BinO2020c; BinO2020d; BinO2021a; BinO2022a) (unified in the arXiv: BinO2016b).

15.1 The Steinhaus–Piccard Theorem

The fine topologies, $\mathcal{T}$, we work with are all shift-invariant; that is, every shift is an open mapping (takes $\mathcal{T}$-open sets to $\mathcal{T}$-open sets). Theorem 15.1.1 may in principle be applied to any of them (e.g. the fine topology, $\mathcal{F}$, of potential theory, §8.1). In this section we will apply the result to the Euclidean and density topologies.

Theorem 15.1.1 *Let $\mathbb{R}$ be given a shift-invariant topology $\mathcal{T}$ under which it is a Baire space, and suppose the homeomorphisms $h_n(x) = x + z_n$ satisfy* (cc), *whenever $\{z_n\} \to 0$ is a null sequence (in the Euclidean topology). For S Baire and non-meagre in $\mathcal{T}$, the difference set $S - S$ contains an interval around the origin.*

Proof Suppose not: then for each positive integer n we may select

$$z_n \in (-1/n, +1/n) \setminus (S - S).$$

Since $\{z_n\} \to 0$ (in the Euclidean topology), the Category Embedding Theorem, 10.2.2, applies, and gives an $s \in S$ and an infinite $\mathbb{M}_s$ such that $\{h_m(s) : m \in \mathbb{M}_s\} \subseteq S$. Then for any $m \in \mathbb{M}_s$,

$$s + z_m \in S, \quad \text{i.e. } z_m \in S - S,$$

a contradiction. $\qquad\qquad\square$

235

We deduce two classical theorems: Piccard's Theorem (Pic1939; Pic1942) and Steinhaus' Theorem (Stei1920). We give two proofs of each, both brief.

Theorem 15.1.2 (Piccard's Theorem) *For S Baire and non-meagre in the Euclidean topology, the difference set $S - S$ contains an interval around the origin.*

First Proof Apply Theorem 15.1.1 since, by Theorem 10.3.1, (cc) holds. □

Second Proof Suppose not: then, as before, for each positive integer n we may select $z_n \in (-1/n, +1/n) \setminus (S - S)$. Since $z_n \to 0$, by Theorem 10.3.3, for quasi all $s \in S$ there is an infinite $\mathbb{M}_s$ such that $\{s + z_m : m \in \mathbb{M}_s\} \subseteq S$. Then, for any $m \in \mathbb{M}_s$, $s + z_m \in S$, i.e. $z_m \in S - S$, a contradiction. □

Remark See BinO2010g for a derivation from here of the more general result (due to Pet1950; Pet1951) that for S, T Baire and non-meagre in the Euclidean topology, the difference set $S - T$ contains an interval. See also §15.2.

Theorem 15.1.3 (Steinhaus' Theorem) *For measurable S of positive measure, the difference set $S - S$ contains an interval around the origin.*

First Proof Arguing as in the first proof above, by Theorem 10.3.1 (cc) holds and S, being measurable non-null, is Baire non-meagre under $\mathcal{D}$ (Property (6) of Chapter 7, on page 108). □

Second Proof Arguing as in the second proof above, Theorem 4.2.1 applies. □

Just as with the Pettis extension of Piccard's result, so also here, Steinhaus proved that, for S, T non-null measurable, $S - T$ contains an interval. This too may be derived from the CET (§10.2); see BinO2010b.

Unlike some of the results in earlier chapters, these results extend to topological groups. See, e.g., Com1984, Th. 4.6, p. 1175 for the positive statement, and the closing remarks for a negative one.

15.2 The Common Basis Theorem

In this section we capture a further key similarity (their topological 'common basis', adapting a term from logic) between the Baire and measure cases. Recall (RogJ1980, p. 460) the usage in logic, whereby a set B is a *basis* for a class C of sets whenever any member of C contains an element in B.

Theorem 15.2.1 (Common Basis Theorem; BinO2010b) *For V, W Baire non-meagre in the line $\mathbb{R}$, equipped with either the Euclidean or the density topology, there is $a \in \mathbb{R}$ such that $V \cap (W + a)$ contains a non-empty open set modulo meagre sets common to both, up to translation. In fact, in both cases, up to translation, the two sets share a Euclidean $\mathcal{G}_\delta$ subset which is non-meagre in the Euclidean case and non-null in the density case.*

Proof In the Euclidean case for V, W Baire non-meagre we may suppose that $V = I \backslash M_0 \cup M_0'$ and $W = J \backslash M_1 \cup M_1'$, where I, J are open intervals and M_i and M_i' are meagre. Take $V_0 = I \backslash M_0$ and $W_0 = J \backslash M_1$. For v, w in V_0, W_0, put $a := v - w$. Thus $v \in I \cap (J + a)$. So $I \cap (J + a)$ differs from $V \cap (W + a)$ by a meagre set. Since $M_0 \cup M_0'$ may be expanded to a meagre $\mathcal{F}_\sigma$ set M, we deduce that $I \backslash M$ and $J \backslash M$ are non-meagre $\mathcal{G}_\delta$-sets.

In the density case, for V, W measurable non-null let V_0 and W_0 be the sets of density points of V and W. For v, w in V_0, W_0, put $a := v - w$. Then $v \in T := V_0 \cap (W_0 + a)$, and so T is non-null and v is a density point of T. So for $T_0 := \phi(T)$ (the density points of T), $T \backslash T_0$ is null, so T_0 differs from $V \cap (W + a)$ by a null set. Evidently T_0 contains a non-null closed, hence $\mathcal{G}_\delta$, subset (as T_0 is measurable non-null, by regularity of Lebesgue measure). $\square$

This leads to a strengthening of Theorem KBD, 10.3.3, which concerns two sets rather than one.

Theorem 15.2.2 *For V, W Baire non-meagre/measurable non-null in $\mathbb{R}$, there is $a \in \mathbb{R}$ such that $V \cap (W + a)$ is Baire non-meagre/measurable non-null and for any null sequence $z_n \to 0$ and quasi all (almost all) $t \in V \cap (W + a)$ there exists an infinite $\mathbb{M}_t$ such that*

$$\{t + z_m : m \in \mathbb{M}_t\} \subset V \cap (W + a).$$

Proof In either case applying the Common Basis Theorem, 15.2.1, for some a, the set $T := V \cap (W + a)$ is Baire non-meagre/measurable non-null. We may now apply Theorem KBD, 10.3.3, to the set T. Thus for almost all $t \in T$ there is an infinite $\mathbb{M}_t$ such that

$$\{t + z_m : m \in \mathbb{M}_t\} \subset T \subset V \cap (W + a). \qquad \square$$

This result concerning mutual inclusion of a non-negligible subset motivates a further strengthening.

Definitions Let S be shift-compact.

1. Call T *commutual* with S if for every null sequence $z_n \to 0$ there is $u \in S \cap T$ and infinite $\mathbb{M}_u$ such that

$$\{u + z_m : m \in \mathbb{M}_u\} \subset S \cap T.$$

Thus S is commutual with T and both are shift-compact.

Call T *weakly commutual* with S if for every null sequence $z_n \to 0$ there is $s \in S$ and infinite $\mathbb{M}_s$ such that

$$\{s + z_m : m \in \mathbb{M}_s\} \subset T.$$

Thus T is shift precompact.

2. Call S *self-commutual* (up to reflected translation) if for some $a \in \mathbb{R}$ and some $T \subset S$, S is commutual with $a - T$.

3. Call S *weakly self-commutual* (up to reflected translation) if for some $a \in \mathbb{R}$ and some $T \subset S$, S is weakly commutual with $a - T$.

As an immediate corollary of Theorems 15.2.1 or 15.2.2, taking $V = S$, $W = -S$, we may now formulate the following.

Theorem 15.2.3 (Self-Commutuality Theorem)   *Suppose that S is Baire non-meagre/measurable non-null. Then S is self-commutual.*

Proof   Fix a null sequence $z_n \to 0$. If S is Baire non-meagre/measurable non-null, then so is $-S$; thus we have for some a that $T := S \cap (a - S)$ is likewise Baire non-meagre/measurable non-null and so for quasi all (almost all) $t \in T$ there is an infinite $\mathbb{M}_t$ such that

$$\{t + z_m : m \in \mathbb{M}_t\} \subset T \subset S \cap (a - S),$$

as required.   □

Commutuality is the additional feature needed to establish the Semi-Group Theorem.

Theorem 15.2.4 (General Semi-Group Theorem)   *If S, T are shift-compact in $\mathbb{R}$ with T (weakly) commutual with S, then $S - T$ contains an interval about the origin. Hence if S is shift-compact and (weakly) self-commutual, then $S + S$ contains an interval. Hence, if additionally S is a semi-group, then S contains an infinite half-line.*

Proof   For S, T (weakly) commutual, we claim that $S - T$ contains $(0, \delta)$ for some $\delta > 0$. Suppose not: then for each positive n there is z_n with

$$z_n \in (0, 1/n) \backslash (S - T).$$

Now $-z_n$ is null, so there is s in S and infinite $\mathbb{M}_s$ such that

$$\{s - z_m : m \in \mathbb{M}_t\} \subset T.$$

For any m in $\mathbb{M}_t$ pick $t_m \in T$ so that $s - z_m = t_m$; then we have

$$s - z_m = t_m \quad \text{so} \quad z_m = s - t_m,$$

contradicting $z_m \notin S - T$. Thus, for some $\delta > 0$, we have $(0, \delta) \subset S - T$.

For S self-commutual, say S is commutual with $T := a - S$, for some a; then $a + (0, \delta) \subset a + (S - T) = a + S - (a - S) = S + S$, i.e. $S + S$ contains an interval. $\qquad\square$

By the Common Basis Theorem, replacing T by $-T$, we obtain as an immediate corollary of Theorems 15.2.3 and 15.2.4 a new proof of three classical results: an extension of the Steinhaus and of the Piccard Theorems (cf. Theorem 15.1.2), and the Classical Semi-Group Theorem, 15.2.6.

The following result is due to Steinhaus (Stei1920) in the measure case, cf. Bec1960, and to Pettis (Pet1950; Pet1951) in the Baire case, cf. Kom1971.

Theorem 15.2.5 (Sum–Set Theorem) *If S, T are Baire non-meagre/measurable non-null in $\mathbb{R}$, then $S + T$ contains an interval.*

The following classical result is due to Hille and Phillips (HillP1957, Th. 7.3.2) (cf. BecCS1958, Th. 2; Bec1960) in the measurable case, and to Bingham and Goldie (BinGo1982a; BinGo1982b) in the Baire case; see BGT1987, Cor. 1.1.5. For a combinatorial form, see BinO2011a.

Theorem 15.2.6 (Classical Semi-Group Theorem) *For an additive Baire (measurable) semi-group S of $\mathbb{R}$, the following are equivalent:*

(i) $S \supseteq (s, \infty)$ *or* $S \supseteq (-\infty, -s)$, *for some s,*
(ii) S *is non-meagre (non-null).*

15.3 Measure Subcontinuity and Interior Points

We begin by stating the Steinhaus–Weil Theorem in its simplest form (Stei1920; for the line, see Wei1940, §11, p. 50; for a Polish locally compact group, see Gros1989).

Theorem 15.3.1 (Steinhaus–Weil) *In a locally compact Polish group G with (left) Haar measure η_G, for non-null Borel B, $B^{-1}B$ (and likewise BB^{-1}) contains a neighbourhood of the identity.*

The context we work with in this and the next sections, unless otherwise stated, is that groups and spaces are to be assumed separable. This both simplifies the exposition and emphasizes that we need only the Axiom of Dependent

Choices (DC – 'what is needed to make induction work' – see Chapter 16), rather than the Axiom of Choice (AC) (again see Chapter 16); cf. BinO2019a. For comments concerning non-separable settings, see the unified arXiv version (BinO2016b, §8.1) of our four-part series on Theorem 15.3.1.

In this general context, the interior-point property of the measure-theoretically 'non-negligible' set B of the theorem is referred to as the *Steinhaus–Weil property*, which encompasses the category variant due to Piccard (Pic1939; Pic1942) and Pettis (Pet1950), cf. BinO2021a, Cor. 2′ and Th. 1B (by reference, when appropriate, to the *quasi-interior* of a set – the largest open set equivalent to it modulo a meagre set). This important result has many ramifications; for example, it is basic to the theory of regular variation – see, e.g., BGT1987, Th. 1.1.1.

The results below hinge on work of Solecki (Sol2006) on amenability at 1 and on an amendment of Fuller's concept of subcontinuity (see below). These are aimed at freeing up the classical dependency on local compactness and the corresponding standard (Haar) reference measure. To the best of our knowledge such aims for topological groups were last undertaken by Xia (Xia1972, Ch. 3), where the emphasis is on (relative) quasi-invariance; cf. BinO2016b, §7.2, a topic we pursued in the related paper, BinO2019c (cf. Bog1998, p. 64) with tools developed here. For background on invariance of measures, see Kha1998. For background on topological vector spaces and their negligible sets, see Bog2018 or BogS2017.

For G a topological group with (admissible) metric d (briefly: metric group), denote by $\mathcal{M}(G)$ the family of regular σ-finite Borel measures on G, with $\mathcal{P}(G) \subseteq \mathcal{M}(G)$ the probability measures (Kec1995, §17E; Par1967), by $\mathcal{P}_{\text{fin}}(G)$ the larger family of finitely additive regular probability measures (cf. Bin2010; Myc1979), and by $\mathcal{M}_{\text{sub}}(G)$ submeasures (monotone, finitely subadditive set functions μ with $\mu(\emptyset) = 0$). Here *regular* is taken to imply both *inner* regularity (inner approximation by compact subsets, also called the *Radon* property, as in Bog2007b, §7.1 and Schw1973), and *outer* regularity (outer approximation by open sets). We recall that a σ-finite Borel measure on a metric space is necessarily outer regular (Bog2007b, Th. 7.1.7; Kall2002, Lemma 1.34; cf. Par1967, Th. II.1.2 albeit for a probability measure) and, when the metric space is complete, inner regular (Bog2007b, Th. 7.1.7; cf. Par1967, Ths. II.3.1 and 3.2). When G is locally compact we denote Haar measure by η_G or just η. For X metric, we denote by $\mathcal{K} = \mathcal{K}(X)$ the family of compact subsets of X (the *hyperspace* of X, where we view it as a topological space under the Hausdorff metric, or the Vietoris topology). For $\mu \in \mathcal{M}(G)$ we write $_g\mu(\cdot) := \mu(g\cdot)$ and $\mu_g(\cdot) := \mu(\cdot g)$; by $\mathcal{M}(\mu)$ we denote the μ-measurable sets of G and by $\mathcal{M}_+(\mu)$ those of finite positive measure and write $\mathcal{K}_+(\mu) := \mathcal{K}(G) \cap \mathcal{M}_+(\mu)$. For G a

Polish group, recall that $E \subseteq G$ is *universally measurable*, which we write in brief as $E \in \mathcal{U}(G)$, if E is measurable with respect to every measure $\mu \in \mathcal{P}(G)$. For background, see, e.g., Kec1995, §21D; cf. Fre2002, 434D, 432; Sho1984; these form a σ-*algebra*. Examples are analytic subsets (see, e.g., RogJD1980, Part 1, §2.9; Kec1995, Th. 21.10; Fre2002, 434Dc) and the σ-algebra that they generate. Beyond these are the provably Δ_2^1 sets of FenN1973 (§3.3) – see §3.3, cf. BinO2019a. The latter are sets whose $\mathbf{\Delta}_2^1$ character can be proved from the standard axiom system ZF (see Chapter 16) with the Axiom of Dependent Choice(s) in place of the Axiom of Choice. See our survey article BinO2019a.

Recall that E is *left Haar null*, $E \in \mathcal{HN}$, as in Solecki (Sol2005; Sol2006; Sol2007) (following Chr1972; Chr1974; see also ShiT1998) if there are $B \in \mathcal{U}(G)$ covering E and $\mu \in \mathcal{P}(G)$ with

$$\mu(gB) = 0 \qquad (g \in G).$$

So if $B \in \mathcal{U}(G)$ is not left Haar null, then for each $\mu \in \mathcal{P}(G)$ there is compact $K = K_\mu \subseteq B$ and $g \in G$ with

$$_g\mu(K) > 0.$$

The question then arises whether there is also $\delta > 0$ with $_g\mu(Kt) > 0$ for *all* $t \in B_\delta$, for $B_\delta = B_\delta(1_G)$ the open δ-ball centred at 1_G: a *right-sided* property *complementing* the earlier *left-sided* property (of nullity, or otherwise). If this is the case for some μ, then (see Corollary 15.3.7) $1_G \in \text{int}(K^{-1}K) \subseteq \text{int}(E^{-1}E)$; indeed, one has

$$K \cap Kt \in \mathcal{M}_+(_g\mu) \qquad (t \in B_\delta) \qquad\qquad (*M)$$

('M for measure', cf. $(*B)$ below , 'B for Baire'), which implies (Lemma 15.3.2):

$$B_\delta \subseteq \text{int}(K^{-1}K) \subseteq \text{int}(E^{-1}E);$$

cf. Kem1957; Kucz1985, Lemma 3.7.2; BinO2010f, Th. K; BinO2018b, Th. 1(iv). As this clearly forces local compactness of G (see Lemma 15.3.2), for the more general context we weaken the 'complementing right-sided property' to hold only *selectively*: on a subset (cf. $B_\delta^\Delta(\mu)$ on p. 244) of B_δ of the form

$$\{z \in B_\delta : |\mu(Kz) - \mu(K)| < \varepsilon\}.$$

We are guided by the close relation between the measure-theoretic *Steinhaus–Weil-like* property $(*M)$ and its category version

$$K \cap Kt \in \mathcal{B}_+(\tau), \qquad\qquad (*B)$$

where the latter term $\mathcal{B}_+(\tau)$ refers to non-meagre *Baire* sets of τ, a refinement of the ambient topology $\mathcal{T}_G = \mathcal{T}_d$ of G, the latter conveniently taken to be generated by a *left*-invariant metric $d = d_L^G$ with associated group-norm (see Chapter 6). We refer to the (left)-invariance of $\mathcal{B}_+(\tau)$ (under translation) as the (left) *Nikodym property* of τ.

Here, in the context of a metric or Polish group G, we study continuity properties of the maps $m_K : t \mapsto \mu(Kt)$ in the light of theorems of Solecki (Sol2006) and of converses to Theorem 15.3.1 (see the Simmons–Mospan Theorem, 15.5.5) and related results. The key here is Fuller's notion of subcontinuity, as applied to the function $m_K(t)$ at $t = 1_G$. This yields a fruitful interpretation of Solecki's notion of *amenability at* 1_G via *selective subcontinuity* and linkage to *shift-compactness* (see Theorem 15.3.8). Since commutative Polish groups are amenable at 1 (Sol2006, Th. 1(ii)), this widens the field of applicability of shift-compactness to non-Haar null subsets of these, as in BinO2017, and leads to a conjecture (see Remarks preceding Theorem 15.3.8) as to whether $\mathcal{HN}$ comprises the negligible sets of some refinement topology of $\mathcal{T}_d$.

We frequently refer for background to the extended commentaries and associated extensive bibliography of BinO2016b, the unified arXiv version of our four-part series on Theorem 15.3.1.

We begin with the promised adaptation of *subcontinuity* (of functions) due to Fuller (Ful1968) (for which see Remark 4 below) to the context of measures. See also Fort1949. We focus on the *right-sided* version of the concept. Subcontinuity is a natural auxiliary in the quest for fuller forms of continuity: as one instance, see Bouz1996 for the step from separate to joint continuity; as another (classic) instance, note that a subcontinuous set-valued map with closed graph (yet another relative of upper semi-continuity) is continuous – see HolN2012 for an extensive bibliography. Here its relevance to the Steinhaus–Weil Theorem (which is relatively new – BinO2020c) yields Theorems 15.3.4 and 15.3.8, linking *amenability at* 1 with *shift-compactness*, for which see Theorem 15.3.8. Thus subcontinuity passes between local compactness and the pathology of invariance associated with non-local compactness: see Oxt1946 and DieS2014, Ch. 10.

Definitions (BinO2018b)   For $\mu \in \mathcal{P}_{\text{fin}}(G)$, and (compact) $K \in \mathcal{K}(G)$, noting that $\mu_\delta(K) := \inf\{\mu(Kt) : t \in B_\delta\}$ is weakly decreasing in δ, put

$$\mu_-(K) := \sup_{\delta > 0} \inf\{\mu(Kt) : t \in B_\delta\},$$

and, for $\mathbf{t} = \{t_n\}$ a *null sequence*, i.e. with $t_n \to 1_G$,

$$\mu_-^{\mathbf{t}}(K) := \liminf_{n \to \infty} \mu(Kt_n).$$

Then

$$0 \le \mu_-(K) \le \mu(K) = \inf_{\delta > 0} \sup\{\mu(Kt) : t \in B_\delta\}$$

by BinO2020c, Prop. 1 (that $t \to Kt$ is upper semi-continuous). We say that a null sequence **t** is *non-trivial* if $t_n \ne 1_G$ infinitely often. Define as follows:

(i) μ is *translation-continuous* (*'continuous'* or *'mobile'*) if $\mu(K) = \mu_-(K)$ for all $K \in \mathcal{K}(G)$;

(ii) μ is *maximally discontinuous* at $K \in \mathcal{K}(G)$ if $0 = \mu_-(K) < \mu(K)$;

(iii) μ is *subcontinuous* if $0 < \mu_-(K) \le \mu(K)$ for all $K \in \mathcal{K}_+(\mu)$;

(iv) μ is *(selectively) subcontinuous* at $K \in \mathcal{K}_+(\mu)$ along **t** if $\mu_-^{\mathbf{t}}(K) > 0$.

Remarks 1. $m_K(t) := \mu(Kt)$ is *continuous* if μ is continuous, since $m_K(st) = m_{Ks}(t)$ and Ks is compact whenever K is compact; for directional continuity of measures in linear spaces, see Bog2010, §3.1. In LiuR1968 (cf. LiuRW1970; Gow1970; Gow1972), a Radon measure μ on a space X, on which a group G acts homeomorphically, is called *mobile* if $t \mapsto \mu(Kt)$ is continuous for all $K \in \mathcal{K}(X)$.

2. For G locally compact (i) holds for μ the left Haar measure η_G, and also for $\mu \ll \eta_G$ (absolutely continuous with respect to η_G).

3. A measure μ singular with respect to Haar measure is maximally discontinuous for its support: this is at the heart of the analysis offered by Simmons (and independently, much later by Mospan) – see Corollary 15.3.7.

4. *Subcontinuity*, in the sense of Ful1968, of a map $f : G \to (0, \infty)$ requires that, for every $t_n \to t \in G$, there is a subsequence $t_{m(n)}$ with $f(t_{m(n)})$ convergent in the range (i.e. to a positive value). The distinguished role of null sequences emerges below in the *Subcontinuity Theorem*, 15.3.4. Null sequences should be viewed here as a stepwise selection of an 'asymptotic direction' (or even, as suggested by Tomasz Natkaniec, arcwise), since one may apply the Hartman–Mycielski Embedding Theorem, 12.1.4, justifying the phrase *'along* **t***'* in (iv) above, and allowing (iv) to be interpreted as a *selective subcontinuity* in 'direction' **t**. The analogous selective concept in a linear space is 'along a vector' as in Bog2010, §3.1.

5. *Selective versus uniform subcontinuity.* Definition (iii) is equivalent to demanding, for $K \in \mathcal{K}_+(\mu)$, that *any* null sequence $\mathbf{t} = \{t_n\}$ have a subsequence $\mu(Kt_{m(n)})$ bounded away from 0; then (iii) may be viewed as demanding 'uniform subcontinuity': selective subcontinuity along *each* **t** for all $K \in \mathcal{K}_+(\mu)$.

6. *Left- versus right-sided versions.* Writing $\tilde{\mu}(E) := \mu(E^{-1})$ with E Borel in G for the *inverse measure* captures versions associated with right-sided translation such as $\tilde{\mu}_-$ and

$$\tilde{\mu}^{\mathbf{t}}_-(K) := \liminf_{n\to\infty} \mu(t_n K).$$

Definition We will say that μ is *symmetric* if $\mu = \tilde{\mu}$; then B is null if and only if B^{-1} is null for B a Borel set, or $B \in \mathcal{U}(G)$.

In Lemma 15.3.2 it suffices for μ to be a bounded, regular submeasure which is *supermodular:*

$$\mu(E \cup F) \geq \mu(E) + \mu(F) - \mu(E \cap F) \qquad (E, F \in \mathcal{U}(G));$$

recall, however, from Bog2007a, 1.12.37 the opportunity to replace, for any $K \in \mathcal{K}(G)$, a supermodular submeasure μ by a dominating $\mu' \in \mathcal{M}_{\mathrm{fin}}(G)$, i.e. with $\mu'(K) \geq \mu(K)$.

For $K \in \mathcal{K}_+(\mu)$ and $\delta, \Delta > 0$, put

$$B_\delta^\Delta = B_\delta^{K,\Delta}(\mu) := \{z \in B_\delta : \mu(Kz) > \Delta\},$$

which is monotonic in $\Delta : B_\delta^\Delta \subseteq B_\delta^{\Delta'}$ for $0 < \Delta' \leq \Delta$. Note that $1_G \in B_\delta^\Delta$ for $0 < \Delta < \mu(K)$.

The specialization below to a mobile measure (see Definition (i) above) may be found in Gow1970 and Gow1972 (cf. BinO2018b, Th. 2.5).

Lemma 15.3.2 *Let $\mu \in \mathcal{P}_{fin}(G)$ for G a metric group. For $K \in \mathcal{K}_+(\mu)$, if $\mu^{\mathbf{t}}_-(K) > 0$ for some non-trivial null sequence $\mathbf{t}$, then, for $\Delta \geq \mu^{\mathbf{t}}_-(K)/4 > 0$, there is $\delta > 0$ with $t_n \in B_\delta^\Delta$ for all large enough n and*

$$\Delta \leq \mu(K \cap Kt) \qquad (t \in B_\delta^\Delta),$$

so that

$$K \cap Kt \in \mathcal{M}_+(\mu) \qquad (t \in B_\delta^\Delta). \tag{$*$}$$

In particular,

$$K \cap Kt \neq \emptyset \qquad (t \in B_\delta^\Delta),$$

or, equivalently,

$$B_\delta^\Delta \subseteq K^{-1}K, \tag{$**$}$$

so that B_δ^Δ has compact closure. A fortiori, if $\mu_-(K) > 0$, then $\delta, \Delta > 0$ may be chosen with $\Delta < \mu_-(K)$ and $B_\delta \subseteq B_\delta^\Delta$ so that $(*)$ and $(**)$ hold with B_δ replacing B_δ^Δ, and, in particular, G is locally compact.

Proof For the first part fix a null sequence $\mathbf{t}$ and $K \in \mathcal{K}_+(\mu)$ with $\mu^{\mathbf{t}}_-(K) > 0$; take any $\Delta \geq \mu^{\mathbf{t}}_-(K)/4 > 0$, and, as above, write B_δ^Δ for $B_\delta^{K,\Delta}$. Then, for $\mu(Kt) > 2\Delta \geq \mu^{\mathbf{t}}_-(K)/2$ and $\delta > 0$ arbitrary, $t \in B_\delta^\Delta$; and so $t_n \in B_\delta^\Delta$ for all large enough n (since also $t_n \in B_\delta$ for all large enough n). So $B_\delta^\Delta(K)\backslash\{1_G\}$ is non-empty for **t** non-trivial.

Put $H_t := K \cap Kt \subseteq K$. By outer regularity of μ, choose $U = U(\Delta, K)$ open with $K \subseteq U$ and $\mu(U) < \mu(K) + \Delta$. By upper semi-continuity of $t \mapsto Kt$, we may now fix $\delta = \delta(\Delta, K) > 0$ so that $KB_\delta \subseteq U$. For $t \in B_\delta^\Delta$, by finite additivity of μ, since $2\Delta < \mu(Kt)$,

$$2\Delta + \mu(K) - \mu(H_t) \leq \mu(Kt) + \mu(K) - \mu(H_t) = \mu(Kt \cup K)$$
$$\leq \mu(U) \leq \mu(K) + \Delta.$$

Comparing the ends gives

$$0 < \Delta \leq \mu(H_t) \qquad (t \in B_\delta^\Delta).$$

For $t \in B_\delta^\Delta$, as $K \cap Kt \in \mathcal{M}_+(\mu)$, take $s \in K \cap Kt \neq \emptyset$; then $s = kt$ for some $k \in K$, so $t = k^{-1}s \in K^{-1}K$. Conversely, $t \in B_\delta^\Delta \subseteq K^{-1}K$ yields $t = k^{-1}k'$ for some $k, k' \in K$; then $k' = kt \in K \cap Kt$.

By the compactness of $K^{-1}K$, B_δ^Δ has compact closure.

As for the final assertions, if $\mu_-(K) > 0$, now take $\Delta := \mu_-(K)/2$. Then $\inf\{\mu(Kt) : t \in B_\delta\} > \Delta$ for all small enough $\delta > 0$, and so in particular $\mu(Kt) > \Delta$ for $t \in B_\delta$, i.e. $B_\delta \subseteq B_\delta^\Delta$. So the argument above applies for small enough $\delta > 0$ with B_δ in lieu of B_δ^Δ, just as before. Here the compactness of $K^{-1}K$ now implies local compactness of G itself. $\qquad\square$

As an immediate and useful corollary, we have

Lemma 15.3.3 *Suppose* $\mu \in \mathcal{P}_{\text{fin}}(G)$, *with* G *a metric group,* **t** *any null sequence and* $K \in \mathcal{K}(G)$. *Then, if* $\mu_-^{\mathbf{t}}(K) > 0$, *there is* $m \in \mathbb{N}$ *with*

$$0 < \mu_-^{\mathbf{t}}(K)/4 < \mu(K \cap Kt_n) \qquad (n > m). \tag{$*'$}$$

In particular,

$$t_n \in K^{-1}K \qquad (n > m). \tag{$**'$}$$

Proof Apply Lemma 15.3.2 to obtain $\Delta, \delta > 0$; for $t \in B_\delta^\Delta$, we have $\mu(Kt) > \Delta$, so, as above, $t_n \in B_\delta^\Delta$ for all large enough n. $\qquad\square$

This permits a connection with left Haar null sets $\mathcal{HN}$ introduced above. Recall that a group G is *amenable at* 1 (Sol2006) (see below for the origin of this term) if, given $\mu := \{\mu_n\}_{n\in\mathbb{N}} \subseteq \mathcal{P}(G)$ with $1_G \in \text{supp}(\mu_n)$ (the support of μ_n), for $n \in \mathbb{N}$ there are σ and σ_n in $\mathcal{P}(G)$ with $\sigma_n \ll \mu_n$ satisfying

$$\sigma_n * \sigma(K) \to \sigma(K) \qquad (K \in \mathcal{K}(G)).$$

In view of Theorem 15.3.4, we term σ (or $\sigma(\mu)$ if context requires) a *selective measure* and the measures σ_n, if needed, as *associated measures* (corresponding to the sequence $\{\mu_n\}_{n\in\mathbb{N}}$).

Solecki explains (Sol2006, end of §2) the use of the term 'amenability at 1' as a localization (via the restriction that all supports contain 1_G) of a *Reiter-like condition* (Pat1988, Prop. 0.4) which characterizes amenability: for $\mu \in \mathcal{P}(G)$ and $\varepsilon > 0$, there is $\nu \in \mathcal{P}(G)$ with

$$|\nu * \mu(K) - \nu(K)| < \varepsilon \qquad (K \in \mathcal{K}(G)).$$

Lemma 15.3.2 and the next several results disaggregate Solecki's Interior-point Theorem (Sol2006, Th. 1(ii)) (Corollary 15.3.6), shedding more light on it and in particular connecting it to shift-compactness (Theorem 15.3.4). Indeed, we see that interior-point theorem itself as an 'aggregation' phenomenon. Theorem 1 of BinO2020d identifies subgroups with a 'disaggregation' topology, refining $\mathcal{T}_G$ by using sets of the form $B_\delta^{K,\Delta}(\sigma)$, the measures σ being provided in our first result, a result motivated by Solecki (Sol2006, Th. 1(ii)):

Theorem 15.3.4 (Subcontinuity Theorem) *For G Polish and amenable at 1_G and* $\mathbf{t}$ *a null sequence, there is $\sigma = \sigma(\mathbf{t}) \in \mathcal{P}(G)$ such that, for each $K \in \mathcal{K}_+(\sigma)$, there is a subsequence $\mathbf{s} = \mathbf{s}(K) := \{t_{m(n)}\}$ with*

$$\lim_n \sigma(Kt_{m(n)}) = \sigma(K) \qquad (n \in \mathbb{N}), \quad so \quad \sigma_-^{\mathbf{s}}(K) > 0.$$

Proof For $\mathbf{t} = \{t_n\}$ null, put $\mu_n := 2^{n-1} \sum_{m\geq n} 2^{-m} \delta_{t_m^{-1}} \in \mathcal{P}(G)$; then $1_G \in \mathrm{supp}(\mu_n) \supseteq \{t_m^{-1} : m > n\}$. By definition of amenability at 1_G, in $\mathcal{P}(G)$ there are σ and $\sigma_n \ll \mu_n$, with $\sigma_n * \sigma(K) \to \sigma(K)$ for all $K \in \mathcal{K}(G)$. For $n \in \mathbb{N}$ choose $\alpha_{mn} \geq 0$ with $\sum_{m\geq n} \alpha_{mn} = 1$ ($n \in \mathbb{N}$) and with $\sigma_n := \sum_{m\geq n} \alpha_{mn} \delta_{t_m^{-1}}$.

Fix $K \in \mathcal{K}_+(\sigma)$ and θ with $0 < \theta < 1$. As K is compact, $\sigma_n * \sigma(K) \to \sigma(K)$; then, without loss of generality,

$$\sigma_n * \sigma(K) > \theta\sigma(K) \qquad (n \in \mathbb{N}).$$

Then, for each $n \in \mathbb{N}$,

$$\sup\{\sigma(Kt_m) : m \geq n\} \cdot \sum_{m\geq n} \alpha_{mn} \geq \sum_{m\geq n} \alpha_{mn}\sigma(Kt_m) > \theta\sigma(K).$$

So for each n there is $m = m(\theta) \geq n$ with

$$\sigma(Kt_m) > \theta\sigma(K).$$

Now choose $m(n) \geq n$ inductively so that $\sigma(Kt_{m(n)}) > (1-2^{-n})\sigma(K)$; then, by the upper semi-continuity of the maps $t \mapsto Kt$ and $t \mapsto \sigma(Kt)$ (cf. BinO2020c, Prop. 1), $\lim_n \sigma(Kt_{m(n)}) = \sigma(K)$: σ is subcontinuous along $\mathbf{s} := \{t_{m(n)}\}$ on K. □

Remark   The selection above of the subsequence **s** mirrors the role of 'admissible directions' which we will encounter, in §15.7, in Cameron–Martin theory (BinO2019c, §2; BinO2016b, §8.2).

We are now able to deduce Solecki's interior-point theorem in a slightly stronger form, which asserts that the sets B_δ^Δ reconstruct the open sets of G using the compact subsets of a 'non-negligible set', as follows. We recall that $\mathcal{K}(X)$ denotes the family of compact subsets of X; below we use the notation $\delta(\Delta, K)$ established in the proof of Lemma 15.3.2.

Theorem 15.3.5 (Aggregation Theorem)   *For G Polish and amenable at 1_G, if $E \in \mathcal{U}(G)$ is not left Haar null and setting*

$$\hat{E} := \bigcup_{\Delta>0,\, g\in G,\, \mathbf{t}} \left\{ B_{\delta(gK,\Delta)}^{gK,\Delta}(\sigma(\mathbf{t})) : K \in \mathcal{K}(E),\, 0 < \sigma(\mathbf{t})(gK)/4 \right.$$
$$\left. \leq \Delta < \sigma(\mathbf{t})(gK) \right\},$$

then

$$1_G \in \mathrm{int}(\hat{E}) \subseteq \hat{E} \subseteq E^{-1}E.$$

In particular, for E open, $1_G \in \mathrm{int}(\hat{E})$.

Proof   Suppose otherwise; then, as in Lemma 15.3.2, for $g \in G$, any null sequence **z**, compact $K \subseteq E$ with $0 < \sigma(\mathbf{z})(gK)/4 \leq \Delta$ and $\delta = \delta(gK, \Delta)$, we have

$$B_\delta^{gK,\Delta}(\sigma(\mathbf{z})) \subseteq (gK)^{-1}gK = K^{-1}K \subseteq E^{-1}E,$$

so that $\hat{E} \subseteq E^{-1}E$. Next, suppose there is for each n,

$$t_n \in B_{1/n} \backslash \hat{E}.$$

Consider $\sigma = \sigma(\mathbf{t})$. As E is not left Haar null, there is g with $\sigma(gE) > 0$. Choose compact $K \subseteq gE$ with $\sigma(K) > 0$. Then, with $h := g^{-1}$ and $H := hK \subseteq E$, it follows that $\sigma(K) = \sigma(gH) = \sigma^s(gH) > 0$ for some subsequence $\mathbf{s} = \{t_{m(k)}\}$. So, again as above and as in Lemma 15.3.2, with $\Delta := \sigma(gH)/4$ for some $\delta = \delta(K, \Delta) > 0$,

$$B_\delta^{gH,\Delta}(\sigma(\mathbf{t})) \subseteq (gH)^{-1}gH = H^{-1}H \subseteq E^{-1}E.$$

Choose n with $n > 1/\delta$. Then $t_n \in B_\delta$ for all $m > n$; so for infinitely many k,

$$t_{m(k)} \in B_\delta^{gH,\Delta}(\sigma(\mathbf{t})) \subseteq \hat{E},$$

contradicting $t_n \notin \hat{E}$. As for the final assertion, for E open, D countable and dense, $G \subseteq \bigcup_{d\in D} dE$, so, for any $\mu \in \mathcal{P}(G)$ (in particular for σ), $\mu(dE) > 0$ for some $d \in D$, and so E is not left Haar null.   □

The immediate consequence from Sol2006, Th. 1(ii), is

Corollary 15.3.6 (Solecki's Interior-Point Theorem)   *For G Polish and amenable at 1_G, if $E \in \mathcal{U}(G)$ is not left Haar null, then $1_G \in \text{int}(E^{-1}E)$.*

Corollary 15.3.7   *For G a Polish group, if $E \in \mathcal{U}(G)$ is not left Haar null and is in $\mathcal{M}_+(\mu)$ for some subcontinuous $\mu \in \mathcal{P}_{\text{fin}}(G)$, then, for some $\delta > 0$,*

$$B_\delta \subseteq \text{int}(E^{-1}E).$$

In particular, this inclusion holds for some $\delta > 0$ in a locally compact group G, for any Baire non-meagre set E.

Proof   The first assertion is immediate from Lemma 15.3.2. As for the second, for a non-meagre Baire set E, if $\tilde{E}$ is the quasi-interior and $K \subseteq \tilde{E}$ is compact with non-empty interior, then $\eta_G(K) > 0$. Since η is subcontinuous, there is $\delta > 0$ with

$$Kt \cap K \neq \emptyset \qquad (\|t\| < \delta),$$

and so

$$\tilde{E}t \cap \tilde{E} \neq \emptyset \qquad (\|t\| < \delta).$$

Then $U := (Et)\widetilde{} \cap \tilde{E} \neq \emptyset$, since $(Et)\widetilde{} = \tilde{E}t$ (the Nikodym property of the *usual* topology of G). So since U is open and non-meagre, also $Et \cap E \neq \emptyset$, and so again (**) holds.   □

The next result establishes the embeddability by (left-sided) translation of an appropriate subsequence of a given *null* sequence into a given target set that (like-sidedly) is non-left Haar null. We establish this, first announced in MilleO2012, in relation to the ideal $\mathcal{HN}$ of *left Haar null sets*. (It is a σ-ideal for Polish G in the presence of amenability at 1: Sol2006, Th 1(i).) This leaves open the 'converse question' of the existence of a refinement topology for which $\mathcal{HN}$ is the associated notion of *negligibility*; this seems plausible under the *continuum hypothesis* (CH) if one restricts attention only to Borel sets in $\mathcal{HN}$ and their subsets by lifting a result concerning $\mathbb{R}$ in CieJ1995, Cor. 4.2, to G – see also the Remark 1 following our next result.

Theorem 15.3.8 (Shift-Compactness Theorem for $\mathcal{HN}$)   *For G Polish and amenable at 1_G, if $E \in \mathcal{U}(G)$ is not left Haar null and z_n is null, then there are $s \in E$ and an infinite $\mathbb{M} \subseteq \mathbb{N}$ with*

$$\{sz_m : m \in \mathbb{M}\} \subseteq E.$$

Indeed, this holds for quasi all $s \in E$, i.e. off a left Haar null set.

Proof Put $t_n := z_n^{-1}$, which is null. With $\sigma = \sigma(\mathbf{t})$ as in the Subcontinuity Theorem, 15.3.4, since E is not left Haar null, there is g with $\sigma(gE) > 0$. For this g, put $\mu := {}_g\sigma$. Fix a compact $K_0 \subseteq E$ with $\mu(K_0) > 0$ and then, passing to a subsequence of $\mathbf{t}$ as necessary (by Theorem 15.3.4), we may assume that $\mu_-^{\mathbf{t}}(K_0) > 0$. Choose inductively a sequence $m(n) \in \mathbb{N}$, and decreasing compact sets $K_n \subseteq K_0 \subseteq E$ with $\mu(K_n) > 0$ such that

$$\mu(K_n \cap K_n t_{m(n)}) > 0.$$

To check the inductive step, suppose K_n is already defined. As $\mu(K_n) > 0$, by the Subcontinuity Theorem there is a subsequence $\mathbf{s} = \mathbf{s}(K_n)$ of $\mathbf{t}$ with $\mu_-^{\mathbf{s}}(K_n) > 0$. By Lemma 15.3.2, there is $k(n) > n$ such that $\mu(K_n \cap K_n s_{k(n)}) > 0$. Putting $t_{m(n)} = s_{k(n)}$ and $K_{n+1} := K_n \cap K_n t_{m(n)} \subseteq K_n$ completes the inductive step.

By compactness, select s with

$$s \in \bigcap\nolimits_{m \in \mathbb{N}} K_m \subseteq K_{n+1} = K_n \cap K_n t_{m(n)} \qquad (n \in \mathbb{N});$$

choosing $k_n \in K_n \subseteq K$ with $s = k_n t_{m(n)}$ gives $s \in K_0 \subseteq E$, and

$$s z_{m(n)} = s t_{m(n)}^{-1} = k_n \in K_n \subseteq K_0 \subseteq E.$$

Finally take $\mathbb{M} := \{m(n) : n \in \mathbb{N}\}$.

As for the final assertion, we follow the idea of the Generic Completeness Principle (§4.1) (but with $\mathcal{U}(G)$ for $\mathcal{B}a$ there): define

$$F(H) := \bigcap\nolimits_{n \in \mathbb{N}} \bigcup\nolimits_{m > n} H \cap H t_m \qquad (H \in \mathcal{U}(G));$$

then $F : \mathcal{U}(G) \to \mathcal{U}(G)$ and F is monotone ($F(S) \subseteq F(T)$ for $S \subseteq T$); moreover, $s \in F(H)$ if and only if $s \in H$ and $s z_m \in H$ for infinitely many m. We are to show that $E_0 := E \backslash F(E)$ is left Haar null. Suppose otherwise. Then renaming g and K_0 as necessary, without loss of generality both $\mu(E_0) > 0$ and $K_0 \subseteq E_0$ (and $\mu(K_0) > 0$). But then, as above, $\emptyset \neq F(K_0) \cap K_0 \subseteq F(E) \cap E_0$, a contradiction, since $F(E) \cap E_0 = \emptyset$. $\qquad \square$

Remarks   1. In the setting of Theorem 15.3.8 any non-empty open set U is not left Haar null (as $\{dU : d \in D\}$ with D countable dense covers G), hence neither is $U \backslash H$ for $H \in \mathcal{HN}$. So the (Hashimoto ideal) topology generated by such sets includes $\mathcal{HN}$ among its negligible sets.

2. The special *abelian* case of Theorem 15.3.8 was independently established by Banakh and Jabłońska in BanaJ2019. A similar result extends to the Haar-meagre sets of Darji (Darj2013); cf. Jab2015. See also EleV2017, EleN2020 and BinO2016b, §8.9. We discuss these results in §§15.7 and 15.8.

Corollary 15.3.9 *For G Polish and amenable at 1_G and z_n null, there is $\mu \in \mathcal{P}(G)$ such that for $K \in \mathcal{K}_+(\mu)$,*

$$K \cap K z_m^{-1} \in \mathcal{M}_+(\mu) \quad \text{for infinitely many } m \in \mathbb{N}$$

if and only if for μ-quasi all $s \in K$ there is an infinite $\mathbb{M} \subseteq \mathbb{N}$ with

$$\{s z_m : m \in \mathbb{M}\} \subseteq K.$$

Proof We will refer to the function F of the preceding proof. First proceed as in the proof of Theorem 15.3.4, taking $t_n := z_n^{-1}$ and $g = 1_G$ (so that $\mu = \sigma$). Fix K with $\mu(K) > 0$. For the forward direction, continue as in the proof of Theorem 15.3.4 with $K_0 = K$ and observe that the proof above needs only that $s_{k(n)} \in K_n^{-1} K_n$ occurs infinitely often whenever $\mu(K_n) > 0$. This yields the desired conclusion that $\mu(K \backslash F(K)) = 0$. For the converse direction, suppose that $\mu(F(K)) > 0$. Since for each $n \in \mathbb{N}$,

$$F(K) \subseteq \bigcup_{m>n} K \cap K t_m,$$

we have $\mu(K \cap K t_m) > 0$ for some $m > n$; so

$$K \cap K t_m \in \mathcal{M}_+(\mu) \quad \text{for infinitely many } m. \qquad \square$$

Remark With E as in the Shift-Compactness Theorem, 11.1.1, if $z_n \in B_{1/n} \backslash E^{-1} E$, then z_n is null; so, for some $s \in E$, $s z_m \in E$ for infinitely many m. Then, for any such m,

$$z_m \in E^{-1} E,$$

contradicting the choice of z_m. So $1_G \in \text{int}(E^{-1} E)$, i.e. E has the Steinhaus–Weil property, as before.

The following sharpens a result due (for Lebesgue measure on $\mathbb{R}$) to Mospan (Mosp2005) by providing the converse below; it is antithetical to Lemma 15.3.2 (and so to Theorem 15.3.4).

Proposition 15.3.10 (Mospan Property) *For G a metric group, $\mu \in \mathcal{P}_{\text{fin}}(G)$ and compact $K \in \mathcal{K}_+(\mu)$:*

(i) *if $1_G \notin \text{int}(K^{-1} K)$, then $\mu_-(K) = 0$, i.e. μ is maximally discontinuous; equivalently, there is a null sequence $t_n \to 1_G$ with $\lim_n \mu(K t_n) = 0$;*

(ii) *there is a null sequence $t_n \to 1_G$ with $\lim_n \mu(K t_n) = 0$, and there is a compact $C \subseteq K$ with $\mu(K \backslash C) = 0$ with $1_G \notin \text{int}(C^{-1} C)$.*

Proof The first assertion follows from Lemma 15.3.2. For the converse, as in Mosp2005, suppose that $\mu(K t_n) = 0$ for some sequence $t_n \to 1_G$. By passing to a subsequence, we may assume that $\mu(K t_n) < 2^{-n-1}$. Put $D_m :=$

$K \setminus \bigcap_{n \geq m} K t_n \subseteq K$; then $\mu(K \setminus D_m) \leq \sum_{n \geq m} \mu(K t_n) < 2^{-m}$, so $\mu(D_m) > 0$ provided $2^{-m} < \mu(K)$. Now choose compact $C_m \subseteq D_m$, with $\mu(D_m \setminus C_m) < 2^{-m}$. So $\mu(K \setminus C_m) < 2^{1-m}$. Also $C_m \cap C_m t_n = \emptyset$, for each $n \geq m$, as $C_m \subseteq K$; but $t_n \to 1_G$, so the compact set $C_m^{-1} C_m$ contains no interior points. Hence, by Baire's Theorem, neither does $C^{-1} C$, since $C = \bigcup_m C_m$, which differs from K by a null set. □

Proposition 15.3.11 *A (regular) Borel measure μ on a locally compact metric topological group G has the Steinhaus–Weil property if and only if either*

(i) *for each $K \in \mathcal{K}_+(\mu)$, the map $m_K : t \to \mu(Kt)$ is subcontinuous at 1_G; or*
(ii) *for each $K \in \mathcal{K}_+(\mu)$, there is no sequence $t_n \to 1_G$ with $\mu(Kt_n) \to 0$.*

Remark This is immediate from Proposition 15.3.10 (cf. Mosp2005).

15.4 The Simmons–Mospan Converse

The converse to the Steinhaus–Weil Theorem for a locally compact group G identifies exactly when a Borel measure μ on G guarantees that 1_G is an interior point of TT^{-1} for any non-μ-null Borel T. Simmons (Sim1975) showed in 1975 that the measure μ has to be absolutely continuous with respect to Haar measure on G. This (and other conditions – cf. Proposition 15.3.10) were investigated independently by Mospan (Mosp2005) in 2005. The result follows from their use of the Fubini Null Theorem, 9.5.1, and the Lebesgue decomposition theorem (Hal1950, §32, p. 134, Th. C), but here we stress the dependence on the Fubini Null Theorem and on left μ-inversion. We revert to the Weil left-sided convention and associated KK^{-1} usage.

We begin with some definitions which stress the one-sided nature of translation in a non-commutative group and the relation to the switching of sides under inversion.

Definitions (i) For $\mu \in \mathcal{M}(G)$, say that $N \subseteq G$ is *left μ-null*, $N \in \mathcal{M}_0^L(\mu)$, if it is contained in a universally measurable set $B \subseteq G$ such that

$$\mu(gB) = 0 \qquad (g \in G).$$

Thus a set $S \subseteq G$ is *left Haar null* (Sol2006 after Chr1974) if it is contained in a universally measurable set $B \subseteq G$ that is left μ-null for some $\mu \in \mathcal{M}(G)$.

(ii) For $\mu \in \mathcal{M}(G)$, say that $N \in \mathcal{M}_0^L(\mu)$ is *left invertibly μ-null*, $N \in \mathcal{M}_0^{L\text{-inv}}(\mu)$, if

$$N^{-1} \in \mathcal{M}_0^L(\mu).$$

(iii) For $\mu, \nu \in \mathcal{M}(G)$, we say ν is *left absolutely continuous* with respect to μ (i.e. $\nu <^{\mathrm{L}} \mu$) if $\nu(N) = 0$ for each $N \in \mathcal{M}_0^{\mathrm{L}}(\mu)$, and likewise for the invertibility version: $\nu <^{\mathrm{L\text{-}inv}} \mu$.

(iv) For $\mu, \nu \in \mathcal{M}(G)$, we say ν is *left singular* with respect to μ (on B), i.e. $\nu \perp^{\mathrm{L}} \mu$ (on B), if B is a support of ν and B is in $\mathcal{M}_0^{\mathrm{L}}(\mu)$, and likewise $\nu \perp^{\mathrm{L\text{-}inv}} \mu$.

We will take for granted the one-sided Lebesgue decomposition theorem:

Theorem 15.4.1 (Theorem LD)   *For G a Polish group, $\mu, \nu \in \mathcal{M}(G)$, there are $\nu_{\mathrm{a}}, \nu_{\mathrm{s}} \in \mathcal{M}(G)$ with*

$$\nu = \nu_{\mathrm{a}} + \nu_{\mathrm{s}} \text{ with } \nu_{\mathrm{a}} <^{\mathrm{L}} \mu \text{ and } \nu_{\mathrm{s}} \perp^{\mathrm{L}} \mu,$$

and likewise, there are $\nu_{\mathrm{a}}', \nu_{\mathrm{s}}' \in \mathcal{M}(G)$ with

$$\nu = \nu_{\mathrm{a}}' + \nu_{\mathrm{s}}' \text{ with } \nu_{\mathrm{a}} <^{\mathrm{L\text{-}inv}} \mu \text{ and } \nu_{\mathrm{s}} \perp^{\mathrm{L\text{-}inv}} \mu.$$

See BinO2020d for a pedestrian proof from DC, and its more rapid alternative classical proof from AC.

Proposition 15.4.2 (Local Almost Nullity)   *For G a Polish group, $\mu \in \mathcal{M}(G)$, $V \subseteq G$ open and $K \in \mathcal{K}(G) \cap \mathcal{M}_0^{\mathrm{L\text{-}inv}}(\mu)$, so that $K, K^{-1} \in \mathcal{M}_0^{\mathrm{L}}(\mu)$, we have, for any $\nu \in \mathcal{M}(G)$, $\nu(tK) = 0$ for μ almost all $t \in V$, and likewise $\nu(Kt) = 0$.*

Proof   For ν left invertibly μ-absolutely continuous, the conclusion is immediate from the definition (iii) above; for general ν this will follow from Theorem LD, 15.4.1, above, once we have proved the corresponding singular version of the assertion: that is the nub of the proof.

Thus, suppose that $\nu \perp^{\mathrm{L\text{-}inv}} \mu$ on K. For $t \in V$ let $t = uw$ be any expression for t as a group product of $u, v \in G$, and note that $\mu(uK^{-1}) = 0$, as $K^{-1} \in \mathcal{M}_0^{\mathrm{L}}(\mu)$. Let H be the set

$$\bigcup\nolimits_{t \in V} (\{t\} \times tK),$$

here viewed as a union of vertical t-sections. We next express it as a union of u-horizontal sections and apply the Fubini Null Theorem, 9.5.1.

Since $u = tk = uwk$ is equivalent to $w = k^{-1}$, the u-horizontal sections of H may now be rewritten, eliminating t, as

$$\{(t, u) : uw = t \in V, u \in tK = uwK\} = \{(uw, u) : uw \in V, uw \in uK^{-1}\}.$$

So H may now be viewed as a union of u-horizontal sections as

$$\bigcup\nolimits_{u \in G} (V \cap (uK^{-1}) \times \{u\}),$$

all of these u-horizontal sections being μ-null. By Theorem 9.5.1, μ-almost all vertical t-sections of H for $t \in V$ are ν-null. As the assumptions on K are symmetric, the right-sided version follows. □

The result here brings to mind the Dodos Dichotomy Theorem (Dod2004a, Th. A) for *abelian* Polish groups G: if an analytic set A is witnessed as Haar null under one measure $\mu \in \mathcal{P}(G)$, then either A is Haar null for quasi all $\nu \in \mathcal{P}(G)$ or else it is non-Haar null for quasi all such ν, i.e. if $A \in \mathcal{M}_0(\mu)$ (omitting the unnecessary superscript L), then either $A \in \mathcal{M}_0(\nu)$ for quasi all such ν or $A \notin \mathcal{M}_0(\nu)$ for quasi all such ν (with respect to the Prokhorov–Lévy metric in $\mathcal{P}(G)$; Dud1989, 11.3, cf. 9.2). Indeed, (Dod2004b, Prop. 5) when A is σ-compact, A is Haar null for quasi all $\nu \in \mathcal{P}(G)$. The result is also reminiscent of Amb1947, Lemma 1.1.

Before stating the Simmons–Mospan specialization to the Haar context and also to motivate one of the conditions in its subsequent generalizations, we cite (and give a direct proof of) the following known result (equivalence of Haar measure η and its inverse $\tilde{\eta}$), encapsulated in the formula

$$\tilde{\eta}(K) := \eta(K^{-1}) = \int_K d\eta(t)/\Delta(t),$$

exhibiting the direct connection between η and $\tilde{\eta}$ via the (positive) modular function Δ (HewR1979, 15.14, or Hal1950, §60.5f). Such an equivalence holds more generally between any two probability measures when one is left and the other right quasi-invariant – see Xia1972, Cor. 3.1.4; this is related to a theorem of Mackey (Mac1957, cf. BinO2016b, §8.16). As will be seen from the proof, in Lemma 15.4.3, there is no need to assume the group is separable: a compact metrizable subspace (being totally bounded) is separable.

Lemma 15.4.3 *In a locally compact metrizable group G, for K compact, if $\eta(K) = 0$, then $\eta(K^{-1}) = 0$, and, by regularity, so also for K measurable.*

Proof Fix a compact K. As K is compact, the modular function Δ of G is bounded away from 0 on K, say by $M > 0$; furthermore, K is separable, so pick $\{d_n : n \in \mathbb{N}\}$ dense in K. Then, for any $\varepsilon > 0$, there are two (finite) sequences $m(1), \ldots, m(n) \in \mathbb{N}$ and $\delta(1), \ldots, \delta(n) > 0$ such that $\{B_{\delta(i)} d_{m(i)} : i \leq n\}$ covers K and

$$M \sum_{i \leq n} \eta(B_{\delta(i)}) \leq \sum_{i \leq n} \eta(B_{\delta(i)}) \Delta(d_{m(i)}) = \sum_{i \leq n} \eta(B_{\delta(i)} d_{m(i)}) < \varepsilon.$$

Then

$$\sum_{i \leq n} \eta(d_{m(i)}^{-1} B_{\delta(i)}) = \sum_{i \leq n} \eta(B_{\delta(i)}) \leq \varepsilon/M.$$

But $\{d_{m(i)}^{-1} B_{\delta(i)} : i \leq n\}$ covers K^{-1} by the symmetry of the balls B_δ (by the symmetry of the norm); so, as $\varepsilon > 0$ is arbitrary, $\eta(K^{-1}) = 0$.

As for the final assertion, if $\eta(E^{-1}) > 0$ for some measurable E, then $\eta(K^{-1}) > 0$ for some compact $K^{-1} \subseteq E^{-1}$, by regularity; then $\eta(K) > 0$, and so $\eta(E) > 0$. □

Proposition 15.4.2 and Lemma 15.4.3 immediately give (cf. Saks1964, III.11; Mosp2005; BartFF2018, Th. 7):

Theorem 15.4.4 *For G a locally compact group with left Haar measure η and ν a Borel measure on G, if the set S is η-null, then, for η-almost all t,*

$$\nu(tS) = 0.$$

In particular, this is so for S the support of a measure ν singular with respect to η.

This in turn allows us to prove the locally compact separable case of the Simmons–Mospan Theorem, 15.4.5 (Sim1975, Th. 1; Mosp2005, Th. 7, later rediscovered in the abelian case in BartFF2018, Th. 10).

Theorem 15.4.5 (The Simmons–Mospan Converse) *In a locally compact Polish group, a Borel measure has the Steinhaus–Weil property if and only if it is absolutely continuous with respect to Haar measure.*

Proof If μ is absolutely continuous with respect to Haar measure η, then μ, being invariant, is subcontinuous, and Lemma 15.3.3 gives the Steinhaus–Weil property. Otherwise, decomposing μ into its singular and absolutely continuous parts with respect to η, choose K a compact subset of the support of the singular part of μ; then $\mu(K) > \mu_-(K) = 0$, by Theorem 15.4.4, and so 15.3.10 (Converse part) applies. □

The next result is motivated by work of Simmons (Sim1975, Lemma) and by BartFF2018, Th. 8.

Proposition 15.4.6 *For G a Polish group, $\mu, \nu \in \mathcal{M}(G)$ and $\nu \perp^{\text{L-inv}} \mu$ concentrated on a compact left invertibly μ-null set K, there is a Borel $B \subseteq K$ such that $K \backslash B$ is ν-null and both BB^{-1} and $B^{-1}B$ have empty interior.*

Proof As we are concerned only with the subspace $KK^{-1} \cup K^{-1}K$, without loss of generality the group G is separable. By Proposition 15.4.2 $Z := \{x : \nu(xK) = 0\}$ is dense and so also

$$Z_1 := \{x : \nu(K \cap xK) = 0\},$$

since $\nu(K \cap xK) \leq \nu(xK) = 0$, so that $Z \subseteq Z_1$. Take a denumerable dense set $D \subseteq Z_1$ and put

$$S := \bigcup_{d \in D} K \cap dK.$$

Then $\nu(S) = 0$. Take $B := K \backslash S$. If $\emptyset \neq V \subseteq BB^{-1}$ for some V open and $d \in D \cap V$, then, for some $b_1, b_2 \in B \subseteq K$,

$$d = b_1 b_2^{-1} : \qquad b_1 = db_2 \in K \cap dK \subseteq S,$$

a contradiction, since B is disjoint from S. So $(K \backslash S)(K \backslash S)^{-1}$ has empty interior. A similar argument based on

$$T := \bigcup\nolimits_{d \in D} Kd \cap K$$

ensures that also $(K \backslash S \backslash T)^{-1} (K \backslash S \backslash T)$ has empty interior.   □

Remark   A non-separable generalization is pursued in BinO2020d.

15.5 The Steinhaus Property AA^{-1} versus AB^{-1}

We clarify below the relation between two versions of the Steinhaus interior points property: the simple (sometimes called 'classical') version concerning sets AA^{-1} and the composite, more embracing one, concerning sets AB^{-1}, for sets from a given family $\mathcal{H}$. The latter is connected to a strong form of metric transitivity: Kominek (Kom1988) showed, for a general separable Baire topological group G equipped with an inner-regular measure μ defined on some σ-algebra $\mathcal{M}$, that AB^{-1} has non-empty interior for all $A, B \in \mathcal{M}_+(\mu)$, the sets in $\mathcal{M}$ of positive μ-measure, if and only if for each countable dense set D and each $E \in \mathcal{M}_+(\mu)$ the set $X \backslash DE \in \mathcal{M}_0(\mu)$, the sets in $\mathcal{M}$ of μ-measure zero. This is considered in Theorem 15.5.5. The composite property is thus related to the Smítal property, for which see BartFN2011. Care is required when moving to the alternative property for AB, since the family $\mathcal{H}$ need not be preserved under inversion.

In general the simple property does not imply the composite: Matoušková and Zelený (MatoZ2003) show that in any non-locally compact abelian Polish group there are closed non-(left) Haar null sets A, B such that $A + B$ has empty interior. Jabłońska (Jab2016) has shown that likewise in any non-locally compact abelian Polish group there are closed non-Haar meagre sets A, B such that $A + B$ has empty interior; see also BanaGJ2021. Bartoszewicz, Filipczak and Filipczak (BartFF2018, Ths. 1, 4) analyze the Bernoulli product measure on $\{0, 1\}^{\mathbb{N}}$ with p the probability of the digit 1; see BinO2011a, §8.15. The product space may be regarded as comprising canonical binary digit expansions of the additive reals modulo 1 (in which case the measure is not invariant). Here the (Borel) set A of binary expansions with asymptotic frequency p of the digit 1 has $[0, 1)$ as its difference set if and only if $\frac{1}{4} \leq p \leq \frac{3}{4}$; however, $A + A$

has empty interior unless $p = \frac{1}{2}$ (the base 2 simple-normal numbers case; cf. §14.1).

Below we identify some conditions on a family of sets A with the simple AA^{-1} property which do imply the AB^{-1} property. What follows is a generalization to a group context of relevant observations from BinO2020d from the classical context of $\mathbb{R}$. For a similar approach, see Kha2019.

The motivation for the definition below is that its subject, the space H, is a subgroup of a topological group G from which it inherits a (necessarily) translation-invariant (either-sidedly) topology τ. Various notions of 'density at a point' give rise to 'density topologies' (BinO2018b), which are translation-invariant since they may be obtained via translation to a fixed reference point: early examples, which originate in spirit with Denjoy as interpreted by Haupt and Pauc (HauP1952), were studied intensively in GofW1961, GofNN1961, soon followed by Marti1961, Marti1964 and Mue1965; more recent examples include FilW2011 and others investigated by the Wilczyński school, cf. Wil2002.

Proposition 15.5.1 embraces as an immediate corollary the case $H = G$ with G locally compact and σ the Haar density topology (see BinO2019c). Proposition 15.5.4 proves that Proposition 15.5.1 applies also to the ideal topology (in the sense of LukMZ1986) generated from the ideal of Haar null sets of an abelian Polish group.

We recall that a group H carries a *left semi-topological* structure τ if the topology τ is left invariant (ArhT2008) ($hU \in \tau$ if and only if $U \in \tau$); the structure is *semi-topological* if it is also right invariant, i.e. briefly, τ is translation invariant. The group H is *quasi-topological* under τ if τ is both left and right invariant and inversion is τ-continuous.

Definition   For H a group with a translation-invariant topology τ, call a topology $\sigma \supseteq \tau$ a *Steinhaus refinement* if:

 (i) $\text{int}_\tau(AA^{-1}) \neq \emptyset$ for each non-empty $A \in \sigma$, and
(ii) σ is involutive-translation invariant: $hA^{-1} \in \sigma$ for all $A \in \sigma$ and all $h \in H$.

Property (ii) above (called simply 'invariance' in BartFN2011) apparently calls for only left invariance, but in fact, via double inversion, delivers translation invariance, since $Uh = (h^{-1}U^{-1})^{-1}$; then H under σ is a semi-topological group with a continuous inverse, so a *quasi-topological group*. We address the step from the simple property to the composite in the next result.

Proposition 15.5.1   *If τ is translation-invariant, and $\sigma \supseteq \tau$ is a Steinhaus refinement topology, then* $\text{int}_\tau(AB^{-1}) \neq \emptyset$ *for non-empty $A, B \in \sigma$. In particular, as σ is preserved under inversion, also* $\text{int}_\tau(AB) \neq \emptyset$ *for $A, B \in \sigma$.*

Proof Suppose $A, B \in \sigma$ are non-empty; as $B^{-1} \in \sigma$, choose $a \in A$ and $b \in B$; then, by (ii),

$$1_H \in C := a^{-1}A \cap b^{-1}B^{-1} \in \sigma.$$

By (i), for some non-empty $W \in \tau$,

$$W \subseteq CC^{-1} = (a^{-1}A \cap b^{-1}B) \cdot (A^{-1}a \cap B^{-1}b) \subseteq (a^{-1}A) \cdot (B^{-1}b).$$

As τ is translation invariant, $aWb^{-1} \in \tau$ and

$$aWb^{-1} \subseteq AB^{-1},$$

the latter since, for each $w \in W$, there are $x \in A$, $y \in B^{-1}$ with

$$w = a^{-1}x. \, yb : \qquad awb^{-1} = xy \in AB^{-1}.$$

So $\mathrm{int}_\tau(AB^{-1}) \neq \emptyset$. □

Corollary 15.5.2 *In a locally compact group the Haar density topology is a Steinhaus refinement.*

Proof Property (i) follows from Weil's Theorem since density-open sets are non-null measurable; left translation invariance in (ii) follows from left invariance of Haar measure, while involutive invariance holds, as any measurable set of positive Haar measure has non-null inverse by Lemma 15.4.3 (HewR1979, 15.14; cf. BinO2020d, §2, Lemma H). □

A weaker version, inspired by metric transitivity, comes from applying the following concept.

Definition Say that a group H *acts transitively* on a family $\mathcal{H} \subseteq \wp(G)$ if, for each $A, B \in \mathcal{H}$, there is $h \in H$ with $A \cap hB \in \mathcal{H}$.

Thus a locally compact topological group acts transitively on its non-null Haar measurable subsets (in fact, either-sidedly); this follows from Fubini's Theorem (Hal1950, 36C), via the average theorem (Hal1950, 59.F):

$$\int_G |g^{-1}A \cap B| \, dg = |A| \cdot |B^{-1}| \qquad (A, B \in \mathcal{M})$$

(note that $g = ab^{-1}$ if and only if $g^{-1}a = b$), cf. TomW2016, §11.3 after Th. 11.17.

In MatoZ2003 it is shown that in any non-locally compact abelian Polish group G there exist two non-Haar null sets, $A, B \notin \mathcal{HN}$, such that $A \cap hB \in \mathcal{HN}$ for all h; that is, G there does *not* act transitively on the non-Haar null sets.

Definition   In a quasi-topological group (H, τ), say that a proper σ-ideal $\mathcal{H}$ has the *simple Steinhaus property AA^{-1}* if AA^{-1} has interior points for universally measurable subsets $A \notin \mathcal{H}$ (cf. BartFN2011).

Proposition 15.5.3   *In a group (H, τ) with τ translation-invariant, if H acts transitively on a family of subsets $\mathcal{H}$ with the simple Steinhaus property, then $\mathcal{H}$ has the (composite) Steinhaus property:*

$$\mathrm{int}_\tau(AB^{-1}) \neq \emptyset \text{ for } A, B \in \mathcal{H}.$$

Furthermore, if $\mathcal{H}$ is preserved under inversion, then also

$$\mathrm{int}_\tau(AB) \neq \emptyset \text{ for } A, B \in \mathcal{H}.$$

Proof   For $A, B \in \mathcal{H}$ choose h with $C := A \cap hB \in \mathcal{H}$; then

$$CC^{-1}h = (A \cap hB)(A^{-1} \cap B^{-1}h^{-1}) \subseteq AB^{-1}. \qquad \square$$

Proposition 15.5.4   *If (H, τ) is a quasi-topological group (i.e. τ is invariant with continuous inversion) carrying a left-invariant σ-ideal $\mathcal{H}$ with the Steinhaus property and $\tau \cap \mathcal{H} = \{\emptyset\}$, then the ideal topology σ with basis*

$$\mathcal{B} := \{U \backslash N : U \in \tau, N \in \mathcal{H}\}$$

is a Steinhaus refinement of τ.

In particular, for (H, τ) an abelian Polish group, the ideal topology generated by its σ-ideal of Haar null subsets is a Steinhaus refinement.

Proof   If $U, V \in \mathcal{B}$ and $w \in U \cap V$, choose $M, N \in \mathcal{H}$ and $W_M, W_N \in \tau$ such that $x \in (W_M \backslash M) \subseteq U$ and $x \in (W_N \backslash N) \subseteq V$. Then, as $M \cup N \in \mathcal{H}$,

$$x \in (W_M \cap W_N) \backslash (M \cup N) \in \mathcal{B}.$$

So $\mathcal{B}$ generates a topology σ refining τ. With the same notation, $hU = hW_M \backslash hM \in \sigma$, as $hM \in H$, and $U^{-1} = W_M^{-1} \backslash M^{-1}$. Finally, UU^{-1} has nonempty τ-interior, as $U \notin \mathcal{H}$ and is non-empty.

As for the final assertion concerned with an *abelian* Polish group context, note that if N is Haar null (i.e. $N \in \mathcal{HN}$), then $\mu(hN) = 0$ for some $\mu \in \mathcal{P}(G)$ and all $h \in H$, so $hN \in \mathcal{HN}$ for all $h \in H$. Furthermore, if $A \notin \mathcal{HN}$, then $A^{-1} \notin \mathcal{HN}$: for otherwise, $\mu(hA^{-1}) = 0$ for some $\mu \in \mathcal{P}(G)$ and all $h \in H$; then, taking $\tilde{\mu}(B) = \mu(B^{-1})$ for Borel B, we have $\tilde{\mu}(A) = 0$ and $\tilde{\mu}(hA) = \mu(A^{-1}h^{-1}) = 0$ for all $h \in H$, a contradiction. $\qquad \square$

Remark   A left Haar null set need not be right Haar null: for one example, see ShiT1998, and for more general non-coincidence, see Sol2005, Cor. 6; cf. Dod2009. So the argument in Proposition 15.5.4 does not extend to the family of left Haar null sets $\mathcal{H}\mathcal{N}$ of a *non-commutative* Polish group. Indeed, Solecki (Sol2006, Th. 1.4) showed in the context of a countable product of countable groups that the simpler Steinhaus property holds for $\mathcal{H}\mathcal{N}_{\text{amb}}$ (involving simultaneous left- and right-sided translation – see BinO2021a, §1) if and only if $\mathcal{H}\mathcal{N}_{\text{amb}} = \mathcal{H}\mathcal{N}$.

Next, we reproduce a result from Kom1988, Th. 5. Recall that μ is *quasi-invariant* if μ-nullity is translation invariant. The transitivity assumption (of co-nullity) is motivated by *Smítal's Lemma*, which refers to a countable dense set – see KuczS1976.

Theorem 15.5.5 (Kominek's Theorem)   *If $\mu \in \mathcal{P}(G)$ is quasi-invariant and there exists a countable subset $H \subseteq G$ with HM co-null for all $M \in \mathcal{M}_+(\mu)$, then $\text{int}(AB^{-1}) \neq \emptyset$ for all $A, B \in \mathcal{M}_+(\mu)$.*

Proof   By regularity, we may assume $A, B \in \mathcal{M}_+(\mu)$ are compact, so AB^{-1} is compact. Fix $g \in G$; then, by quasi-invariance, $\mu(gB) > 0$. So by the transitivity assumption, both $G \backslash HgB$ and $G \backslash HA$ are null, and so $HA \cap HgB \neq \emptyset$. Say $h_1 a = h_2 gb$, for some $a \in A$, $b \in B$, $h_1, h_2 \in H$; then $g = h_2^{-1} h_1 ab^{-1}$. As g was arbitrary,

$$G = \bigcup_{h_1, h_2 \in H} h_2^{-1} h_1 AB^{-1}.$$

By Baire's Theorem, as H is countable, $\text{int}(AB^{-1}) \neq \emptyset$.   □

15.6 Borell's Interior-Point Property

For completeness of this overview of the Steinhaus–Weil interior-point property, we offer in brief here the context and statement of a (by now) classical Steinhaus-like result in probability theory; this differs in that the Polish group now specializes to an infinite-dimensional topological vector space and the reference measure is Gaussian, so no longer invariant. We refer to the related paper, BinO2019c, for further details and background literature, and to our generalizations to Polish groups and other reference measures.

For X a locally convex topological vector space, γ a probability measure on the σ-algebra of the cylinder sets generated by the dual space X^* (equivalently, for X separable Fréchet, e.g. separable Banach, the Borel sets), with $X^* \subseteq L^2(\gamma)$: then γ is called *Gaussian* on X ('gamma for Gaussian', following Bog1998) if and only if $\gamma \circ \ell^{-1}$ defined by

$$\gamma \circ \ell^{-1}(B) = \gamma(\ell^{-1}(B)) \qquad \text{for all Borel } B \subseteq \mathbb{R}$$

is Gaussian (normal) on $\mathbb{R}$ for every $\ell \in X^* \subseteq L^2(\gamma)$. For a monograph treatment of Gaussianity in a Hilbert-space setting, see Jan1997. Write $\gamma_h(K) := \gamma(K + h)$ for the translate by h. Recall that *relative quasi-invariance* of γ_h and γ means that, for all compact K,

$$\gamma_h(K) > 0 \quad \text{if and only if} \quad \gamma(K) > 0.$$

This property holds relative to a set of vectors $h \in X$ (termed the *admissible directions*) forming a vector subspace known as the *Cameron–Martin space*, $H(\gamma)$. Then γ_h and γ are equivalent, written $\gamma \sim \gamma_h$, if and only if $h \in H(\gamma)$. Indeed, if $\gamma \sim \gamma_h$ fails, then the two measures are mutually singular, i.e. $\gamma_h \perp \gamma$ (the Hajek–Feldman Theorem – cf. Bog1998, Th. 2.4.5, 2.7.2).

Continuing with the assumption above on X^*, as $X \subseteq X^{**} \subseteq L^2(\gamma)$, one can equip $H = H(\gamma)$ with a norm derived from that on $L^2(\gamma)$. In brief, this is done with reference to a natural covariance under γ obtained by regarding $f \in X^*$ as a random variable and working with its zero-mean version, $f - \gamma(f)$; then, for $h \in H$, we can represent δ_h^γ, i.e. the (shifted) evaluation map defined by $\delta_h^\gamma(f) := f(h) - \gamma(f)$ for $f \in X^*$, as $\langle f - \gamma(f), \hat{h}\rangle_{L^2(\gamma)}$ for some $\hat{h} \in L^2(\gamma)$. (Here, for γ symmetric $\gamma(f) = 0$, so $\delta_h^\gamma = \delta_h$ is the Dirac measure at h.) This is followed by identifying h with $\hat{h}$ (for $h \in H$), and noting that $|h|_H := \|\hat{h}\|_{L^2(\gamma)}$ is a norm on H arising from the inner product

$$(h, k)_H := \int_X \hat{h}(x)\hat{k}(x)\,d\gamma(x).$$

Formally, the construction first requires an extension of the domain of δ_h^γ to X_γ^*, the closed span of $\{x^* - \gamma(x^*) : x^* \in X^*\}$ in $L^2(\gamma)$, a Hilbert subspace in which to apply the Riesz Representation Theorem.

We may now state the Steinhaus-like property due, essentially in this form, to Christer Borell (Borel1976). (Proposition 1 of LeP1973 offers a weaker, 'one-dimensional section' form with the origin an interior point of the difference set relative to each line of H passing through it; we may call it the H-radial form by analogy with the $\mathbb{Q}$-radial form (Kucz1985, §10.1) of Euclidean spaces: the rational points are indeed an additive subgroup. The alternative term 'algebraic interior point' is also in use, e.g. in the literature of functional equations – cf. Brz1992.)

The following result is due to Borell (Borel1976, Cor. 4.1) – see Bog1998, p. 64.

Theorem 15.6.1 (Borell's Interior-Point Theorem)   *For γ a Gaussian measure on a locally convex topological space X with $X^* \subseteq L^2(\gamma)$, and A any non-null γ-measurable subset A of X, the difference set $A - A$ contains a $|.|_H$-open neighbourhood of 0 in the Cameron–Martin space $H = H(\gamma)$, above. That is, $(A - A) \cap H$ contains an H-open neighbourhood of 0.*

This follows from the continuity in h of the density of γ_h with respect to γ (Bog1998, Cor. 2.4.3), as given in the *Cameron–Martin–Girsanov formula*:

$$\exp\left(\hat{h}(x) - \frac{1}{2}\|\hat{h}\|^2_{L^2(\gamma)}\right), \tag{CM}$$

where $\hat{h}$ 'Riesz-represents' h, i.e. $x^*(h) = \langle x^*, \hat{h}\rangle$, for $x^* \in X^*$, as above. Thus here a modified Steinhaus Theorem holds: the *relative interior-point theorem*.

15.7 Haar-Meagre Sets in an Abelian Polish Group

We begin with Darji's definition, in an abelian Polish group X, of the *Haar-meagre* sets, which we denote $\mathcal{HM}_X$, and his theorem that these sets form a σ-ideal and that they are meagre. Both results rely on the Kuratowski–Ulam Theorem. We then proceed to give Jabłońska's proof of a Steinhaus–Weil–Piccard theorem for non-Haar-meagre sets and an open homomorphism theorem for 'Darji-measurable' functions. (A set is *Darji-measurable* if it is the union of a Borel set and a Haar-meagre set.)

Definition (Darj2013)   In a normed group X, say that $A \subseteq X$ is *Haar meagre* if, for some Borel set B with $A \subseteq B \subseteq X$, there is a continuous function $f\colon K \to B$ with K compact metric such that $f^{-1}(B + x)$ is meagre for each $x \in X$. Such a function f will be termed a *witnessing function*.

Remarks   1. Evidently for K compact and f a witnessing function $f(K)$ is compact and this covers $B \cap (B + x)$.

2. For X a *locally compact* group and M meagre, suppose, without loss of generality, M to be a meagre $\mathcal{F}_\sigma$. Take any U open in X with compact closure and put $K := \bar{U}$. As $(x + M)$ is meagre for any x, as also is $K \cap (x + M)$, we may take $f = \mathrm{id}_K$ (with $\mathrm{id}_K(x) \equiv x$ for $x \in K$) to witness that any meagre M in X is Haar meagre. Thus Darji's definition extends to the non-locally compact context an analogue of Haar null sets.

3. In a non-locally compact Polish group compact sets (and, by implication, σ-compact sets) are Haar meagre. Indeed, for such a set K, here again id_K is a witnessing function.

4. For a witnessing function $f\colon K \to B$ as above, any $k \in K$ and any norm open ball U in X with $f(k) \in U$, take $H := \overline{f^{-1}(U)}$. Then H is non-meagre in K, with $k \in H$, and the range of f on H is in the closed ball. Consider the translate $g\colon H \to B$ with

$$g(h) := f(h) - f(k), \text{ for } h \in H;$$

then $g^{-1}(B+x) = H \cap f^{-1}(B - f(k) + x)$ is meagre for any x. Thus the range of the witnessing function can be contained in an arbitrarily small ball around 0_X.

Theorem 15.7.1 (Darj2013)   *The family of Haar-meagre sets in X, $\mathcal{HM}$, is a σ-ideal.*

Proof   It suffices to consider a sequence of Haar meagre Borel sets A_n and associated witnessing functions $f_n : K_n \to A_n$ with the range of f_n lying in the ball of radius 2^{-n} around 0_X. Define $f : \prod_n K_n \to X$ by

$$f(\{k_n\}_n) = \sum_n f_n(k_n),$$

which is continuous. Since, for $A = \bigcup_j A_j$,

$$f^{-1}(A + x) = \bigcup_j f^{-1}(A_j + x),$$

it suffices to show that each set $f^{-1}(A_j + x)$ is meagre. To this end, fix j and view $\prod_n K_n$ as the product of K_j and $\prod_{n \neq j} K_n$. For any fixed $\{k_n\}_{n \neq j}$, the following set is meagre in K_j:

$$\left\{ a_j \in K_j : f_j(a_j) + \sum_{n \neq j} f_n(k_n) \in A_j + x \right\} = f_j^{-1}\left(A_j + x - \sum_{n \neq j} f_n(k_n) \right).$$

But the set $f^{-1}(A_j + x)$ is Borel (as A_j is Borel and f is continuous), so has the Baire property, and

$$f^{-1}(A_j + x) = \left\{ \{k_n\} : \sum_n f_n(k_n) \in A_j + x \right\}.$$

We may thus apply the Kuratowski–Ulam Theorem, 9.5.2, to conclude that $f^{-1}(A_j + x)$ is meagre, as claimed.   □

Theorem 15.7.2 (Darj2013)   *Haar-meagre sets are meagre: $\mathcal{HM} \subseteq \mathcal{M}$.*

Proof   For Borel $A \in \mathcal{HM}$ and $f : K \to A$ a witnessing function, put

$$\Sigma_A := \{(x, y) \in X \times K : f(y) \in (A - x)\}.$$

As A is Borel, Σ_A is also Borel and so has the Baire property. Now, for each $x \in X$,

$$\{y \in K : (x, y) \in \Sigma_A\} = f^{-1}(A - x),$$

so is meagre in K by assumption. Thus, by the Kuratowski–Ulam Theorem, Σ_A is meagre in $X \times K$. Applying the Kuratowski–Ulam Theorem again, but now the other way about, there is, in particular, $y \in K$ such that

$$B = \{x : (x, y) \in \Sigma_A\}$$

is meagre in X. But

$$B = \{x : f(y) \in (A - x)\} = \{x : x \in A - f(y)\} = A - f(y).$$

That is, $A - f(y)$ is meagre in X. However, in a Polish group, meagre sets are translation invariant; so A is meagre in X. □

Theorem 15.7.3 (Jab2015) *In an abelian Polish group, for Borel $A \notin \mathcal{HM}_X$,*

$$0 \in \text{int}(A - A).$$

Proof Suppose otherwise. Then, for some $A \notin \mathcal{HM}_X$, the origin 0 is a member but not an interior point of the set

$$\Delta(A) := \{x \in X : A \cap (A + x) \notin \mathcal{HM}_X\} \subseteq A - A;$$

indeed, $x \in A - A$ if and only if $A \cap (A + x) \neq \emptyset$. By the Birkhoff–Kakutani Theorem, 6.1.5, X may be equipped with a group-norm $\|. \|$. So we may choose $z_n \to 0$ with $\|z_n\|_X \leq 2^{-n}$ and $z_n \notin \Delta(A)$. Take

$$A_0 = A \backslash \bigcup_n A \cap (A + z_n).$$

Then A_0 is not Haar meagre, since $A \cap (A + z_n) \in \mathcal{HM}_X$ for each n. Give the Cantor group $K = \{0, 1\}^{\mathbb{N}}$ the group-norm:

$$\|k\| = \sum_n 2^{-n} k_n,$$

so that $\|e_n\| = 2^{-n}$ for the natural basis vectors, and define

$$g_z(k) = k \cdot z = \sum_n k_n z_n.$$

Then

$$\|g_z(k)\|_X \leq \sum_n \|k_n z_n\| = \sum_n k_n \|z_n\| \leq \sum_n k_n 2^{-n} = \|k\|.$$

Thus $g: K \to X$ is continuous and so there exists y with $g^{-1}(A_0 + y)$ not meagre in K. By the Piccard–Pettis Theorem, 9.2.1, for some $\delta > 0$,

$$B_0^K(\delta) \subseteq g^{-1}(A_0 + y) -_K g^{-1}(A_0 + y).$$

Choose n with $\|e_n\| < \delta$; then, for some $a, b \in g^{-1}(A_0 + y)$,

$$e_n = a -_K b.$$

Thus $a = b +_K e_n$; according as $b_n = 0$ or 1, working in X:

$$g_z(a) - g_z(b) = (1 + b_n) z_n - b_n z_n = \pm z_n.$$

That is, $\pm z_n \in (A_0 + y) - (A_0 + y) = A_0 - A_0$, a contradiction, since $z_n + a_0 = a_0' \in A_0$ for some $a_0, a_0' \in A_0$. Yet $A_0 \cap (A_0 + z_n)$ is disjoint from A_0. □

We may now apply the preceding theorem and the fact that any abelian Polish group is not Haar meagre to establish the next theorem, after the following definition.

Definition Call a set D-*measurable* if and only if it is the union of a Borel set and a Haar-meagre set. ('D for Darji'.)

Since the Borel sets form a σ-algebra and the Haar-meagre sets are a σ-ideal, the D-measurable sets also form a σ-algebra. It is thus natural to speak of D-*measurable maps*, when preimages of D-measurable sets are D-measurable.

Theorem 15.7.4 (Jab2015) *Any D-measurable homomorphism $f : X \to Y$ between abelian Polish groups X and Y is continuous.*

Proof We can work entirely in the closed image Z of f, as this as a subspace will also be a Polish group. To prove continuity at 0_X, consider any open neighbourhood $U = B^Z(\varepsilon)$ of 0_Y. Taking $V = B^Z(\varepsilon/2)$, we have $V - V \subseteq U$, and so, by symmetry of the implied metric,

$$Z = \overline{f(X)} \subseteq f(X) + V.$$

By separability of Z, there are points $x_n \in X$ with

$$f(X) \subseteq Z \subseteq \bigcup_n f(x_n) + V :$$
$$X \subseteq \bigcup_n f^{-1}(V) + x_n;$$

this last holds as f is a homomorphism. As X is not Haar meagre, neither is $f^{-1}(V) + x_n$ for some n (as $\mathcal{H}\mathcal{M}_X$ is a σ-ideal), and so neither is $f^{-1}(V)$. As f is D-measurable, $f^{-1}(V) = B \cup M$ for some Borel B and some $M \in H\mathcal{M}_X$. Since B is not Haar meagre in X, by the preceding theorem, for some $\delta > 0$,

$$B^X(\delta) \subseteq B - B \subseteq f^{-1}(V) - f^{-1}(V) \subseteq f^{-1}(V - V) \subseteq U.$$

This proves continuity.   □

Remark This result will also follow from the fact that non-Haar-meagre sets in an abelian Polish group are shift-compact, our next result. Thus there are meagre sets which, by virtue of being not Haar meagre (such sets exist in non-locally compact groups), are shift-compact.

The following theorem is enuciated for a Borel set A; however, the proof works just as well for any set that is universally Baire.

Theorem 15.7.5 (BanaJ2019) *In an abelian Polish group, any Borel $A \notin \mathcal{H}\mathcal{M}_X$ is shift-compact.*

Proof Suppose otherwise. Then there is a null-sequence $\{z_n\}_n$ such that for any $x \in X$ the set $\{n \in \mathbb{N} : x + z_n \in A\}$ is finite. Passing to a subsequence, if necessary, we can assume that $\|z_n\| < 2^{-n}$ for all n. Taking

$$K_i := \{0\} \cup \{z_m\}_{m \geq i},$$

which is compact (in X), define on $K := \prod_{n \in \mathbb{N}} K_n$ the function

$$g(\{x_n\}) := \sum_{n \in \mathbb{N}} x_n,$$

which is continuous. Since the set A is not Haar meagre and is Borel, there exists a point $x \in X$ such that

$$B := g^{-1}(x + A)$$

is non-meagre and Baire in K (by continuity of g). Hence its (non-empty) quasi-interior $U \subset K$ has $U \cap B$ co-meagre in U. Passing to a basic subset of U, we may, without loss of generality, assume that $U := \{b\} \times \prod_{n \geq J} K_n$, for some $J \in \mathbb{N}$ and some $b \in \prod_{n < J} K_n$. For $m \geq J$, each constituent subset of U of the form $\{b\} \times \{z_m\} \times \prod_{n > J} K_n$ is open in K, and so

$$C_m := \left\{ y \in \prod_{n > J} K_n : \{b\} \times \{z_m\} \times \{y\} \subset U \cap B \right\}$$

is co-meagre in $\prod_{n>J} K_n$. Hence $\bigcap_{m \geq J} C_m$ is co-meagre in $\prod_{n>J} K_n$, and so contains at least one point y for which we have

$$\{b\} \times (K_j \backslash \{0\}) \times \{y\} \subset B.$$

Fix such a point y. For v with $\{v\} = \{b\} \times \{0\} \times \{y\}$ and $m \geq J$, we have, by considering $\{b\} \times \{z_m\} \times \{y\}$, that

$$g(v) + z_m \in g(B) = x + A.$$

Put $u := -x + g(v)$; then $u + z_m \in A$ for all $m \geq J$. So $\{n \in \mathbb{N} : u + z_n \in A\}$ is infinite, contradicting the choice of $(z_n)_{n \in \mathbb{N}}$. $\qquad\square$

15.8 Haar-Null Sets Revisited (the Abelian Case)

We return to the result in Theorem 15.3.8 where we showed that, in a Polish group which is amenable at 1, a set that is not (left) Haar null (in the sense of Christensen and Solecki) is shift-compact. When the group is abelian this can be proved by a method that is highly reminiscent of a similar result concerning sets that are not Haar meagre. The argument here again comes from BanaJ2019.

Theorem 15.8.1 *In an abelian Polish group any non-Haar null universally measurable set is shift-compact.*

Proof (BanaJ2019) Suppose otherwise, and suppose too that A is universally measurable and not Haar null in X, but that there is a null-sequence $(z_n)_{n \in \mathbb{N}}$ such that for each $x \in X$ the set $\{n \in \mathbb{N} : x + z_n \in A\}$ is finite. Passing to a subsequence, without loss of generality we may assume that $\|k_i z_i\| \le 2^{-i}$ for $n \in \mathbb{N}$. Now consider the compact product space of finite discrete spaces

$$\Pi := \prod_{n \in \mathbb{N}} \{0, 1, \dots, 2^n\},$$

endowed with the product measure λ of uniform measures. On Π define the map

$$g(k) = \sum_{i=1}^{\infty} k_i z_i,$$

where we note that k_i are integers. Since $\|k_i z_i\| \le 2^i$, as $\|z_i\| \le 4^{-i}$, the series is convergent and the function g is well defined and continuous.

Since A is not Haar null, there exists $x \in X$ such that $g^{-1}(x + A)$ has positive λ-measure, and so there is a compact $K \subseteq g^{-1}(x + A)$ of positive measure. For every $n \in \mathbb{N}$ consider the subcube

$$\Pi_n := \prod_{i=1}^{n-1} \{0, 1, 2, \dots, 2^i\} \times \prod_{i=n}^{\infty} \{1, 2, \dots, 2^i\} \subseteq \Pi$$

and observe that $\lambda(\Pi_n) \to 1$ as $n \to \infty$. Replacing K by $K \cap \Pi_\ell$ for a sufficiently large ℓ, we can assume that $K \subset \Pi_\ell$. For $m \ge \ell$ define the 'back-tracking' map from Π_n to Π by

$$s_m(y) = \begin{cases} y_i, & i \ne m, \\ y_i - 1, & i = m. \end{cases}$$

Claim *For any compact set $C \subset \Pi_\ell$ with $\lambda(C) > 0$ and $\varepsilon > 0$,*

$$\lambda(C \cap s_m(C)) > (1 - \varepsilon)\lambda(C)$$

for all large enough $m > \ell$.

Proof of Claim By the regularity of the measure λ, the set C has a neighbourhood $O(C) \subset \Pi$ such that $\lambda(O(C) \backslash C) < \varepsilon \cdot \lambda(C)$. For each $c \in C$ there is $m(c)$ such that $s_m(c) \in O(C)$. By the compactness of C, there exists $k \ge \ell$ such that, for any $m \ge k$, the image $s_m(C) \subseteq O(C)$. Hence

$$\lambda(s_m(C) \backslash C) \le \lambda(O(C) \backslash C) < \varepsilon \cdot \lambda(C)$$

and so

$$\lambda(s_m(C) \cap C) = \lambda(s_m(C)) - \lambda(s_m(C) \backslash C) > \lambda(C) - \varepsilon \cdot \lambda(C) = (1 - \varepsilon)\lambda(C),$$

and the claim is proved.

Returning to the proof of the theorem, using the claim we can choose an increasing sequence of integers $(m(k))_{k \in \mathbb{N}}$ with $m_0 > \ell$ such that the set

$$K_\infty := \bigcap\nolimits_{k \in \mathbb{N}} s_{m(k)}(K)$$

has positive measure and so contains a point $b := (b(i))_{i \in \mathbb{N}}$. It follows that, for each $k \in \mathbb{N}$, $b_k := s_{m(k)}^{-1}(b) \in K \subset g^{-1}(x + A)$. Now $b_k(m(k)) = b(m(k)) + 1$ and so

$$g(b) + z_{m(k)} = \sum\nolimits_{n \neq m(k)} b(n)z_n + (b(m(k)) + 1)z_{m(k)} = g(b_k) \in x + A.$$

So $-x + g(b) + z_{m(k)} \in A$ for all $k \in \mathbb{N}$, which contradicts the choice of the sequence $(z_n)_{n \in \mathbb{N}}$. $\qquad\square$

16

Axiomatics of Set Theory

16.1 The Three Elephants in the Room

We summarize here briefly some of the contents of our 2019 survey, 'Set Theory and the Analyst' (BinO2019a). We begin with the three 'elephants in the room'. (An 'elephant in the room' is something which is all too obviously there, but which no one wants to mention.)

The Canonical Status of the Reals

The first of the three elephants in BinO2019a, p. 5, concerns *the canonical status of the reals* $\mathbb{R}$. The rationals, $\mathbb{Q}$, are canonical from any point of view. The reals are canonical (so justifying the use of the definite article here) *from many points of view, including uniqueness as a complete Archimedean ordered field up to isomorphism, but not including cardinality*. The *Continuum Hypothesis* (CH)

$$2^{\aleph_0} = \aleph_1 \tag{CH}$$

(see, e.g., Kec1995, 16D; the notation on the left derives from binary expansion of the reals), that the cardinality of $\mathbb{R}$ is the smallest uncountable cardinal, is and will always remain just that, a hypothesis. For, as Cohen (Coh1963) showed, both CH and its negation are consistent with the 'ordinary rules of mathematics', Zermelo–Fraenkel set theory (ZF).

So one can choose which cardinality of the reals one wishes to work with (indeed, 'which real line one wishes to work with'). As Solovay (Solo1965) puts it, 'it (the cardinality of the reals) can be anything it ought to be'. The only constraint is that its cofinality be uncountable, by a result of König of 1905 – see, e.g., Kun1983, §10.40; Kun2011, I.13.12.

If one drops the Archimedean requirement, other possibilities emerge. For such 'super-real' fields, see DalW1996.

Impossibility of Proving Consistency of Set Theory

The third elephant (BinO2019a, p. 7) was disposed of in 1931 by Gödel's incompleteness theorems (God1940): first, the existence of sentences that can be neither proved nor disproved; second, the impossibility of a rich enough axiom system being able to provide a proof of its own consistency (rich enough to accommodate arithmetic).[1] Thus, neither ZF nor ZFC (ZF plus the Axiom of Choice, AC) can be proved consistent in ZF, resp. ZFC. These apparently disturbing results are facts of life, with which one must live in mathematics.

Which Sets of Reals Can We Use?

There remains the second elephant (BinO2019a, p. 6): which *sets* of reals are available? Here (perhaps not surprisingly in view of the above) the answer depends crucially on what *axioms of set theory* one assumes, hence this chapter. At one extreme, one can assume AC and work with ZFC, giving a maximal supply of sets to work with. A first advantage of this choice is that it yields the Vitali example of a non-measurable set (see, e.g., Rud1966, §2.22; SteS2005, pp. 24–25), albeit non-constructively.

At the other extreme, one could assume, instead of AC, the axiom of Dependent Choice(s) (DC). (This is what is needed to make mathematical induction work, a minimal requirement for a useful axiom system in mathematics.) One could augment this with LM (the axiom that all sets are *Lebesgue-measurable*) and/or PB (that all sets have the *property of Baire*).

Intermediate between these is PD – that all sets in the *projective* hierarchy (§3.2) are *determined* (§2.4). This determinacy is in the sense that in the relevant Banach–Mazur game, one player has a winning strategy. See also BinO2019a, §7.2. For the connection between large cardinals (below in §16.3) and PD, see JudSp1997.

16.2 Hypotheses and Axioms

Generalized Continuum Hypothesis

Much stronger than CH is the generalized continuum hypothesis (GCH): writing κ^+ for the successor cardinal of an infinite cardinal κ,

$$2^\kappa = \kappa^+ \qquad \text{for every infinite cardinal } \kappa. \qquad \text{(GCH)}$$

Gödel proved (God1940) that GCH is consistent with ZF. Foreman and Woodin (ForW1991) showed that 'the generalized continuum hypothesis can fail everywhere'.

---

[1] There are consistent systems capable of proving their own consistency, but with restricted arithmetic power, see, e.g., Hilbert's axiomatization of arithmetic, denoted A_0 in Kneebone (Kne1963).

Gödel's Axiom of Constructibility, V = L

Here (see, e.g., BinO2019a, §2) one may begin with von Neumann's 1923 definition of the natural numbers $\mathbb{N}$ starting naturally with 0:

$$0 = \emptyset; \quad 1 = \{0\}; \quad 2 := \{0, 1\}; \quad 3 := \{0, 1, 2\}; \quad \cdots ;$$
$$n + 1 : = n \cup \{n\} := n \cup \{0, 1, \ldots, n - 1\}.$$

The von Neumann scheme can be naturally extended: once the class of ordinals $\alpha \in On$ (below) is established, it yields the *cumulative hierarchy* V_α, introduced inductively by

$$V_{\alpha+1} := \wp(V_\alpha), \qquad V_\lambda := \bigcup \{V_\alpha : \alpha < \lambda\},$$

with $\wp(V_\alpha)$ the power set of V_α, where λ is a limit ordinal. The 'class of all sets' is then

$$V := \bigcup \{V_\alpha : \alpha \in On\},$$

known as the von Neumann universe, which one may view as the 'Cantor universe'. (Here we recall the dual framework of formal set theory which distinguishes *sets* admitted by the axioms and *classes* defined by arbitrary 'properties', expressed in the formal language of set theory; this is the established formalization aimed at avoiding Russell's paradox. Classes may fail to be sets. Thus an ordinal is a set well ordered, transitively, by membership; however, since the class of ordinals is well ordered by membership, it fails to be a set, or it would be a member of itself; that would contradict the ZF axiom of well-foundedness of membership.)

The 'class of all constructible sets', L, is defined similarly. For a set S, write $cl(S)$ for the closure of S under the Gödel operations which include operations admitted by the ZF axioms as leading from sets to sets (such as the union of two sets: for these, see, e.g., Jec1973, §3.4). Define

$$L_0 := 0, \qquad L_\alpha := \bigcup_{\beta < \alpha} L_\beta \qquad \text{if } \alpha \text{ is a limit ordinal,}$$
$$L_{\alpha+1} := \wp(L_\alpha) \cap cl(L_\alpha \cup \{L_\alpha\}),$$
$$L : = \bigcup_{\alpha \in On} L_\alpha.$$

This gives the *constructible hierarchy L*, with $L \subset V$.

Gödel's *Axiom of Constructibility* (God1940) is

$$V = L$$

'all sets are constructible'. It is stronger than both GCH and AC; see, e.g., Dev1977, §1. For a monograph treatment, see Dev1984.

Without further axioms, one cannot say whether a given set has a non-constructible subset; see the section on zero-sharp in §16.3.

The class L, as a subclass of V, is said to be an *inner model* of set theory. Gödel's contribution thus opened the door to the study of more general inner models, with L the minimal such model. Thus one may expand L to include sets not in L to enable their study. In the simplest case one takes a *transitive* set x (one on which membership is a transitive relation) and starts a *relative constructible hierarchy* (relative to x) with $L_0 = x$, applying at each stage α the Gödel operations to $L_\alpha(x)$ to obtain the class $L(x)$ (see, e.g., Dra1974, p. 149). Two examples of interest below are $L(x)$, for $x \subseteq \omega$ (so x here is interpreted via its characteristic function as a real number), and $L(\mathbb{R})$, where again the members of $\mathbb{R}$ are interpreted as subsets of ω. The axiom system ZF holds in the latter inner model, but AC fails as the model has all sets of reals Lebesgue-measurable and Baire: see Woo1988. These examples adjoin to L sets from the low levels of V. Maximal inner models, known as *core models*, adjoin to L structures whose elements extend all the way up to below some 'large' cardinal (for which see below in §16.3), enabling the study of strong axioms of infinity. For instance, adjoin an ultrafilter U, presumed to exist on a large enough cardinal κ, yielding a class denoted by $L[U]$. Here the transfinite induction again mimics Gödel's but constructs $L_{\alpha+1}[U]$ by reference to the set $U \cap L_\alpha[U]$ (see, e.g., Dra1974, p. 151). For a textbook account, see Zem2002 or the later *Handbook of Set Theory* from 2010 (ForK2010).

Martin's Axiom

A topological space X is said to satisfy the *countable chain condition* (ccc) if every disjoint family of its open sets is countable. Thus $\mathbb{R}$ satisfies ccc (as does any separable space); connected to this is *Souslin's hypothesis* (SH) that every (non-empty) complete, dense, linear order without first or last element is order isomorphic with $\mathbb{R}$, the study of which gave rise to *Martin's Axiom* (MA). Its statement is that (writing $\mathfrak{c}$ for the cardinality of the continuum) in every non-empty compact space satisfying ccc, the intersection of fewer than $\mathfrak{c}$ open dense sets is non-empty. In this context the statement thus asserts Baire's Theorem as extending from countable to fewer than $\mathfrak{c}$ intersections. The Continuum Hypothesis (CH) is equivalent to the same statement but without the condition ccc. So,

CH Implies MA.

Regarding consistency, it is known that if ZFC is consistent, it remains consistent after adding any one of the three additional axioms CH, MA and MA + ¬CH (MA and the negation of CH).

The last of these axioms, MA + ¬CH, implies SH; on the other hand, $V = L$ (in fact a much weaker combinatorial hypothesis, implied by this axiom) contradicts SH, by constructing a counterexample.

Martin's Axiom appears in the Martin–Solovay paper (MartS1970). It has been extremely influential, *as a substitute for and weakening of the CH*. For a monograph treatment of its many and fruitful consequences, see Fre1984; see also Weis1984.

Countability, Category and Measure

As above, Martin's Axiom may be regarded as freeing (so far as possible) Baire category from countability. There is a measure-theoretic counterpart to this: *tau-additivity* – see Fre2003 for a monograph treatment. A measure μ is *tau-additive* (or τ-additive) if, for any increasing (perhaps uncountable) family of sets, $(A_\lambda)_{\lambda < \kappa}$ say,

$$\sup\nolimits_{\lambda < \kappa} \mu(A_\lambda) = \mu(\sup\nolimits_{\lambda < \kappa} A_\lambda), \quad \text{i.e.} \quad = \mu\left(\bigcup_{\lambda < \kappa} A_\lambda\right).$$

The concept of tau-additivity should be compared with the more restrictive property of a measure μ being *κ-additive* where for any disjoint family $\{A_\lambda : \lambda < \eta\}$ with $\eta < \kappa$

$$\mu\left(\bigcup\nolimits_{\lambda < \eta} A_\lambda\right) = \sum\nolimits_{\lambda < \eta} \mu(A_\lambda).$$

16.3 Ordinals and Cardinals

First, recall the ordinal numbers (§16.2; see, e.g., Dev1993, Ch. 3). The first ordinal is 0, the second is $1 := \{0\}$, and the $(n + 1)$th is $n := \{0, 1, \ldots, n - 1\}$ (as in the von Neumann definition of $\mathbb{N}$ above). Then the first infinite ordinal is $\omega := \{0, 1, 2, \ldots, n, \ldots\}$, the second is $\omega + 1 := \{0, 1, 2, \ldots, n, \ldots, \omega\}$, etc. We write

$$\alpha + 1 := \alpha \cup \{\alpha\},$$

the *successor ordinal* of α. Other ordinals are called *limit ordinals*.

Next, recall the cardinal numbers (or just cardinals) $\aleph_\alpha$, indexed by the ordinals, using Cantor's notation. Call two sets *equipotent* if there is a bijection between them. This is an equivalence relation on any given collection of sets. Extending the order relation on the natural numbers $\mathbb{N}$ to infinite sets involves the Schröder–Bernstein Theorem: if two sets have injections from each to the other, there is a bijection between them. (For proof, see, e.g., BirM1953, or

Cohn1989, both algebra texts; AC is not needed here, although DC is, cf. Dev1993, pp. 77–78.) The *cardinals* are the *initial ordinals* α for which there is no surjection $f : \beta \to \alpha$ for an ordinal $\beta < \alpha$. This introduces a second ordinal-inspired notation for cardinals, due to von Neumann, replacing aleph with omega so that

$$\aleph_\alpha = \omega_\alpha \qquad (\alpha \in On),$$

with $\omega_0 = \omega$. The finite cardinals n are just the finite ordinals, the number of points in an n-point set; the smallest infinite cardinal is $\aleph_0$, the cardinality of $\mathbb{N}$ (countable); the uncountable cardinals are $\aleph_1, \aleph_2, \ldots ..$ We regard any ordinal (including any cardinal) as the set of its predecessors, as above. In brief: *ordinals are absolute, cardinals are relative* (cf. BinO2019a, p. 12), the latter depending on what surjections are admitted on a given 'putative' cardinal.

There is a further (fourth) 'elephant in the room' here: the concept of a cardinality attaching to sets presents a call on AC to choose a cardinal and a bijection.

The Continuum Hypothesis may thus be written

$$\mathfrak{c} = \aleph_1. \tag{CH}$$

Measurable Cardinals

We quote the following result of Ulam (Ula1930): if a finite countably additive measure μ is defined on all subsets of a set X of cardinality $\aleph_1$, and vanishes on all singletons, then μ vanishes identically.

Ulam's Theorem leads to the related concepts of measurable cardinal and Ulam number. A cardinal κ is called *real measurable* (in brief, *measurable*) if there is some topological space of cardinality κ and a κ-additive probability measure μ defined on *all* its subsets and vanishing on singletons, *non-measurable* otherwise. More restrictively, if the probability measure here takes only the value 0 and 1, κ is called *two-valued measurable* (so two-valued non-measurability is less restrictive than (real) non-measurability). Non-measurable cardinals are called *Ulam numbers*. (In measure theory, measurable sets are the 'well-behaved' sets; here, it is the non-measurable cardinals that are the 'well-behaved' cardinals. So we will say 'two-valued non-measurable' rather than 'not two-valued measurable', etc.) The countable cardinal $\aleph_0$ is non-measurable. Any cardinal less than a non-measurable one is non-measurable, so the non-measurable cardinals form an 'initial interval' in the cardinals. By the Ulam–Tarski Theorem, the immediate successor of a non-measurable is non-measurable, and the supremum of non-measurably many non-measurables is non-measurable (see, e.g., Bog2007a, Th. 1.12.42 or Fed1969, 2.1.6).

Existence of Measurable Cardinals

Write M for the axiom that there exists a measurable cardinal, ZFM for ZF + M. In ZFM, non-constructible sets exist (Sco1961). So Gödel's Axiom of Constructibility $V = L$ is *incompatible with the existence of a measurable cardinal* (LevS1967, p. 240). Unfortunately, this casts no light on CH. If ZFM is consistent, then so are both ZFM + CH and ZFM + ¬CH (LevS1967, Th. 1). Similarly for ZFM and Souslin's hypothesis SH (above) and its negation, and ZFM + all projective sets are Lebesgue-measurable (LevS1967, p. 236).

Classes of Cardinals

Call a cardinal κ *inaccessible* if the class of all smaller cardinals has no maximal element, and no set comprising cardinals less than κ of cardinality less than κ has supremum κ. Furthermore, $2^\lambda < \kappa$ for all cardinals $\lambda < \kappa$ (the latter condition referred to as strong inaccessibility). So if measurable cardinals exist, the smallest one is inaccessible, and so all are. We quote (see, e.g., Bog2007a, 1.12(x)):

(a) If κ is two-valued non-measurable, so is 2^κ.

(b) $\aleph_1$ is non-measurable, being the successor of the non-measurable $\aleph_0$.

(c) $\mathfrak{c}$ is two-valued non-measurable.

This follows from (a). If further one assumes Martin's Axiom, then:

(MA) $\mathfrak{c}$ is (real) non-measurable.

Ramsey Cardinals

The branch of infinite combinatorics relevant here, *partition calculus*, derives from Erdős and Rado; see, e.g., BinO2019a, §4.1; Dra1974, Chs. 7, 8. Here, as there, it is convenient to use Erdős's notation: write

$$\kappa \to (\alpha)_2^{<\omega}$$

as shorthand for: if the class $[\kappa]^{<\omega}$ of finite subsets of κ is partitioned into two classes (using two colours, say), then there is a monochromatic ('homogeneous') subset of κ of order type (isomorphism type, of linearly ordered sets) α. Ramsey's Theorem (Rams1930) is then written $\omega \to (\omega)_2^2$.

For κ measurable, one has further

$$\kappa \to (\kappa)_2^{<\omega}$$

(see BinO2019a, §4.2), the definition of a *Ramsey cardinal*. For accounts of Ramsey theory, see also, e.g., Kec1995, §19; Bol1998, Ch. VI; HinS1998, Chs. 5, 18.

Supercompact Cardinals

We need three further kinds of large cardinal (see, e.g., BinO2019a, §4.3; Mit2010, pp. 1487–1488). A cardinal κ is *supercompact* if it is λ-supercompact for all $\lambda \geq \kappa$; here κ is λ-supercompact if there is a (necessarily non-trivial) elementary embedding $j = j_\lambda \colon V \to M$ with M a transitive class, such that j has critical point κ, and $M^\lambda \subseteq M$, i.e. M is closed under arbitrary sequences of length λ. Under AC, without loss of generality, $j(\kappa) > \lambda$.

Strong Cardinals

For κ a cardinal and $\lambda > \kappa$ an ordinal, κ is said to be λ-*strong* if for some transitive inner model (transitive class containing the ordinals), M say, there is an elementary embedding (elementary, as below), $j_\lambda \colon V \to M$ with critical point κ, $j_\lambda(\kappa) \geq \lambda$, and

$$V_\lambda \subseteq M.$$

Furthermore, κ is said to be a *strong cardinal* if it is λ-strong for all ordinals $\lambda > \kappa$.

Elementarity here concerns the requirement that any sentence referring to a finite number of elements of the structure $\langle V, \in \rangle$ holds in the structure if and only if the same sentence, but rewritten so as to refer to the j images of the said finite set, holds in $\langle M, \in \rangle$ (BelS1969, Ch. 4).

This notion may be relativized to subsets S to yield the concept of λ-S-*strong* by requiring, in place of the inclusion above, only that

$$j(S) \cap V_\lambda = S \cap V_\lambda;$$

one says that j 'preserves' S up to λ.

Woodin Cardinals

This leads to our final definition. The cardinal δ is a *Woodin cardinal* (Woo1999) if δ is strongly inaccessible, and, for each $S \subseteq V_\delta$, there exists a cardinal $\theta < \delta$ which is λ-S-strong for every $\lambda < \delta$. (So the last of these three embedding-based definitions calls for more 'preservation' than the second, but less than the first.)

The *consistency strength* (see the next paragraph) of various extensions of the standard axioms, ZFC, by the addition of further axioms, may then be compared (perhaps even assessed on a well-ordered scale) by determining which canonical large-cardinal hypothesis will suffice to create a model for the proposed extension. Thus, for κ supercompact, V_κ is a model for the sentence asserting that a strong cardinal exists (abbreviated, using satisfaction $\models$ as below, to $V_\kappa \models \exists \mu['\mu$ is strong']) which places supercompact 'above' strong, in the sense that the assumption of the existence of a supercompact cardinal is stronger than the assumption of the existence of a strong cardinal (indeed, also of a strong cardinal below a supercompact). Likewise, for κ strong, V_κ is a model in which the sentence asserting the existence of a measurable cardinal holds (abbreviated to $V_\kappa \models \exists\mu[\text{‘}\mu$ is measurable’]), placing measurability ‘below’ strong, in the same sense. (Below that in turn is the existence of a Ramsey cardinal, as above.)

Consistency Strength

We have just met measurable and inaccessible cardinals and ZFC, whose consistency we assume. But this cannot imply the existence of an inaccessible cardinal κ, because of Gödel's incompleteness theorem (see, e.g., BinO2019a, p. 16), as otherwise ZF would provide a proof of its own consistency, since all the axioms of ZF would be provably satisfied within the set V_κ. So to go further here, one needs a stronger axiom of infinity (stronger than that which admits $\mathbb{N}$ as a set) – or *axioms*. These involve *large cardinals* (see, e.g., Dra1974 or BinO2019a, §§1,4,7,10). The axioms that we will use are comparable in their logical strength (are ordered by implication). Using ‘>’ as shorthand for ‘existence of needs a stronger assumption than’, the categories of large cardinal we use can be ranked, or compared, as follows (BinO2019a, p. 17):

> supercompact > Woodin > strong > measurable > Ramsey.

Zero Sharp

For background on $0^\sharp$ and other sharps, see, e.g., Mit2010; BinO2019a, p. 21; Dra1974, §8.5 and the brief account below. As existence of Ramsey cardinals is in the least restrictive position in the diagram above, we note (again, see, e.g., Mit2010):

(i) The existence of $0^\sharp$ is an immediate consequence of that of a Ramsey cardinal.

(ii) ‘Assuming the existence of a large cardinal (a Ramsey cardinal is much more than enough), it can be shown to be consistent that $0^{\sharp\alpha}$ exists for all ordinals α.’

To describe $0^\sharp$ requires first some notation.

Recall Gödel's use of natural numbers to code formulas of the standard formal language of set theory (involving the symbols = and $\in$). We write $\ulcorner\varphi\urcorner$ for the number coding the formula φ. For a transitive set or class M and formula φ with no more than n *free* variables, these to be denoted $x_1, \ldots, x_n$, we write $M \models \varphi[a_1, \ldots, a_n]$ to mean that φ is satisfied (holds) in M when x_1 is interpreted as referring to a_1, x_2 as $a_2, \ldots$ (where $a_1, \ldots, a_n$ are in M). For M a set, the relation of satisfaction can be defined formally by induction on the complexity of the formula φ (starting with $M \models (x_1 \in x_2)[a, b]$ if and only if

$a \in b$); we note the need to go outside M to describe what goes on in M, when eliminating the operation of quantifiers, as in $M \models \exists x (x \in x_1)[b]$ if and only if there exists $a \in M$ with $a \in b$.

Define f on ω by $f(i) = \omega_i$ and, following Dra1974, p. 257, put

$$0^\# := \{ \ulcorner \varphi \urcorner : L \models \varphi[f] \},$$

so that any free variables in any formula φ are assigned interpretations from the sequence f.

Despite Tarski's Theorem on the undefinability of truth in a structure within itself, this may be a well-defined subset of ω assuming, for instance, the existence of a cardinal κ for which the partition relation $\kappa \to (\omega_1)^{<\omega}$ holds, e.g. if κ is measurable or a Ramsey cardinal. In such circumstances, the structures L_λ for λ an uncountable cardinal constitute an elementary chain, allowing passage from 'language to meta-language', so that one can define truth in L in terms of satisfaction in L_λ for large enough λ.

The phrase '$0^\#$ exists' is used to describe circumstances that enable satisfaction in L to be definable. An insightful result due to Kunen (Kun1970) gives as a necessary and sufficient condition the existence of a non-trivial elementary embedding of L to itself (one that moves ordinals above a certain critical point).

A similar definition, allowing relativization to a parameter $x \subseteq \omega$, can be given for $x^\#$, in which L is replaced by $L(x)$.

The proof of Kunen's characterization of the existence of $0^\#$ involving a further equivalence to the existence of a closed unbounded class of L-indiscernibles (for which see our survey paper) may be found in several textbooks, including those of Jech (Jec1997), Devlin (Dev1984) and Kanamori (Kan2003). Indeed, the definition we give, after Drake, relies on a specific sequence, f, of indiscernibles.

Epilogue: Topological Regular Variation

Recall from the Preface that the book grew out of our work on Topological Regular Variation, then our intended title for the book, now that of this Epilogue, and the use of the Prologue and Epilogue on regular variation as a framing device for our main text, Chapters 2 to 16.

The text below may be viewed as a brief summary of (or, overview of and introduction to) the relevant papers in our corpus:

BinO2009b: generic regular variation;
BinO2009c: the index theorem;
BinO2009d: foundations;
Ost2010: topological dynamics;
BinO2010b: regular variation without limits;
BinO2010c; BinO2010d; BinO2010e: the topological regular variation trilogy;
BinO2013: Steinhaus theory and regular variation;
BinO2014: Beurling aspects.

The significant contribution regular variation plays in the classical probability theory of random variables naturally guided the research when the context more recently expanded to finite-dimensional vector spaces (see MeeS2001) and further to function spaces such as $C[0, 1]$ and the Skorokhod space $D[0, 1]$ of $\mathbb{R}^d$-valued càdlàg functions on $[0, 1]$ (see HulLM2005). It was realized by Bajšanski and Karamata as early as 1969 (in BajK1969) that some of their foundational work on regular variation can in fact be conducted in a group-theoretic framework working with functions $h: G \to H$ between topological groups G and H. Their investigation was brief[1] and with the sole exception of Bal1973 that point of view was not pursued much further for three decades.

---

[1] The notion of convergence in BajK1969 was based on filters; we have espoused a sequential approach.

Our purpose here is to show how the mathematical panoply assembled in the preceding chapters can upgrade the classical theory and how its algebraic stance provides a hitherto unknown lens through which to see and simplify classical arguments. The principal tool marries the Category Embedding Theorem with the Effros Theorem. The first theorem embeds into any non-negligible set S, quasi always in S, any *null sequence* ψ_n of homeomorphisms (i.e. homeomorphisms converging to the identity), $\psi_n \to id_X$ meaning that, for q.a. $s \in S$,

$$\{\psi_n(s) : m \in \mathbb{M}_s\} \subseteq S$$

for some infinite set $\mathbb{M}_s \subseteq \mathbb{N}$. The second theorem provides, for any convergent sequence $z_n \to z_0$, a null sequence of homeomorphisms with $\psi_n(z_0) = z_n$. We may call such a sequence an *Effros null sequence*.

Uniform Convergence Theorem
The foundation stone of classical regular variation, and so the primary object of study for its topological counterpart, is the Uniform Convergence Theorem (UCT). This assumes in its multiplicative form, as we recall from the Prologue (§P.8), a *regularly varying* function $h \colon \mathbb{R} \to \mathbb{R}$; that is, one with a well-defined limit (its kernel):

$$k(t) := \lim_{x \to \infty} h(tx)/h(x) \text{ for } t \in \mathbb{R}_+ := (0, \infty),$$

and concludes uniform convergence as t ranges over compact sets, but subject to h being (Lebesgue) measurable or Baire (having the Baire property). The multiplicative version of regular variation hints already that it is natural to study a normed group X (i.e. a group with a group-norm, as in Chapter 6) and a *Baire* map $h \colon X \to H$ with H also a normed group. The expression tx is then interpreted in the context of a group action $\varphi(t, x)$ on X provided by the group $\mathcal{H}(X)$ of norm-bounded autohomeomorphisms of X under the supremum group-norm:

$$\|\alpha\|_\infty = \|\alpha\|_{\mathcal{H}} =: \sup_{x \in X} \|\alpha(x)\|_X.$$

For connections to the compact-open topology on the homeomorphisms (Eng1989, §3.4 or Kur1968, Ch. 4, §44), see Mille O2012. See also Dij2005. The action may be restricted to some subgroup of homeomorphisms of interest, $T \subseteq \mathcal{H}(X)$; for instance, T might comprise a group of *translations* of X, say the left translations $\lambda_x \colon u \mapsto xu$ for $x, u \in X$. However, taken in as wide a context as the premise of UCT allows (evidently, with H a normed group, with identity element e_H), X itself may more simply be a topological space on which a topological group T acts *transitively,* thus making X *homogeneous.* (See also Ost2010 for a more general context.) An associated definition of regular variation gives meaning to *slow variation* in the following sequential convergence:

$$h(\varphi_n(x))h(\varphi_n(z_0))^{-1} \to e_H \qquad (x \in X).$$

The task is then to deduce that this convergence holds uniformly in x, i.e. for x ranging over any compact $K \subseteq X$. Here z_0 is a distinguished (fixed) *reference point* or *null point*. What actions φ_n should we allow? Following Doob's convention we write $\alpha_\circ = (\alpha_n)_n$ for a sequence of elements and demand that it be a (pointwise) *divergent sequence*: for any $M > 0$, ultimately all n satisfy

$$d^X(\alpha_n(x), x) \geq M \text{ for all } x.$$

Shift-compactness appearing in proofs of the UCT in the main text required us to transfer an arbitrary convergent sequence $u_n \to u$ in the given compact set K to a 'null sequence'; here we take sequences $z_n \to z_0$ as our canonical null sequences, in view of the distinguished null point. The transfer is enabled by assuming T-homogeneity of X: we choose $\sigma \in \mathcal{H}(X)$ with $\sigma z_0 = u$ so that, with $z_n := \sigma^{-1}u_n$, we have $u_n = \sigma z_n \to \sigma z_0$. Next, the *Effros property*, as mentioned above, supplies a sequence of homeomorphisms converging (in the supremum norm of $\mathcal{H}(X)$) to the identity, say $\psi_n \to \mathrm{id}_X$, with $z_n = \psi_n(z_0)$. It emerges that the Category Embedding Theorem (CET) of Chapter 10 is applicable to any sequence converging to the identity. Thus one is able to deduce from CET, for any non-meagre Baire set S, that for quasi all $s \in S$ (in particular for some one s) and corresponding $\mathbb{M}_s \subseteq \mathbb{N}$,

$$\{\psi_n(s) : m \in \mathbb{M}_s\} \subseteq S.$$

Hence $\alpha_\circ$-*slowly varying* functions $h \colon X \to H$ involve taking all $\varphi_\circ$ of the form

$$\varphi_n = \alpha'_n = \alpha_n \sigma \psi_n.$$

These *perturbed shifts* are also divergent. We thus demand, for all $x \in X$ and all perturbed shifts α'_n of α_n, that

$$h(\alpha'_n(x))h(\alpha'_n(z_0))^{-1} \to e_H.$$

The UCT is then derivable in the following general form from CET:

Theorem E.1 *Suppose $\alpha_\circ$ in $\mathcal{H}(X)$ divergent, T is a subgroup, and X is non-meagre, T-transitive and T-Effros. If $h \colon X \to H$ is Baire and α-slowly varying, then, for compact $K \subseteq X$,*

$$h(\alpha_n(x))h(\alpha_n(z_0))^{-1} \to_{\text{uniformly}} e_H \qquad (x \in K).$$

For the proof, see BinO2010c. (There we refer to 'unconditional' slow variation emphasizing with the qualifier that the assumed convergence of slow variation involves all $\alpha_\circ'$ arising from a given divergent sequence $\alpha_\circ$.)

UCT on the L_1 Algebra of a Locally Compact Metric Group G

In the preceding contexts derivation of the UCT relied on transitive action. But one can sometimes have an alternative approach. Here we have G a locally compact metric group and work on $L_1(G)$ with its natural action of convolution; however, this action need not be transitive. Let G be equipped with a (left) Haar measure η. Take the domain and range of regularly varying functions to be $X = L_1(G, \eta)$, regarded as the Banach algebra of η-integrable functions $x: G \to \mathbb{R}$ under convolution. Thus

$$\|x\|_1 = \int |x(g)| \, d\eta(g).$$

The group G defines a natural action on X, namely $*: G \times X \to X$, where

$$(g * x)(t) := x(g^{-1}t) \qquad (g, t \in G, x \in X).$$

Definitions (1) Call the map $h: X \to X$ *slowly varying* (with respect to the net $\mathbf{x} := \{x_\delta\}$) if

$$\lim_\delta \|h(g * x_\delta) - h(x_\delta)\|_1 = 0, \quad \text{for each } g \in G. \tag{SV}$$

(2) Say that $h: X \to X$ is *Baire relative to convolution* if the maps $h_x: G \to X$ defined by

$$h_x(g) := h(g * x) \qquad (g \in G, x \in X)$$

are Baire for all x off a meagre set, the *exceptional set*, E_h, of h.

(As always, the map $h: X \to X$ is Baire if h^{-1} takes open sets to sets with the Baire property.)

(3) For Baire $h: X \to X$, say that h is slowly varying *with respect to regular nets* if it is slowly varying with respect to nets $\{x_\delta\}$ with all $x_\delta \notin E_h$; as in (2) above, here E_h is the meagre exceptional set corresponding to h.

It emerges that for G separable and $h: X \to X$ Baire, there does exist a meagre set E_h in X such that h_x is Baire for each x in the co-meagre set $R := X \backslash E_h$; when h is continuous E_h is empty. Under these circumstances we have:

Theorem E.2 (UCT for $L_1(G)$; BinO2013) *For G a locally compact metric group with Haar measure η and $X = L_1(G, \eta)$:*

(i) *for G separable (i.e. σ-compact), if h: X → X is Baire and slowly varying with respect to regular nets, then the convergence in* (SV) *is uniform on compacts;*

(ii) *for general G, uniform convergence in* (SV) *holds for h continuous and slowly varying with respect to arbitrary nets.*

Here the weaker assumption on the action extracts a price: slow variation is defined relative to regular nets, i.e. nets consisting of points avoiding a specified meagre set – in order to secure the Baire property for the maps h_x.

For G again a locally compact metric group with Haar measure η and $X = L_1(G)$, denote the non-negative probability densities by

$$P(G) := \{y \in L_1(G) : y \geq 0, \int y \, d\eta = 1\},$$

and say that the (weak) *Reiter condition* (Pat1988, Prop. 0.4) holds for a net $\{x_\delta\}$ in $P(G)$ if

$$\|g * x_\delta - x_\delta\|_1 \to 0 \text{ for each } g \in G. \tag{R}$$

Taking $h(x) := x$ which is continuous, one may deduce in our context that the condition (R) then holds uniformly on compact sets, that is:

$$\| g * x_\delta - x_\delta \|_1 \to 0 \text{ on compact sets of } g \in G. \tag{UR}$$

This is a significant fact (cf. Pat1988, Prop. 4.4) in amenability theory, as the Reiter condition (R) is equivalent to amenability. (An invariant mean can be extracted as any weak* cluster-point of the net $\{\hat{x}_\nu\}$, where $\hat{x}$ represents x in the second dual $L_1(G)'' = L_\infty(G)'$; for background on nets, see Eng1989, §1.1.6, or Kel1955, Ch. 2.)

Characterization Theorem

The UCT is linked to slow variation of a Baire function, $h: X \to H$, whose kernel is at its simplest: $k(t) \equiv e_H$. Aiming for a general kernel, the immediate question is its existence and characterization relative to the group of actions T on X. Here it is helpful that the domain of the action is a group T, since

$$S := \{t \in T : \text{ there exists a limit } \lim_{x \to \infty} h(tx)/h(x)\}$$

is a subgroup, and so the Subgroup Dichotomy Theorem (Chapter 9) can be called into play. We thus have:

Theorem E.3 (Characterization Theorem; cf. BGT1987, Th. 1.4.1)   *Let X and H be normed groups, $T \subseteq \mathcal{H}(X)$ a connected non-meagre subgroup, Baire* *under the (right) norm topology, acting on the group X and let* $h \colon X \to H$ *be Baire. If the limit*

$$k(t) = \partial_X h(t) := \lim_{x \to \infty} h(tx)h(x)^{-1}$$

exists on a non-meagre subset of T, then $\partial_X h(t)$ *exists on all of T and is a continuous homomorphism from T to H:*

$$k(st) = k(s)k(t) \qquad (s, t \in T).$$

For the proof, see BinO2010d, §2. For the role of homomorphisms arising in the associated functional equations of probability theory, see Ost2017. The continuity assertion above is linked to continuity properties of Baire functions described in the Discontinuity Theorem and the Banach–Neeb Theorem, for which see Chapter 12. Matters are simpler when the groups in question are normed spaces. We recall the following result (cf. Ban1932, 1.3.4, p. 40, albeit for 'Baire-measurable' functions; Meh1964).

Theorem E.4 (Banach–Mehdi Theorem) *An additive Baire function between complete normed vector spaces is continuous, and so linear, provided the image space is separable.*

For a proof, see BinO2009c, §2 and also Chapter 12. For a locally compact Hausdorff space T, in the case of $C_0(T)$, in the context of the space of real-valued continuous functions with compact support: from this last result, the standard Riesz representation theorem (Rud1966, Ch. 6) and reference to $\Phi := \{\varphi_u : u \in C_0(T)\}$ the group of shift homeomorphisms $\varphi_u(x) = x + u$, we have:

Theorem E.5 *Let* $X = C_0(T)$ *with T a locally compact Hausdorff space. For* Φ *the group of shift homeomorphisms, if* $h \colon X \to \mathbb{R}$ *is Baire* Φ-*regularly varying with distinguished null point the zero of X, then, for some measure* μ *on T, we have*

$$k(x) := \partial_\Phi h(x) = \int_T x(t)\, d\mu(t).$$

Real flows

A further case in which a representation is achievable is the context of *real flows* where for T one takes either of $T = \mathbb{R}$, the additive reals, or $T = \mathbb{R}_+$, the multiplicative reals, to act on a space X. Here flow homogeneity occurs. For the proof see BinO2009c, §3.

Theorem E.6 (Index Theorem) *For a (real) flow* φ *on X and kernel k of* $h \colon X \to H$, *there is a flow on H associated with the kernel k*

$$\varphi_k(t, z) = z\rho(t) \text{ for } z \in H, t \in \mathbb{R}$$

with ρ multiplicative (on H) $\mathbb{R}_+ \to H$ such that

$$k(\varphi(t, x)) = \varphi_k(t, k(x)), \quad \textit{i.e. } k(tx) = t * k(x).$$

For X, H complete with H separable, there exists $\rho \in H$ such that

$$\varphi_k(t, z) = z + \rho t.$$

Illustrative Example   For $X = \mathbb{R}^2 = C(\{0, 1\})$, the vector flow in direction u given by

$$\varphi(t, x) = \varphi_{tu}(x) = (x_0 + tu_0, x_1 + tu_1),$$

and $h(x) = \alpha_0 x_0 + \alpha_1 x_1$, we compute its kernel to be $k(x) = h(x)$. Indeed,

$$h(\varphi(t, x)) - h(\varphi(t, 0)) = \alpha_0(x_0 + tu_0) + \alpha_1(x_1 + tu_1) - (\alpha_0 u_0 t + \alpha_1 u_1 t)$$
$$= \alpha_0 x_0 + \alpha_1 x_1 = k(x).$$

So, for $u = e_0 = (1, 0)$, writing $\varphi_{\text{horiz}}(x) := (x_0 + 1, x_1)$,

$$k(\varphi_{\text{horiz}}(t, x)) = \alpha_0(x_0 + t) + \alpha_1 x_1 = k(x) + \alpha_0 t :$$
$$\rho_{\text{horiz}} = \alpha_0.$$

Similarly $\rho_{\text{vert}} = \alpha_1$ and $\rho_u = \alpha_0 u_0 + \alpha_1 u_1 = (\rho_{\text{horiz}}, \rho_{\text{vert}}) \cdot u$.

Calculus of Regular Variation: The Differential Modulus

Our main concern in this section is with products of regularly varying functions. In the classical context of the real line, it is obvious that the product of two regularly varying functions is regularly varying. This is also true in the current context of functions $h: X \to H$ when the group H is abelian and the metric is invariant. To see this observe that a regularly varying function is the product of a multiplicative function and a slowly varying function.

What may be said if H is non-commutative? It emerges that while φ-slowly varying functions have a group structure, identifying what happens when taking a product of φ-regularly varying functions, f_1 and f_2 say, requires the intervention of a modular function akin to that used in switching between left and right Haar measures. The need for a switching 'differential modulus' (differential, since kernels are derivatives at infinity) follows from the *First Factorization Theorem*, E.8, which permits us to write $f_i = k_i h_i$ for corresponding kernels k_i and slowly varying factors h_i, yielding:

$$f_1 f_2 = k_1 h_1 k_2 h_2.$$

So h_1 needs to switch position around k_2.

Indeed, it has to be appreciated that our definition of regular variation opted for *division on the right*, so to be fair the question should address one-sided multiplication (in fact on the left, see below). To guess at the answer, take $X = H$ and focus on the special case of two multiplicative functions $k(x)$ and $K(x)$ with $K(x) = x$. If the (pointwise) product $k(x)K(x)$ were to be regularly varying, one would expect it to be multiplicative, and the latter property is equivalent to

$$k(x)k(y)xy = k(xy)xy = k(xy)K(xy) = k(x)xk(y)y, \quad \text{i.e. } k(y)x = xk(y),$$

for all $x, y \in H$. This asserts that each value $k(y)$ commutes with each element x in the group H. From this one conjectures that the range of k must lie in the centre $Z(H)$ of the group H. (We recall that the subgroup $Z(H) = \{a \in H : ah = ha$ for all $h \in H\}$ is the *algebraic centre*.)

The conjecture is upheld as correct by the *Centrality Theorem*, E.13, but the argument requires the normed group H to have a bi-invariant metric:

Definition A metric d^H on H is *bi-invariant* if $d^H(ax, ay) = d^H(x, y) = d^H(xa, ya)$ for $a, x, y \in H$.

Bi-invariance is equivalent, as Klee (Kle1952) shows, to the existence of a metric possessing what we will term *Klee's property*:

$$d(ab, xy) \leq d(a, x) + d(b, y), \quad \text{i.e. } \|ab(xy)^{-1}\| \leq \|ax^{-1}\| + \|by^{-1}\|.$$

The bi-invariance property acts as a replacement for commutativity, and is exactly the condition which allows a proper development of the calculus of regularly varying functions, mimicking the non-commutative development of the Haar integral (see, e.g., DieS2014). Note that a Klee group is a topological group; cf. BinO2010g, Th. 3.4.

We now make our *blanket assumptions* that:

(i) the norm on H has the Klee property above;
(ii) h is φ-regular varying when the limit function below (the kernel) exists:

$$k(x) = \partial_\varphi h(x) := \lim_{n \to \infty} h(\varphi_n(x))h(\varphi_n(e_X))^{-1};$$

(iii) $\varphi_\circ$ is a fixed and, as earlier, divergent sequence of homomorphisms in $\mathcal{H}(X)$; and that functions g, h, k will always map from X to H.

For the proofs of results cited below, see BinO2010e.

Proposition E.7 (Group Structure of *SV*) *If $h: X \to H$ is φ-slowly varying, then the involutory mapping $h^{-1}: x \to h(x)^{-1}$ is φ-slowly varying. Hence the product of two φ-slowly varying functions is φ-slowly varying.*

Theorem E.8 (First Factorization Theorem) *If $h: X \to H$ is φ-regularly varying, then, for $k = \partial_\varphi h(t)$,*

(i) $k(t)$ is φ-regularly varying and $k(t) = \partial_\varphi k(t)$;

(ii) $\bar{h}(t) := k(t)^{-1} h(t)$ is φ-slowly varying. Thus $h(t)$ is the left product of its limit function with a slowly varying function $\bar{h}$:

$$h(t) = \partial_\varphi h(t) \bar{h}(t).$$

Theorem E.9 (Second Factorization Theorem)   *If g is φ-regularly varying and h is slowly varying, then $g(t)h(t)$ is regularly varying with limit $\partial_\varphi g$.*

It emerges that in the non-commutative case the set of points a of convergence of a sequence of conjugates $\{h_n a h_n^{-1}\}$ is well structured. The choice of sign in the notation below is motivated by the *Modular Flow Theorem*, E.11, to be established shortly.

Theorem E.10 (Asymptotic Conjugacy Theorem)   *For $h_\circ$ an arbitrary sequence in H, the two sets of convergence defined by*

$$D_\pm(h_\circ) = \{a : h_n^{\pm 1} a h_n^{\mp 1} \text{ is convergent}\}$$

are subgroups and are closed if H is complete. They support corresponding (asymptotically inner) automorphisms between themselves:

$$A_\pm(h_\circ, a) := \lim h_n^{\pm 1} a h_n^{\mp 1} \in D_\mp(h_\circ) \text{ for } a \in D_\pm(h_\circ).$$

These automorphisms are mutual inverses:

$$A_\pm(h_\circ, A_\mp(h_\circ, a)) = a \text{ for } a \in D_\pm(h_\circ).$$

Definition   A function $h \colon X \to H$ generates a sequence $h_n =: h(\varphi_n(e_X))$ which we may denote by $h_\circ$ without ambiguity. We may now define the *forward and backward moduli* of h relative to φ by

$$\Delta_\pm(h, a) = A_\pm(h_\circ, a) = \lim h_n^{\pm 1} a h_n^{\mp 1} \text{ for } a \in D_\pm(h_\circ).$$

We shall say that $h \colon X \to H$ is *modular* if all points of H are points of convergence for $h_\circ$; that is, $D_+(h_\circ) = H$, so that one may thus say that H is *asymptotically invariant* for $h_\circ$. Put $H_\circ := \{h \in H^X : h \text{ is modular}\}$ and set $1_H(x) \equiv e_H$.

Theorem E.11 (Modular Flow Theorem)   *$H_\circ$ is a group with identity element 1_H, and Δ_+ is an $H_\circ$-flow on H with*

$$\Delta_+(gh, a) = \Delta_+(g, \Delta_+(h, a)) \text{ and } \Delta_+(1_H, a) = a,$$
$$\Delta_+(h, \Delta_-(h, a)) = a.$$

Theorem E.12 (Left Product Theorem) *Suppose that g, h are φ-regularly varying with kernel functions k and K, and with g modular. Then gh is φ-regularly varying with kernel (limit)*

$$k(x)\Delta_+(g, K(x)).$$

Theorem E.13 (Centrality Theorem) *Suppose g is φ-regularly varying and modular with kernel function k. Then, for every φ-regularly varying function h and for all x, y, each element $\Delta_-(h, k(y))$ commutes with each element $\partial_\varphi h(x)$. In particular, k is central, i.e. the range $\{k(x) : x \in X\}$ is a subset of the centre $Z(H)$.*

For H complete under its norm, modularity of h follows from the following summability criterion for $h_n = h(\varphi_n(e_X))$:

$$\sum_{n=1}^{\infty} \|h_{n+1}h_n^{-1}\| < \infty;$$

this is a strengthened form of one of Kendall's sequential conditions for regular variation, namely $\|h_{n+1}h_n^{-1}\| \to 0$ (see §P.8 in the Prologue). Closeness of modularity to centrality is indicated by the next result (in which H need not be abelian).

Theorem E.14 *For H complete and $h: X \to H$ φ-slowly varying and h satisfying the summability condition:*

(i) *for k central, kh and hk are modular;*
(ii) *kh is modular if and only if k is central.*

Through an Algebraic Lens: Beurling Regular Variation Reappraised
Recall from the Prologue (§P.6) the definition of Beurling slow variation:

$$\varphi(x + t\varphi(x))/\varphi(x) \to 1, \qquad \text{as } x \to \infty \qquad (t \in \mathbb{R}) \qquad \text{(BSV)}$$

for $\varphi: \mathbb{R} \to \mathbb{R}_+$ with $\varphi(x) = o(x)$; recall also its role in the Beurling Tauberian Theorem (§P.7). This prompted the study (BinO2014) of Beurling φ-regular variation, in which kernels are defined by

$$k(t) = \lim_{x \to \infty} f(x + t\varphi(x))/f(x).$$

Allowing $\varphi(x) = O(x)$ and adjusting (BSV) to have a more general limit gains a wider remit, embracing classical as well as Beurling regular variation:

$$\varphi(x + t\varphi(x))/\varphi(x) \to \eta(t) \qquad \text{as } x \to \infty \qquad (t \in \mathbb{R}).$$

Here φ is said to be *self-equivarying* if the convergence is locally uniform (which will be the case for Baire φ; see Ost2015a). Then it emerges that the limit η satisfies the Gołąb–Schinzel equation

$$\eta(u + v\eta(u)) = \eta(u)\eta(v). \tag{GS}$$

Imposing positivity implies that η is continuous and of the form $\eta(t) = 1 + \rho t$ for some $\rho \geq 0$. Significantly, the form of the equation prompts a binary operation, termed the *Popa operation* (see Pop1965, cf. Jav1968, the first two papers in which it was studied):

$$a \circ_\eta b := a + b\eta(a).$$

This induces a group structure on the set $\mathbb{G}_\eta := \{x \in \mathbb{R} : \eta(x) \neq 0\}$ and allows η to be viewed as a homomorphism. By comparison, for fixed x, the binary operation of 'addition accelerated by a factor $\varphi(x)$':

$$a \circ_\varphi b = a + b\varphi(x)$$

may properly be interpreted as an *asymptotic action* in which

$$\eta_x(t) := \varphi(x \circ_\varphi t)/\varphi(x) \to \eta(t).$$

This asymptotic view enables a transparent study of regular variation in which the symbolism

$$f(x \circ_\varphi t)/f(x) \to k(t) \qquad \text{as } x \to \infty \qquad (t \in \mathbb{R})$$

is guided through the algebraic lens of topological regular variation. For details see BinO2016a; BinO2020a. The Popa group structure yielded a simplification to the hardest theorem of Regular Variation (see §P.6 on Quantifier Thinning in the Prologue) and allowed a proper understanding of the *Goldie equation* at the heart of that theorem. See Ost2016b; BinO2022b; BinO2022c; BinO2022d.

Cocycles and Coboundaries

It is just and fitting for the Epilogue to turn the algebraic lens briefly on the regular variation of the Prologue, only to discover the inevitable common features: the use in algebraic topological dynamics of *cocycle* terminology, for which see p. 289 (cf. Ellis' monograph, Elli1969, and its review in MR0267561). We take up this theme from the vantage point of the following.

Definition Let $\varphi = \{\varphi_n\}$ be a divergent sequence. Say that $h\colon X \to H$ is φ-*regularly varying*, or if context permits, just *Fréchet-regularly varying*, if for some function $k(\cdot) = \partial_\varphi h(\cdot)$ and, for each t,

$$h(\varphi_n(t)h(\varphi_n(z_0))^{-1} \to k(t).$$

See also BinO2010g, Section 5 for an integrated treatment in the context of normed groups. For the background and relation here between Gateaux, Fréchet, and Hadamard differentiability, see BinO2009c, §3, Remark.

Proposition E.15 (Concatenation Formula)   *If h is φ-regularly varying for the distinguished point $z = z_0$, then for any w the corresponding Fréchet limit $k_w(x) = \lim_n h(\varphi_n(x))h(\varphi_n(w))^{-1}$ exists and*

$$k_z(x) = k_z(w)k_w(x).$$

Definition   Let $\mathcal{H}_0 = \{\varphi \in \mathcal{H}(X) : \varphi(z_0) = z_0\}$ be the stabilizer subgroup (of the distinguished null point). Note that this is conjugate to the stabilizer of any other point of the (homogeneous) space X. Thus, for σ, τ in $\mathcal{H}(X)$ with $\sigma(z_0) = \tau(z_0)$, we have $\sigma^{-1}\tau \in \mathcal{H}_0$. We will regard two homeomorphisms σ, τ in $\mathcal{H}(X)$ that are $\mathcal{H}_0$-equivalent (i.e. both in the same coset of $\mathcal{H}_0$, e.g. $\tau \in \sigma\mathcal{H}_0$) as equal. Whenever convenient, we will denote by σ_x the unique homeomorphism (up to equivalence) taking z_0 to x. This is particularly useful when G is a topological group, where the canonical choice is

$$\sigma_u(g) = \lambda_u(g) = ug,$$

as we then have $\sigma_u\sigma_v = \sigma_{uv}$. The following result justifies the use of $\mathcal{H}_0$-equivalence.

Proposition E.16   *If h is strongly φ-regularly varying and σ is a bounded homeomorphism with $\sigma(z_0) = z_0$, then the corresponding Fréchet limit function satisfies $k(\sigma(t)) = k(t)$.*

For our next result we need to observe that in the context of real-valued functions the equation

$$\sigma_f(t, x) := f(tx)f(x)^{-1} .$$

defines a *cocycle*, since

$$\sigma_f(st, x) = \sigma_f(t, x)\sigma_f(s, tx) = \frac{f(tx)}{f(x)}\frac{f(stx)}{f(tx)}.$$

A cocycle $\sigma(t, x)$ is a *coboundary* if for some f,

$$f(tx) = f(x)\sigma(t, x).$$

Thus in particular the kernel k of a regularly varying function f gives rise to a coboundary, as in the next result, which refers to Chapter 11 and its Effros properties, Theorems 11.1.2 and 11.2.1.

Theorem E.17 (Continuous Coboundary Theorem)   *Suppose that X is a Baire space with the Effros Crimping property, §11.3 (as in the UCT above). Let H be a topological group such that the pair (X, H) enables Baire continuity. If $h: X \to H$ is Baire regularly varying with limit function k, then k is Baire, has the coboundary property*

$$k(\sigma_y) = k_{\sigma(x,y)}k(\sigma_x),$$

equivalently,

$$k(\sigma_x\sigma_y) = k(\sigma_x)k(\sigma_y)$$

and is continuous.

Recall that 'enabling Baire continuity' is discussed in Chapter 13. For the proof, see BinO2021a.

This leads us to our final result, which plays a thematic role throughout the theory.

Corollary E.18 (Continuous Homomorphism Theorem)   *Suppose that $h: X \to H$ is a Baire regularly varying function defined on a Baire topological group X with values in the topological group H, and that the pair (X, H) enables Baire continuity. If h has a limit function (kernel) k, then k is a continuous homomorphism, i.e.*

$$k(xy) = k(x)k(y).$$

References

[AarGM1970a] Aarts, J.M., de Groot, J. and McDowell, R.H. Cocompactness. *Nieuw Arch. Wisk.* **18** (1970), 2–15. 57

[AarGM1970b] Aarts, J.M., de Groot, J. and McDowell, R.H. Cotopology for metrizable spaces. *Duke Math. J.* **37** (1970), 291–295. 57, 135, 136

[AarL1974] Aarts, J.M. and Lutzer, D.J. Completeness properties designed for recognizing Baire spaces. *Dissertationes Math. (Rozprawy Mat.)* **116** (1974), 48. 37

[AczD1989] Aczél, J. and Dhombres, J. *Functional Equations in Several Variables.* Encycl. Math. Appl. **31**. Cambridge University Press, 1989. 8

[AlbM2016] Alberti, G. and Marchese, A. On the differentiability of Lipschitz functions with respect to measures in the Euclidean space. *Geom. Funct. Anal.* **26** (2016), 1–66. 228

[AlbGI2017] Albeverio, S., Garko, I., Ibragim, M. and Torbin, G. Non-normal numbers: Full Hausdorff dimensionality vs zero dimensionality. *Bull. Sci. Math.* **141** (2017), 1–19. 226

[AloS2008] Alon, N. and Spencer, J.H. *The Probabilistic Method*, 3rd ed. Wiley, 2008 (2nd ed. 2000, 1st ed. 1992). 37

[Amb1947] Ambrose, W. Measures on locally compact topological groups. *Trans. Am. Math. Soc.* **61** (1947), 106–121. 253

[AnaL1985] Anantharaman, R. and Lee, J.P. Planar sets whose complements do not contain a dense set of lines. *Real Anal. Exchange* **11** (1985–1986), 168–179. 156

[Anc1987] Ancel, F.D. An alternative proof and applications of a theorem of E.G. Effros. *Michigan Math. J.* **34** (1987), 39–55. 172, 173, 176, 178, 181, 182, 196, 216

[Arh1963] Arhangelskii, A.V. On a class of spaces containing all metric and all locally compact spaces. *Soviet Math. Dokl.* **151** (1963), 751–754. 49, 165

[Arh1995] Arhangelskii, A.V. Paracompactness and metrization: The method of covers in the classification of spaces. In A.V. Arhangelskii (ed), *General Topology III.* Encyclopaedia Math. Sci. **51**. Springer, 1995, 1–70. 49

[ArhM1998] Arhangelskii, A.V. and Malykhin, V.I. Metrizability of topological groups (in Russian). *Vestnik Moskov. Univ. Ser. I Mat. Mekh.* **91** (1996), 13–16; English translation in *Moscow Univ. Math. Bull.* **51** (1996), 9–11. 86

[ArhR2005] Arhangelskii, A.V. and Reznichenko, E.A. Paratopological and semi-topological groups versus topological groups. *Topol. Appl.* **151** (2005), 107–119. 109

[ArhT2008] Arhangelskii, A. and Tkachenko, M. *Topological Groups and Related Structures.* Atlantis Press, 2008. 256

[Avi1998] Avigad, J. An effective proof that open sets are Ramsey. *Arch. Math. Logic* **37** (1998), 235–240. 122

[Bae1969] Baernstein, A. A non-linear Tauberian theorem in function theory. *Trans. Am. Math. Soc.* **146** (1969), 87–105. 5

[Bae1974] Baernstein, A. A generalization of the $\cos \pi \lambda$ theorem. *Trans. Am. Math. Soc.* **193** (1974), 181–197. 6

[BagW1997] Bagaria, J. and Woodin, H.W. Δ_n^1-sets of reals. *J. Symbol. Logic* **62** (1997), 1379–1428. 222, 231

[Bai1899] Baire, R. Thèse: Sur les fonctions de variable réelle. *Ann. di Math.* **3** (1899), 1–123. 34, 36

[Bai1909] Baire, R. Sur la représentation des fonctions discontinues (2me partie). *Acta Math.* **32** (1909), 97–176. 34

[BajK1969] Bajšanski, B. and Karamata, J. Regular varying functions and the principle of equicontinuity. *Publ. Ramanujan Inst.* **1** (1969), 235–246. 278

[Bal1973] Balkema, A.A. *Monotone Transformations and Limit Laws.* Mathematical Centre Tracts **45**. Mathematisch Centrum, 1973. v+170 pp. 278

[Ban1920] Banach, S. Sur l'équation fonctionelle $f(x + y) = f(x) + f(y)$. *Fund. Math.* **1** (1920), 123–124. Reprinted in *Collected Works* **I**, 47–48, PWN, 1967 (Commentary by H. Fast, p. 314). 8, 190

[Ban1930] Banach, S. Über additive Maßfunktionen in abstrakten Mengen. *Fund. Mathe.* **15** (1930), 97–101.

[Ban1931] Banach, S. Über metrische Gruppen. *Studia Math.* **III** (1931), 101–113. Reprinted in *Collected Works* **II**, 401–411, PWN, 1979. vii+254 pp. 184

[Ban1932] Banach, S. *Théorie des Opérations Linéaires.* Monog. Mat. **I** (1932). Reprinted in *Collected Works* **II**, PWN, 1979. 21, 162, 176, 184, 186, 187, 189, 190, 191, 232, 283

[BanT1924] Banach, S. and Tarski, A. Sur la décomposition des ensembles de points en parties respectivement congruents. *Fund. Math.* **6** (1924), 244–277. 223

[Bana2004] Banakh, T. Cardinal characteristics of the ideal of Haar null sets. *Comment. Math. Univ. Carolin.* **45** (2004), 119–137. 234

[BanaGJ2021] Banakh, T., Głab, S., Jabłońska, E. and Swaczyna, J. Haar-$\mathcal{I}$ sets: Looking at small sets in Polish groups through compact glasses. *Diss. Math.* **564** (2021), 1–105. 255

[BanaJ2019] Banakh, T. and Jabłońska, E. Null-finite sets in topological groups and their applications. *Israel J. Math.* **230** (2019), 361–386. 249, 264, 265, 266

[Bar1955] Bartle, R.G. Implicit functions and solutions of equations in groups. *Math. Z.* **62** (1955), 335–346. 86

[BartFF2018] Bartoszewicz, A., Filipczak, M. and Filipczak, T. On supports of probability Bernoulli-like measures. *J. Math. Anal. Appl.* **462** (2018), 26–35. 254, 255

[BartFN2011] Bartoszewicz, A., Filipczak, M. and Natkaniec, T. On Smítal properties. *Topol. Appl.* **158** (2011), 2066–2075. 255, 256, 258

[Barto1984] Bartoszyński, T. Additivity of measure implies additivity of category. *Trans. Am. Math. Soc.* **28** (1984), 209–213. 230, 232

[Barto2000] Bartoszyński, T. A note on duality between measure and category. *Proc. Am. Math. Soc.* **128** (2000), 2745–2748. 138, 228

[Barto2010] Bartoszyński, T. Invariants of measure and category. In M. Forman and A. Kanamori (eds), *Handbook of Set Theory*, vol. 1. Springer, 2010, 491–556. 232

[BartoGJ1993] Bartoszyński, T., Goldstern, M., Judah, H. and Shelah, S. All meager filters may be null. *Proc. Am. Math. Soc.* **117** (1993), 515–521. 228

[BartoJS1993] Bartoszyński, T., Judah, H. and Shelah, S. The Cichoń diagram. *J. Symbol. Logic* **58** (1993), 401–423. 138

[BartoJ1995] Bartoszyński, T. and Judah, H. *On the Structure of the Real Line.* A.K. Peters, 1995. 154, 222, 232, 233, 234

[Barw1977] Barwise, J. (ed). *Handbook of Mathematical Logic.* North-Holland, 1977. 301

[Bary1964] Bary, N.K. *A Treatise on Trigonometric Series*, vols. I, II. Pergamon, 1964. 120, 138

[Bau1975] Bauer, F. Aspects of modern potential theory. In *Proc. Int. Conf. Math. Vancouver,* 1974, vol., 41–51. Canadian Mathematical Congress, 1975. 115, 118

[Bec1960] Beck, A. A note on semi-groups in a locally compact group. *Proc. Am. Math. Soc.* **11** (1960), 992–993. 239

[BecCS1958] Beck, A., Corson, H.H. and Simon, A.B. The interior points of the product of two subsets of a locally compact group. *Proc. Am. Math. Soc.* **9** (1958), 648–652. 239

[BeckK1996] Becker, H. and Kechris, A.S. *The Descriptive Set Theory of Polish Group Actions.* London Math. Soc. Lect. Note Ser. **232**. Cambridge University Press, 1996. 233

[BelS1969] Bell, J.L. and Slomson, A.B. *Models and Ultraproducts: An Introduction.* North-Holland, 1969. Reprinted by Dover, 2006. 62, 275

[Ber1963] Berge, C. *Topological Spaces, Including a Treatment of Multi-valued Functions, Vector Spaces and Convexity.* Oliver and Boyd, 1963. Reprinted by Dover, 1997. 54

[BerghHW1997] Bergelson, V., Hindman, N. and Weiss, B. All-sums sets in (0,1] – Category and measure. *Mathematika* **44** (1997), 61–87. 74, 76, 79

[Bert1996] Bertoin, J. *Lévy Processes.* Cambridge Tracts in Math. **121**. Cambridge University Press, 1996. 12

[Berz1975]   Berz, E. Sublinear functions on $\mathbb{R}$. *Aequat. Math.* **12** (1975), 200–206. 232

[BeuD1959] Beurling, A. and Deny, J. Dirichlet spaces. *Proc. Nat. Acad. Sci. USA* **45** (1959), 208–215. 120

[Bin1981]   Bingham, N.H. Tauberian theorems and the central limit theorem. *Ann. Probab.* **9** (1981), 221–231. 14, 17

[Bin1984a] Bingham, N.H. Tauberian theorems for summability methods of random-walk type. *J. London Math. Soc.* (2) **30** (1984), 281–287. 16

[Bin1984b] Bingham, N.H. On Valiron and circle convergence. *Math. Z.* **186** (1984), 273–286. 15

[Bin1988]   Bingham, N.H. Tauberian theorems for Jakimovski and Karamata–Stirling methods. *Mathematika* **35** (1988), 216–224. 16

[Bin2007]   Bingham, N.H. Regular variation and probability: The early years. *J. Comput. Appl. Math.* **200** (2007), 357–363. 3

[Bin2010]   Bingham, N.H. Finite additivity versus countable additivity. *Elect. J. His. Probab. Stat.* **6** (2010), 35p. 240

[Bin2014]   Bingham, N.H. The worldwide influence of the work of B. V. Gnedenko. *Theory Probab. Appl.* **58** (2014), 17–24. 2

[Bin2015a] Bingham, N.H. On scaling and regular variation. *Publ. Inst. Math. Beograd* (NS) **97** (111) (2015), 161–174. 22

[Bin2015b] Bingham, N.H. Hardy, Littlewood and probability. *Bull. London Math. Soc.* **47** (2015), 191–201. 3

[Bin2019]   Bingham, N.H. Riesz means and Beurling moving averages. In P.M. Barrieu (ed). *Risk and Stochastics*: *Ragnar Norberg Memorial Volume*. World Scientific, 2019, 159–172; arXiv:1502.07494. 17

[BinF2010] Bingham, N.H. and Fry, J.M. *Regression: Linear models in Statistics.* Springer Undergraduate Series in Mathematics (SUMS), Springer, 2010. 23

[BinG2015] Bingham, N.H. and Gashi, B. Logarithmic moving averages. *J. Math. Anal. Appl.* **421** (2015), 1790–1802. 17

[BinG2017] Bingham, N.H. and Gashi, B. Voronoi means, moving averages and power series. *J. Math. Anal. Appl.* **449** (2017), 682–696. 17

[BinGo1982a] Bingham, N.H. and Goldie, C.M. Extensions of regular variation, I. Uniformity and quantifiers. *Proc. London Math. Soc.* **44** (1982), 473–496. 10, 145, 239

[BinGo1982b] Bingham, N.H. and Goldie, C.M. Extensions of regular variation, II. Representations and indices. *Proc. London Math. Soc.* **44** (1982), 497–534. 11, 12, 239

[BinGo1983] Bingham, N.H. and Goldie, C.M. On one-sided Tauberian conditions. *Analysis* **3** (1983), 159–188. 15, 17

[BGT1987]   Bingham, N.H., Goldie, C.M. and Teugels, J.L. *Regular Variation,* Encycl. Math. Appl. **27**, Cambridge University Press, 1987 (2nd ed. 1989). 6, 19, 64, 78, 80, 81, 82, 140, 141, 144, 145, 239, 240, 282

[BinI1997] Bingham, N.H. and Inoue, A. The Drasin–Shea–Jordan theorem for Fourier and Hankel transforms. *Quart. J. Math.* **48** (1997), 279–307. 5

[BinI1999] Bingham, N.H. and Inoue, A. Ratio Mercerian theorems with applications to Hankel and Fourier transforms. *Proc. London Math. Soc.* **79** (1999), 626–648. 5

[BinI2000a] Bingham, N.H. and Inoue, A. Abelian, Tauberian and Mercerian theorems for arithmetic sums. *J. Math. Anal. Appl.* **250** (2000), 465–493. 6

[BinI2000b] Bingham, N.H. and Inoue, A. Tauberian and Mercerian theorems for systems of kernels. *J. Math. Anal. Appl.* **252** (2000), 177–197. 7, 8, 17

[BinJJ2020] Bingham, N.H., Jabłońska, E., Jabłoński. W. and Ostaszewski, A.J. On subadditive functions bounded above on a large set. *Results Math.* **75**:58 (2020), 12p. xiii

[BinO2008] Bingham, N.H. and Ostaszewski, A.J. Generic subadditive functions. *Proc. Am. Math. Soc.* **136** (2008), 4257–4266. 63, 123

[BinO2009a] Bingham, N.H. and Ostaszewski, A.J. Very slowly varying functions II. *Colloq. Math.* **116** (2009), 105–117. 176

[BinO2009b] Bingham, N.H. and Ostaszewski, A.J. Beyond Lebesgue and Baire: Generic regular variation. *Colloq. Math.* **116** (2009), 119–138. 63, 73, 137, 141, 146, 163, 278

[BinO2009c] Bingham, N.H. and Ostaszewski, A.J. The index theorem of regular variation. *J. Math. Anal. Appl.* **358** (2009), 238–248. 64, 278, 283, 289

[BinO2009d] Bingham, N.H. and Ostaszewski, A.J. Infinite combinatorics and the foundations of regular variation. *J. Math. Anal. Appl.* **360** (2009), 518–529. 145, 278

[BinO2009e] Bingham, N.H. and Ostaszewski, A.J. Automatic continuity: Subadditivity, convexity, uniformity. *Aequat. Math.* **78** (2009), 257–270. 218

[BinO2009f] Bingham, N.H. and Ostaszewski, A.J. Infinite combinatorics on functions spaces: Category methods. *Publ. Inst. Math. Beograd* (NS) **86** (100) (2009), 55–73. 117, 123

[BinO2010a] Bingham, N.H. and Ostaszewski, A.J. Automatic continuity by analytic thinning. *Proc. Am. Math. Soc.* **138** (2010), 907–919. 10, 11, 22, 145, 146

[BinO2010b] Bingham, N.H. and Ostaszewski, A.J. Regular variation without limits. *J. Math. Anal. Appl.* **370** (2010), 322–338. 53, 62, 63, 65, 236, 237, 278

[BinO2010c] Bingham, N.H. and Ostaszewski, A.J. Topological regular variation. I, Slow variation. *Topol. Appl.* **157** (2010), 1999–2013. 278, 281

[BinO2010d] Bingham, N.H. and Ostaszewski, A.J. Topological regular variation. II, The fundamental theorems. *Topol. Appl.* **157** (2010), 2014–2023. 278, 283

[BinO2010e] Bingham, N.H. and Ostaszewski, A.J. Topological regular variation. III, Regular variation. *Topol. Appl.* **157** (2010), 2024–2037. 278, 285

[BinO2010f] Bingham, N.H. and Ostaszewski, A.J. Kingman, category and combinatorics. In N.H. Bingham and C.M. Goldie (eds), *Probability and Mathematical Genetics*: *Sir John Kingman Festschrift*. London Math.

Soc. Lecture Notes in Mathematics **378**, Cambridge University Press, 2010, 135–168. 71, 72, 81, 241

[BinO2010g] Bingham, N.H. and Ostaszewski, A.J. Normed groups: Dichotomy and duality. *Dissertationes Math.* **472** (2010), 138p. 74, 86, 89, 99, 138, 139, 140, 141, 162, 185, 199, 217, 221, 236, 285, 289

[BinO2010h] Bingham, N.H. and Ostaszewski, A.J. Beyond Lebesgue and Baire II: Bitopology and measure-category duality. *Colloq. Math.* **121** (2010), 225–238. 73, 137, 145

[BinO2011a] Bingham, N.H. and Ostaszewski, A.J. Dichotomy and infinite combinatorics: The theorems of Steinhaus and Ostrowski. *Math. Proc. Cambridge Phil. Soc.* **150** (2011), 1–22. 21, 62, 99, 141, 239, 255

[BinO2011b] Bingham, N.H. and Ostaszewski, A.J. Homotopy and the Kestelman–Borwein–Ditor theorem. *Canadian Math. Bull.* **54** (2011), 12–20. 162

[BinO2013] Bingham, N.H. and Ostaszewski, A.J. Steinhaus theory and regular variation: De Bruijn and after. *Indag. Math.* (N. G. de Bruijn Memorial Issue) **24** (2013), 679–692. 278, 281

[BinO2014] Bingham, N.H. and Ostaszewski, A.J. Beurling slow and regular variation. *Trans. London Math. Soc.* **1** (2014), 29–56. 13, 278, 287

[BinO2015] Bingham, N.H. and Ostaszewski, A.J. Cauchy's functional equation and extensions: Goldie's equation and inequality, the Gołąb–Schinzel equation and Beurling's equation. *Aequat. Math.* **89** (2015), 1293–1310. 218

[BinO2016a] Bingham, N.H. and Ostaszewski, A.J. Beurling moving averages and approximate homomorphisms. *Indag. Math.* **27** (2016), 601–633. 17, 62, 63, 288

[BinO2016b] Bingham, N.H. and Ostaszewski, A.J. The Steinhaus–Weil property and its converse: Subcontinuity and amenability. arXiv:1607.00049. (See BinO2020c,d, BinO2021a, BinO2022a.) 228, 235, 240, 242, 247, 249, 253

[BinO2017] Bingham, N.H. and Ostaszewski, A.J. Category–measure duality: Jensen convexity and Berz sublinearity. *Aequat. Math.* **91** (2017), 801–836. 62, 232, 242

[BinO2018a] Bingham, N.H. and Ostaszewski, A.J. Additivity, subadditivity and linearity: Automatic continuity and quantifier weakening. *Indag. Math.* **29** (2018), 687–713. 10, 11, 232

[BinO2018b] Bingham, N.H. and Ostaszewski, A.J. Beyond Lebesgue and Baire IV: Density topologies and a converse Steinhaus theorem. *Topol. Appl.* **239** (2018), 274–292. 104, 234, 241, 242, 244, 256

[BinO2019a] Bingham, N.H. and Ostaszewski, A.J. Set theory and the analyst. *Eur. J. Math.* **5** (2019), 2–48. 229, 240, 241, 268, 269, 270, 273, 274, 275, 276

[BinO2019b] Bingham, N.H. and Ostaszewski, A.J. Variants on the Berz sublinearity theorem. *Aequat. Math.* **93** (2019), 351–369. 232, 234

[BinO2019c] Bingham, N.H. and Ostaszewski, A.J. Beyond Haar and Cameron–Martin: The Steinhaus support. *Topol. App.* **260** (2019), 23–56. 240, 247, 256, 259

[BinO2020a] Bingham, N.H. and Ostaszewski, A.J. General regular variation, Popa groups and quantifier weakening. *J. Math. Anal. Appl.* **483** (2020), 123610. 10, 11, 17, 18, 288

[BinO2020b] Bingham, N.H. and Ostaszewski, A.J. Sequential regular variation: Extensions to Kendall's theorem. *Quart. J. Math.* **71** (2020), 1171–1200. 18, 19

[BinO2020c] Bingham, N.H. and Ostaszewski, A.J. The Steinhaus–Weil property and its converse. I, Subcontinuity and amenability. *Sarajevo Math. J.* **16** (2020), 13–32. 228, 232, 234, 235, 242, 243, 246

[BinO2020d] Bingham, N.H. and Ostaszewski, A.J. The Steinhaus–Weil property and its converse. II, The Simmons–Mospan converse. *Sarajevo Math. J.* **16** (2020), 179–186; 235, 246, 252, 255, 256, 257

[BinO2021a] Bingham, N.H. and Ostaszewski, A.J. The Steinhaus–Weil property and its converse. III, Weil topologies. *Sarajevo Math. J.* **17** (2021), 129–142. 235, 240, 259, 290

[BinO2021b] Bingham, N.H. and Ostaszewski, A.J. Extremes and regular variation. In L. Chaumont and E.A. Kyprianou (eds), *A Lifetime of Excursions through Random Walks and Lévy Processes* (A volume in honour of Ron Doney's 80th birthday). Progr. Prob. **78**, 2021b, 121–137, Birkhäuser. 3, 13

[BinO2022a] Bingham, N.H. and Ostaszewski, A.J. The Steinhaus–Weil property and its converse. IV, Other interior-point properties. *Sarajevo Math. J.* **18** (2022), 203–210. 235

[BinO2022b] Bingham, N.H. and Ostaszewski, A.J. Homomorphisms from functional equations: The Goldie equation II. ArXiv:1910.05816. 288

[BinO2022c] Bingham, N.H. and Ostaszewski, A.J. Homomorphisms from functional equations: The Goldie equation III. ArXiv:1910.05817. 288

[BinO2022d] Bingham, N.H. and Ostaszewski, A.J. The Gołąb–Schinzel and Goldie functional equations in Banach algebras. ArXiv:2105.07794. 288

[BinO2024] Bingham, N.H. and Ostaszewski, A.J. Parthasarathy, shift-compactness and infinite combinatorics. *Indian J. Pure Appl. Math.* **55** (2024), 931–948. 73

[BinS1990] Bingham, N.H. and Stadtmüller, U. Jakimovski methods and almost-sure convergence. In G.R. Grimmett and D.J.A. Welsh (eds), *Disorder in Physical Systems (J.M. Hammersley Festschrift)*. Oxford University Press, 1990, 5–18. 16

[BinT1986] Bingham, N.H. and Tenenbaum, G. Riesz and Valiron means and fractional moments. *Math. Proc. Cambridge Phil. Soc.* **99** (1986), 143–149. 15, 17

[Bir1936] Birkhoff, G. A note on topological groups. *Compos. Math.* **3** (1936), 427–430. 86

[BirM1953] Birkhoff, G. and Mac Lane, S. *A Survey of Modern Algebra*, revised ed., Macmillan, 1953, (1st ed. 1941). 272

[Bla1984] Blass, A. Existence of bases implies the Axiom of Choice. *Contemp. Math.* **31** (1984), 31–33. 20

[Blo1976] Bloom, S. A characterization of *B*-slowly varying functions. *Proc. Am. Math. Soc.* **54** (1976), 243–250. 13

[BluG1968] Blumenthal, R.M. and Getoor, R.K. *Markov Processes and Potential Theory*. Academic Press, 1968. 117

[BobBC1981] Boboc, N., Bucur, Gh. and Cornea, A. *Order and Convexity in Potential Theory: H-cones*. Lecture Notes in Math. **853** (1981), Springer. 118, 120

[Bog1998] Bogachev, V.I. *Gaussian Measures*. Math. Surveys and Monographs **62**, Am. Math. Soc., 1998. 240, 259, 260, 261

[Bog2007a] Bogachev, V.I. *Measure Theory, I*. Springer, 2007. xiii, 24, 25, 28, 29, 30, 31, 33, 73, 104, 126, 232, 244, 273, 274

[Bog2007b] Bogachev, V.I. *Measure Theory, II*. Springer, 2007. 25, 33, 240

[Bog2010] Bogachev, V.I. *Differentiable Measures and the Malliavin Calculus*. Math. Surveys and Monographs **164**, Am. Math. Soc., 2010. 243

[Bog2018] Bogachev, V.I. Negligible sets in infinite-dimensional spaces. *Anal. Math.* **44** (2018), 299–323. 240

[BogS2017] Bogachev, V.I. and Smolyanov, O.G. *Topological Vector Spaces and Their Applications*, Springer, 2017. 240

[BojK1963] Bojanic, R. and Karamata, J. On a class of functions of regular asymptotic behaviour. *Math. Res. Center Tech. Rep.* **436** (1963), Madison WI. Reprinted in Kar2009, pp. 545–569. 9

[Bol1998] Bollobás, B. *Modern Graph Theory*, Springer, 1998. 274

[Bor1989] Border, K.C. *Fixed Point Theorems with Applications to Economics and Game Theory*. Cambridge University Press, 1989. 50

[Bore1909] Borel, E. Les probabilités dénombrables et leurs applications arithmétiques. *Rend. Circ. Mat. Palermo* **27** (1909), 247–271. 225

[Borel1976] Borell, C. Gaussian Radon measures on locally convex spaces. *Math. Scand.* **36** (1976), 265–284. 260

[BorwD1978] Borwein, D. and Ditor, S.Z. Translates of sequences in sets of positive measure. *Canadian Math. Bull.* **21** (1978), 497–498. 74

[Bou1966] Bourbaki, N. *Elements of Mathematics: General Topology*. Parts 1 and 2. Hermann, Paris/Addison-Wesley, 1966. 86

[Bouz1993] Bouziad, A. The Ellis theorem and continuity in groups. *Topol. Appl.* **50** (1993), 73–80. 49, 176, 197

[Bouz1996] Bouziad, A. Every Čech-analytic Baire semitopological group is a topological group. *Proc. Am. Math. Soc.* **124** (1996), 953–959. 197, 198, 242

[BraG1960] Bradford, J.C. and Goffman, C. Metric spaces in which Blumberg's theorem holds. *Proc. Am. Math. Soc.* **11** (1960), 667–670. 40, 127

[Bre1941] Brelot, M. Sur la théorie autonome des fonctions sousharmoniques. *Bull. Sci. Math.* **65** (1941), 72–98. 116

[BreuS2011] Breuer, J. and Simon, B. Natural boundaries and spectral theory. *Adv. Math.* **226** (2011), 4902–4920. 229

[Brez1991] Brezinski, C. *History of Continued Fractions and Padé Approximants*. Springer, 1991. 226

[Bru1971] Bruckner, A.M. Differentiation of integrals. *Am. Math. Monthly* **78** (9) (1971), Part II, ii+51 pp. 104

[Brz1992] Brzdęk, J. Subgroups of the group $\mathbb{Z}_n$ and a generalization of the Gołąb–Schinzel functional equation. *Aequat. Math.* **43** (1992), 59–71. 260

[Bug2012] Bugeaud, Y. *Distribution Modulo One and Diophantine Approximation.* Cambridge Tracts in Math. **193**, Cambridge University Press, 2012. 223

[Buk2011] Bukovský, L. *The Structure of the Real Line.* Monografie Matematyczne (New Series) **71**, Birkhäuser, 2011. 234

[BurBI2001] Burago, D., Burago, Y. and Ivanov, S. *A Course in Metric Geometry.* Graduate Studies in Mathematics **33**, Am. Math. Soc., 2001. 86

[Burg1983a] Burgess, J.P. Classical hierarchies from a modern standpoint. I: C-sets. *Fund. Math.* **115** (1983), 81–95. 62

[Burg1983b] Burgess, J.P. Classical hierarchies from a modern standpoint. II: r-sets. *Fund. Math.* **115** (1983), 97–105. 62

[CabC2002] Cabello Sánchez, F. and Castillo, J.M.F. Banach space techniques underpinning a theory for nearly additive mappings. *Dissertationes Math. (Rozprawy Mat.)* **404** (2002), 73pp.

[CalMS2003] Calude, C.S. and Marcus, S. A topological characterization of random sequences. *Inform. Proc. Lett.* **88** (2003), 245–250. 234

[CaoDP2010] Cao, J., Drozdowski, R. and Piotrowski, Z. Weak continuity properties of topologized groups. *Czechoslovak Math. J.* **60** (2010), 133–148. 197

[CaoM2004] Cao J. and Moors, W.B. Separate and joint continuity of homomorphisms defined on topological groups. *New Zealand J. Math.* **33** (2004), 41–45. 197

[Car1945] Cartan, H. Théorie du potentiel newtonien: énergie, capacité, suites de potentiels. *Bull. Soc. Math. France* **73** (1945), 74–106. 116

[Car1946] Cartan, H. Théorie générale du balayage en potentiel newtonien. *Ann. Inst. Fourier Grenoble* **22** (1946), 221–280. 117

[Cha1933] Champernowne, D. G. The construction of decimals normal in the scale of ten. *J. London Math. Soc.* **8** (1933), 254–260. 225

[ChanM1952] Chandrasekharan, K. and Minakshisundaram, S. *Typical Means.* Oxford University Press, 1952. 17

[CharC2001] Charatonik, J.J. and Charatonik, W.J. The Effros metric. *Topol. Appl.* **110** (2001), 237–255. 176

[CharM1966] Charatonik, J.J. and Maćkowiak, T. Around Effros' theorem. *Trans. Am. Math. Soc.* **298** (1986), 579–602. 173

[Cho1953] Choquet, G. Theory of capacities. *Ann. Inst. Fourier, Grenoble* **5** (1953–54), 131–295. 116

[Cho1969] Choquet, G. *Lectures on Analysis.* Vol. I, Benjamin, 1969. 45

[Chr1972] Christensen, J.P.R. On sets of Haar measure zero in abelian Polish groups. *Israel J. Math.* **13** (1973), 255–260. 241

[Chr1974] Christensen, J.P.R. *Topology and Borel Structure. Descriptive Topology and Set Theory with Applications to Functional Analysis and Measure Theory.* North-Holland, 1974. 47, 241, 251

[Chr1981] Christensen, J.P.R. Joint continuity of separately continuous functions. *Proc. Am. Math. Soc.* **82** (1981), 455–461. 197

[Chu1968] Chung, K.-L. *A Course in Probability Theory.* Academic Press, 1968. (3rd ed., 2001.) 117

[Chu1974] Chung, K.-L. *Elementary Probability Theory with Stochastic Processes.* Springer, 1974. (4th ed., 2003.) 117

[Chu1982] Chung, K.-L. *Lectures from Markov Processes to Brownian Motion.* Grundlehren Math. Wiss. **249**, Springer, 1982. 116, 117, 118

[Chu1995] Chung, K.-L. *Green, Brown and Probability.* World Scientific, 1995. 118

[Cie1997] Ciesielski, K. Set-theoretic real analysis. *J. Appl. Anal.* **3** (1997), 143–190. 234

[CieJ1995] Ciesielski, K. and Jasiński, J. Topologies making a given ideal nowhere dense or meager. *Topol. Appl.* **63** (1995), 277–298. 248

[CieL1990] Ciesielski, K. and Larson, L. The density topology is not generated. *Real Anal. Exchange* **16** (1990/1991), 522–525. 167

[CieLO1994] Ciesielski, K., Larson, L. and K. Ostaszewski, I-*Density Continuous Functions.* Mem. Am. Math. Soc. **107** (1994), no. 515. 167

[Coh1963] Cohen, P. The independence of the continuum hypothesis. *Proc. Natl. Acad. Sci. USA* **50** (1963), 105–110. 268

[Cohn1989] Cohn, P.M. *Algebra, Vol.* 2, 2nd ed. Wiley, 1989. (1st ed. 1977.) 273

[Com1984] Comfort, W.W. *Topological Groups.* In KunV1984, Chapter 24. 94, 140, 236

[ComN1965] Comfort, W.W. and Negrepontis, S. The ring $C(X)$ determines the category of X. *Proc. Am. Math. Soc.* **16** (1965), 1041–1045. 40, 41

[ComN1974] Comfort, W.W. and Negrepontis, S. *The Theory of Ultrafilters.* Grundlehren Math. Wiss. **211**, Springer, 1974. 64

[ConC1972] Constantinescu, C. and Cornea, A. *Potential Theory on Harmonic Spaces.* Grundlehren Math. Wiss. **158**, Springer, 1972. 116, 118

[Conw1990] Conway, J.B. *A Course in Functional Analysis.* 2nd ed. Graduate Texts in Mathematics, **96** Springer, 1990. 182, 183

[Cro1957] Croft, H.T. A question of limits. *Eureka* **20** (1957), 11–13. 18

[CsiE1964] Csiszár, I. and Erdős, P. On the function $g(t) = \limsup_{x \to \infty}(f(x + t) - f(x))$. *Magyar Tud. Akad. Mat. Kutató Int. Közl. A* **9** (1964), 603–606. 145, 170

[Dal1978] Dales, H.G. Automatic continuity: A survey. *Bull. London Math. Soc.* **10**(1978), 129–183. 185

[Dal2000] Dales, H.G. *Banach Algebras and Automatic Continuity.* London Math. Soc. Monog. New Series, **24**, Oxford University Press, 2000. 197

[DalW1987] Dales, H.G. and Woodin, W.H. *An Introduction to Independence for Analysts.* London Math. Soc. Lecture Note Series, **115**. Cambridge University Press, 1987. 232

[DalW1996] Dales, H.G. and Woodin, W.H. *Super-Real Fields. Totally Ordered Fields with Additional Structure.* London Math. Soc. Monog., New Series, **17**, Oxford University Press, 1996. 268

[Dar1875] Darboux, G. Sur la composition des forces en statiques. *Bull. des Sci. Math.* **9** (1875), 281–288. 144

[Darj2013] Darji, Udayan B. On Haar meager sets. *Topol. Appl.* **160** (2013), 2396–2400. 249, 261, 262

[Dav1952] Davies, R.O. Subsets of finite measure in analytic sets. *Nederl. Akad. Wetensch. Proc. Ser. A.* **14** (1952), 488–489. 195

[DavJO1977] Davies, Roy O., Jayne, J.E., Ostaszewski, A.J. and Rogers, C.A. Theorems of Novikov type. *Mathematika* **24** (1977), 97–114. 59

[Deb1986] Debs, G. Points de continuité d'une fonction séparément continue. *Proc. Am. Math. Soc.* **97** (1986), 167–176. 47

[Deb1987] Debs, G. Points de continuité d'une fonction séparément continue II. *Proc. Am. Math. Soc.* **99** (1987), 777–782. 48

[DebSR1987] Debs, G. and Saint Raymond, J. Ensembles d'unicité et d'unicité au sens large, *Ann. Inst. Fourier, Grenoble* **37**(1987) 217–239. 138

[Del1972] Dellacherie, C. *Capacités et Processus Stochastiques*, Ergebnisse Math. **67**, Springer, 1972. 117

[Del1980] Dellacherie, C. Un cours sur les ensembles analytiques, Part II. In RogJD1980, pp. 183–316. 51

[DelM1975–1992] Dellacherie, C. and Meyer, P.-A. *Probabilités et Potentiel*, Ch. I–IV (1975), Ch. V–VIII (1980), Ch. IX–XI (1983), Ch. XII–XVI (1987), Ch. XVII–XXIV (with Maisonneuve, B.) (1992). Hermann, Paris. 118

[Den1915] Denjoy, A. Sur les fonctions dérivées sommable. *Bull. Soc. Math. France* **43** (1915), 161–248. 107

[Dev1973] Devlin, K.J. *Aspects of Constructibility.* Lecture Notes Math. **354**, Springer, 1973. 53, 64

[Dev1977] Devlin, K.J. Constructibility. In Barw1977, p. 453–489. 270

[Dev1984] Devlin, K.J. *Constructibility.* Perspectives Math. Logic, Springer, 1984. 270, 277

[Dev1993] Devlin, K.J. *The Joy of Sets: Fundamentals of Contemporary Set Theory*, 2nd ed. Springer, 1993. (1st ed. 1979.) 272, 273

[DezDD2006] Deza, E., Deza, M.M. and Deza, M. *Dictionary of Distances,* Elsevier, 2006. 86

[DieS2014] Diestel, J. and Spalsbury, A. *The Joys of Haar Measure,* Grad. Studies in Math. **150**, Am. Math Soc., 2014. 242, 285

[Dij2005] Dijkstra, J.J. On homeomorphism groups and the compact-open topology. *Am. Math. Monthly* **112** (2005), 910–912. 279

[Dod2004a] Dodos, P. Dichotomies of the set of test measures of a Haar-null set. *Israel J. Math.* **144** (2004), 15–28. 253

[Dod2004b] Dodos, P. On certain regularity properties of Haar-null sets. *Fund. Math.* **191** (2004), 97–109. 253

[Dod2009] Dodos, P. The Steinhaus property and Haar-null sets. *Bull. Lon. Math. Soc.* **41** (2009), 377–384. 258

[Doo1984] Doob, J.L. *Classical Potential Theory and its Probabilistic Counterpart.* Grundl. Math. Wiss. **262**, Springer, 1984. 115, 116, 117, 118, 119, 120, 123

[DorFN2013] Dorais, F.G., Filipów, R. and Natkaniec, T. On some properties of Hamel bases and their applications to Marczewski measurable functions. *Cent. Eur. J. Math.* **11** (2013), 487–508. 22

[DouF1994] Dougherty, R. and Foreman, M. Banach–Tarski decompositions using sets with the Baire property. *J. Am. Math. Soc.* **7** (1994), 75–124. 138, 224

[Douw1984] van Douwen, E.K. The integers and topology. In K. Kunen and J.E. Vaughan (eds), *Handbook of Set-Theoretic Topology*. North-Holland, 1984, 111–167. 234

[Douw1989] van Douwen, E.K. Fubini's theorem for null sets. *Am. Math. Monthly* **96**(8) (1989), 718–721. 228

[Douw1992] van Douwen, E. A technique for constructing honest locally compact submetrizable examples. *Topol. Appl.* **47** (1992), 179–201. 145

[Dow1947] Dowker, C.H. Mapping theorems for non-compact spaces. *Am. J. Math.* **69** (1947), 200–242. 135

[Dra1974] Drake, F.R. *Set Theory: An Introduction to Large Cardinals*, North-Holland, 1974. 271, 274, 276, 277

[Dras1968] Drasin, D. Tauberian theorems and slowly varying functions. *Trans. Am. Math. Soc.* **133** (1968), 333–356. 4

[Dras2010] Drasin, D. Baernstein's thesis and entire functions with negative zeros. *Mat. Stud.* **34** (2010), 160–167. 5

[DrasS1969] Drasin, D. and Shea, D.F. Asymptotic properties of entire functions extremal for the $\cos \pi \rho$ theorem. *Bull. Am. Math. Soc.* **75** (1969), 119–122. 6

[DrasS1970] Drasin, D. and Shea, D.F. Complements to some theorems of Bowen and Macintyre on the radial growth of entire functions with negative zeros. In H. Shankar (ed), *Mathematical Essays Dedicated to A.J. Macintyre*. Ohio University Press, 1970, 101–121. 4

[DrasS1972] Drasin, D. and Shea, D.F. Pólya peaks and the oscillation of positive functions. *Proc. Am. Math. Soc.* **34** (1972), 403–411. 6

[DrasS1976] Drasin, D. and Shea, D.F. Convolution inequalities, regular variation and exceptional sets. *J. Analyse Math.* **29** (1976), 232–293. 4, 10, 18

[Dud1989] Dudley, R.M. *Real Analysis and Probability*. Cambridge Studies in Advanced Mathematics **74**. Cambridge University Press, 2002. (1st ed. 1989.) 29, 30, 253

[Dug1966] Dugundji, J. *Topology*. Allyn and Bacon, 1966. 35, 85, 176, 197

[Edg1990] Edgar, G.A. *Measure, Topology and Fractal Geometry*. Undergrad. Texts in Math., Springer, 1990. 226

[Edr1969] Edrei, A., Locally Tauberian theorems for meromorphic functions of lower order less than one. *Trans. Am. Math. Soc.* **140** (1969), 309–332. 5

[EdrF1966] Edrei, A. and Fuchs, W.H.J. Tauberian theorems for a class of meromorphic functions with negative zeros and positive poles. In *Contemporary Problems in Anal. Functions* (Proc. Internat. Conf. Erevan, 1965, Russian), 339–358. Nauka, 1966. 4, 5

[Eff1965] Effros, E.G. Transformation groups and C^*-algebras. *Ann. Math.* **81** (1965), 38–55. 172

[EleN2020] Elekes, M. and Nagy, D. Haar null and Haar meagre sets: A survey and new results. *Bull. Lon. Math. Soc.* **52** (2020), 561–619 (arXiv:1606.06607v2) 249

[EleV2017] Elekes, M. and Vidyánsky, Z. Naively Haar null sets in Polish groups. *J. Math. Anal. Appl.* **446** (2017), 193–200. 249

[Ell1974] Ellentuck, E. A new proof that analytic sets are Ramsey. *J. Symbolic Logic* **39**, 163-165. 121

[Elli1953] Ellis, R. Continuity and homeomorphism groups. *Proc. Am. Math. Soc.* **4** (1953). 969–973. 183, 197, 219

[Elli1957] Ellis, R. A note on the continuity of the inverse, *Proc. Am. Math. Soc.* **8** (1957). 372–373. 197

[Elli1969] Ellis, R. *Lectures on Topological Dynamics*, Benjamin, 1969. 288

[EmeFK1979] Emeryk, A., Frankiewicz, R. and Kulpa, W. On functions having the Baire property. *Bull. Acad. Polon. Sci. Sér. Sci. Math.* **27** (1979), 489–491. 185

[Eng1969] Engelking, R. On closed images of the space of irrationals. *Proc. Am. Math. Soc.* **21** (1969), 583–586. 211

[Eng1989] Engelking, R. *General Topology.* Heldermann, 1989. xiii, 25, 26, 27, 28, 35, 36, 41, 44, 56, 58, 85, 86, 120, 136, 153, 157, 162, 165, 193, 194, 208, 214, 215, 279, 282

[ErdKR1963] Erdős, P., Kestelman, H. and Rogers, C.A. An intersection property of sets with positive measure, *Coll. Math.* **11** (1963), 75–80. 168

[Ess1975] Essén, R.R. *The* $\cos \pi \lambda$ *Theorem.* Lecture Notes in Math. **467**, Springer, 1975. 6

[Fal1985] Falconer, K.J. *The Geometry of Fractal Sets.* Cambridge Tracts in Math. **85**, Cambridge University Press, 1985. 226

[FalGO2015] Falconer, K., Gruber, Peter M., Ostaszewski, A.J. and Stuart, Trevor. Claude Ambrose Rogers: 1 November 1920–5 December 2005. *R. Soc. Biogr. Mem.* **61** (2015), 403–435. xiii

[Fed1969] Federer, H. *Geometric Measure Theory.* Grundl. Math. Wiss. **153**, Springer, 1969. 228

[Fel1966] Feller, W. *An Introduction to Probability Theory and Its Applications, Vol. II*, Wiley, 1966. (2nd ed. 1971.) 2

[FenN1973] Fenstad, J.E. and Normann, D. On absolutely measurable sets. *Fund. Math.* **81** (1973/74), 91–98. 62, 241

[FilW2011] Filipczak, M. and Wilczyński, W. Strict density topology on the plane. Measure case. *Rend. Circ. Mat. Palermo*(2) **60** (2011), 113–124. 256

[FisFK2014] Fischer, V., Friedman, S.S. and Khomskii, Y. Cichoń diagram, regularity properties and Δ_3^1 sets of reals. *Arch. Math. Logic* **53** (2014), 695–729. 231

[ForK2010] Foreman, M. and Kanamori, A. (eds). *Handbook of Set Theory.* Springer, 2010. 231, 271, 312, 315, 316

[ForW1991] Foreman, M. and Woodin, W.H. The generalized continuum hypothesis can fail everywhere. *Ann. Math.* **133** (1991), 1–35. 269

[Fort1949] Fort Jr., M.K. A unified theory of semi-continuity. *Duke Math. J.* **16** (1949), 237–246. 242

[Fos1993] Fosgerau, M. When are Borel functions Baire functions? *Fund. Math.* **143** (1993), 137–152. 191

[Fra1982] Frankiewicz, R. On functions having the Baire property, II. *Bull. Acad. Polon. Sci. Sér. Sci. Math.* **30** (1982), 559–560. 185

[FraK1987] Frankiewicz, R. and Kunen, K. Solution of Kuratowski's problem on function having the Baire property, I. *Fund. Math.* **128** (1987), 171–180. 185

[Fre1913] Fréchet, M., Pri la funkcia ekvacio f(x+y) = f(x) + f(y) (in Esperanto). Enseignement Math. **15** (1913), 390–393 21

[Fre1914] Fréchet, M. Sur la notion de différentielle d'une fonction de ligne. *Trans. Am. Math. Soc.* **15** (1914), 135–161. 21

[Fre1980] Fremlin, D.H. *Čech-Analytic Spaces*, (1980). Available from www1 .essex.ac.uk/maths/people/fremlin/n80l08.pdf. 57

[Fre1984] Fremlin, D.H. *Consequences of Martin's Axiom*. Cambridge Tracts in Mathematics **84**, Cambridge University Press, 1984. 272

[Fre1987] Fremlin, D.H. Measure-additive coverings and measurable selectors. *Dissertationes Math. (Rozprawy Mat.)* **260** (1987), 116pp. 185

[Fre2000a] Fremlin, D.H. Universally Kuratowski–Ulam spaces, (2000a). Available from `www1.essex.ac.uk/maths/people/fremlin/preprints.htm`. 149, 158

[Fre2000b] Fremlin, D.H. *Measure Theory, 1. The Irreducible Minimum.* Torres Fremlin, 2000. Available from `www1.essex.ac.uk/maths/people/fremlin/mt1.2011/index.htm`. 25

[Fre2001] Fremlin, D.H. *Measure Theory, 2. Broad Foundations.* Torres Fremlin, 2001. Available from `www1.essex.ac.uk/maths/people/fremlin/mt2.2016/index.htm`. 25, 207

[Fre2002] Fremlin, D.H. *Measure Theory, 3. Measure Algebras.* Torres Fremlin, 2002. Available from `www1.essex.ac.uk/maths/people/fremlin/mt3.2012/index.htm`. 25, 148, 149, 241

[Fre2003] Fremlin, D.H. *Measure Theory, 4. Topological Measure Spaces. Part I, II.* Torres Fremlin, 2003. Available from `www1.essex.ac.uk/maths/people/fremlin/mt4.2013/index.htm`. 25, 207, 272

[Fre2008] Fremlin, D.H. *Measure Theory, 5. Set-Theoretic Measure Theory. Part I, II.* Torres Fremlin, 2008. Available from www1.essex.ac .uk/maths/people/fremlin/mt5.2015/index.htm. 25, 33, 138, 222, 232

[FreG1995] Fremlin, D.H. and Grekas, S. Products of completion regular measures. *Fund. Math.* **147** (1995), 27–37. 154

[FreNR2000] Fremlin, D., Natkaniec, T. and Recław, I. Universally Kuratowski–Ulam spaces. *Fund. Math.* **165** (2000), 239–247. 139, 148, 149, 159, 175, 228

[FriS2020] Friedman, S.D. and Schrittesser, D. Projective measure without projective Baire. *Memoirs Am. Math. Soc.* **267** No. 1298, 2020. 231

[Fro1960] Frolík, Z. Generalizations of the G_δ-property of complete metric spaces. *Czechoslovak Math. J.* **10** (1960), 359–379. 56, 58, 195

[Fro1970] Frolík, Z. Absolute Borel and Souslin sets. *Pacific J. Math.* **32** (1970), 663–683. 56

[FroH1981] Frolík, Z. and Holický, P. Decomposability of completely Suslin-additive families. *Proc. Am. Math. Soc.* **82** (1981), 359–365. 209, 218

[FroN1990] Frolík, Z. and Netuka, I. Čech completeness and fine topologies in potential theory and real analysis. *Expos. Math.* **8** (1990), 81–89. 111

[Fuc1970] Fuchs, L. *Infinite Abelian Groups, I.* Pure and Applied Mathematics **36**, Academic Press, 1970. 142

[Fug1971] Fuglede, B. The quasi topology associated with a countably subadditive set function. *Ann. Inst. Fourier (Grenoble)* **21** (1971), 123–169. 116, 123

[FukOT1994] Fukushima, M., Oshima, Y. and Takeda, M. *Dirichlet Forms and Markov Processes.* Walter de Gruyter, 1994. (1st ed., Fukushima, M., North-Holland, 1980.) 120

[Ful1968] Fuller, R.V. Relations among continuous and various non-continuous functions. *Pacific J. Math.* **25** (1968), 495–509. 242, 243

[Gao2009] Gao, S. *Invariant Descriptive Theory*, CRC Press, 2009. 62

[GerK1970] Ger, R. and Kuczma, M. On the boundedness and continuity of convex functions and additive functions. *Aequat. Math.* **4** (1970), 157–162. 123

[GhoM1985] Ghoussoub, N. and Maurey, B. $\mathcal{G}_\delta$-embeddings in Hilbert space. *J. Funct. Anal.* **61** (1985), 72–97. 219

[GilJ1960] Gillman, L. and Jerison, M. *Rings of Continuous Functions.* Van Nostrand, 1960. (Reprinted in Graduate Texts in Mathematics **43**, Springer, 1976.) 41

[Gli1961] Glimm, J. Locally compact transformation groups. *Trans. Am. Math. Soc.* **101** (1961), 124–138. 172

[GneK1954] Gnedenko, B.V. and Kolmogorov, A.N. *Limit Distributions for Sums of Independent Random Variables.* Addison-Wesley, 1954. 2

[God1940] Gödel, K. *The Consistency of the Axiom of Choice and of the Generalized Continuum Hypothesis.* Ann. Math. Studies **3**, Princeton University Press, 1940. 269, 270

[Gof1950] Goffman, C. On Lebesgue's density theorem. *Proc. Am. Math. Soc.* **1** (1950), 384–388. 112

[Gof1962] Goffman, C. On the approximate limits of a real function. *Acta Sci. Math. (Szeged)* **23** (1962), 76–78. 107

[Gof1975] Goffman, C. Everywhere differentiable functions and the density topology. *Proc. Am. Math. Soc.* **51** (1975), 250. 112

[GofNN1961] Goffman, C., Neugebauer , C.J. and Nishiura, T. Density topology and approximate continuity. *Duke Math. J.* **28** (1961), 497–505. 107, 109, 110, 113, 167, 256

[GofW1961] Goffman, C. and Waterman, D. Approximately continuous transformations. *Proc. Am. Math. Soc.* **12** (1961), 116–121. 107, 110, 167, 256

[GolSW2000] Goldstern, M., Schmeling, J. and Winkler, R. Metric, fractal dimensional and Baire results on the distribution of subsequences. *Math. Nachr.* **219** (2000), 97–108. 226

[Gos1985] Goswami, K.C. Density topology on $\mathbb{R}$ is not a Borel subset of its Stone–Čech compactification. *Indian J. Pure Appl. Math.* **16** (1985), 45–48. 111

[Gow1970] Gowisankaran, C. Radon measures on groups. *Proc. Am. Math. Soc.* **25** (1970), 381–384. 243, 244

[Gow1972] Gowisankaran, C. Quasi-invariant Radon measures on groups. *Proc. Am. Math. Soc.* **35** (1972), 503–506. 243, 244

[GraRS1990] Graham, R.L., Rothschild, B.L. and Spencer, J.H. *Ramsey Theory*, 2nd ed. Wiley, 1990. (1st ed. 1980.) 141

[Gre1828] Green, G. An Essay on the application of mathematical analysis to the theories of electricity and magnetism. T. Wheelhouse, Nottingham, 1828. (Facsimile edition, Wazäta-Melins Aktiebolag, Göteberg, 1958.) 114

[Gro1963] de Groot, J. Subcompactness and the Baire category theorem. *Nederl. Akad. Wetensch. Proc. Ser. A* **66** (1963), 761–767. 135

[Gros1987] Grosse-Erdmann, K.-G. Holomorphe Monster und Universelle Funktionen. *Mitt. Math. Sem. Giessen* **176** (1987). 228

[Gros1989] Grosse-Erdmann, K.-G. An extension of the Steinhaus–Weil theorem. *Colloq. Math.* **57** (1989), 307–317. 239

[Gros1999] Grosse-Erdmann, K.-G. Universal families and hypercyclic operators. *Bull. Am. Math. Soc.* **36** (1999), 345–381. 228

[Gru1984] Gruenhage, G. Generalized metric spaces. In KunV1984, Chapter 10. 44, 45, 49

[Gru1998] Gruenhage, G. Irreducible restrictions of closed mappings. *Topol. Appl.* **85** (1998), 127Ű135.

[Haa1970] de Haan, L. On regular variation and its applications to the weak convergence of sample extremes. *Math. Centre Tracts* **32**, Amsterdam, 1970. 3, 9

[Hal1950] Halmos, P.R. *Measure Theory*, Van Nostrand, 1950. (Reprinted as Graduate Texts in Math. **18**, Springer, 1974.) 24, 25, 75, 148, 251, 253, 257

[Ham1905] Hamel, G. Eine Basis aller Zahlen und die unstetigen Lösungen der Funktionalgleichung $f(x + y) = f(x) + f(y)$. *Math. Ann.* **60** (1905), 459–462. 20

[Han2004] Hand, D. J. *Measurement Theory and Practice: The World through Quantification*. Arnold. 22

[HansT1992] Hansel, G. and Troallic, J.-P. Quasicontinuity and Namioka's theorem. *Topol. Appl.* **46** (1992), 135–149. 197

[Hanse1971] Hansell, R.W. Borel measurable mappings for nonseparable metric spaces. *Trans. Am. Math. Soc.* **161** (1971), 145–169. 39, 185, 208, 209, 217, 218

[Hanse1972] Hansell, R.W. On the nonseparable theory of Borel and Souslin sets. *Bull. Am. Math. Soc.* **78** (1972), 236–241. 205

[Hanse1973a] Hansell, R.W. On the representation of nonseparable analytic sets. *Proc. Am. Math. Soc.* **39** (1973), 402–408. 203, 205

[Hanse1973b] Hansell, R.W. On the non-separable theory of k-Borel and k-Souslin sets. *General Topol. Appl.* **3** (1973), 161–195. 205, 208

[Hanse1974] Hansell, R.W. On characterizing non-separable analytic and extended Borel sets as types of continuous images. *Proc. London Math. Soc.* **28** (1974), 683–699. 193, 208, 209, 210, 211, 218

[Hanse1992] Hansell, R.W. Descriptive topology. In H. Hužek and J. van Mill (eds).. Elsevier, 1992, 275–315. 56, 58, 204, 209, 213, 215, 218

[Hanse1999] Hansell, R.W. Nonseparable analytic metric spaces and quotient maps. *Topol. Appl.* **85** (1998), 143–152. 209, 219

[HanseJR1983] Hansell, R.W., Jayne, J.E. and Rogers, C.A. *K*-analytic sets. *Mathematika* **30**(2) (1983), 189–221. 57, 207

[Har1904] Hardy, G.H. Researches in the theory of divergent series and divergent integrals. *Quart. J. Math.* **35** (1904), 22–66. (Reprinted in *Collected Papers of G. H. Hardy, Volume VI*, Oxford University Press, 1974, 37–84.) 17

[Har1922] Hardy, G.H. A new proof of the functional equation for the zeta-function. *Matematisk Tidsskrift B* (1922), 71–73. Reprinted in *Collected Papers of G. H. Hardy, Volume II*.1, Oxford University Press. 227

[Har1949] Hardy, G.H. *Divergent Series*. Oxford University Press, 1949. 13, 14, 15, 17

[HarL1916] Hardy, G.H. and Littlewood, J.E. Theorems concerning the summability of series by Borel's exponential method. *Rend. Circ. Mat. Palermo* **41** (1916), 36–53. (Reprinted in *Collected Papers of G.H. Hardy, Volume VI*, Oxford University Press, 1974, 609–628.) 16

[HarL1924] Hardy, G.H. and Littlewood, J.E. The equivalence of certain integral means. *Proc. London Math. Soc.* **22** (1924), 60–63. (Reprinted in *Collected Papers of G.H. Hardy, Volume VI*, Oxford University Press, 1974, 677–680.) 12

[HarR1915] Hardy, G.H. and Riesz, M. *The General Theory of Dirichlet's Series*. Cambridge Tracts in Math. **18**, Cambridge University Press, 1915. 17

[HarW2008] Hardy, G.H. and Wright, E.M. *An Introduction to the Theory of Numbers*, 6th ed. (revised D.R. Heath-Brown and J.H. Silverman), Oxford University Press, 2008. 8, 10, 140

[Harm1988] Harman, G. *Metric Number Theory*. LMS Monographs **18**, Oxford University Press, 1988. 225

[HarrKL1990] Harrington, L.A., Kechris, A.S. and Louveau, A. A Glimm–Effros dichotomy for Borel equivalence relations. *J. Amer. Math. Soc.* **3** (1990), 903–928. 173

[Hart1975] Hartman, S. Travaux de W. Sierpiński sur la théorie des ensembles et ses applications IV. Mesure et catégorie. Congruence des ensembles. In Sie1975, 20–25. 137

[HartM1958] Hartman, S. and Mycielski, J. On the imbedding of topological groups into connected topological groups. *Colloq. Math.* **5** (1958), 167–169. 185

[HauP1952] Haupt, O. and Pauc, C. La topologie approximative de Denjoy envisagée comme vraie topologie. *C. R. Acad. Sci. Paris* **234** (1952), 390–392. 107, 167, 256

[Haw1975] Hawkes, J. On the potential theory of subordinators. *Z. Wahrschein.* **33** (1975), 113–132. 118

[HawoM1977] Haworth, R.C. and McCoy, R.C. Baire spaces. *Dissertationes Math.* **141** (1977). 37, 156

[HayP1970] Hayes, C.A. and Pauc, C.Y. *Derivation and Martingales.* Ergebnisse Math. **49**, Springer, 1970. 105

[Haym1964] Hayman, W.K. *Meromorphic Functions.* Oxford University Press, 1964. 5, 6

[Haym1970] Hayman, W.K. Some examples related to the $\cos \pi \rho$ theorem. In H. Shankar (ed), *Mathematical Essays Dedicated to A.J. Macintyre.* Ohio University Press, 1970, 149–170. 6

[Hei1971] Heiberg, C. A proof of a conjecture by Karamata. *Publ. Inst. Math. Beograd* (NS) **12** (26), 41–44. 10

[Hel1969] Helms, L.L. *Introduction to Potential Theory.* Wiley, 1969. 115, 116

[HewR1979] Hewitt, E. and Ross, K.A. *Abstract Harmonic Analysis, I. Structure of Topological Groups, Integration Theory, Group Representations,* 2nd ed. Grundl. Math. Wiss. **115**, Springer, 1979. 173, 191, 253, 257

[Hey1977] Heyer, H. *Probability Measures on Locally Compact Groups.* Ergebnisse Math. **94**, Springer, 1977. 186

[Hil1974] Hildenbrandt, W. *Core and Equilibria of a Large Economy.* Princeton University Press, 1974. 54

[HillP1957] Hille, W. and Phillips, R.S. *Functional Analysis and Semi-Groups.* Am. Math. Soc. Colloq. Publ. **31**, Am. Math. Soc., 1957. 239

[HinS1998] Hindman, N. and Strauss, D. *Algebra in the Stone-Čech Compactification. Theory and Applications.* De Gruyter Expos. Math. **27**, Walter de Gruyter, 1998 (2nd revised and extended ed., de Gruyter, 2012). 64, 274

[Hof1980] Hoffmann-Jørgensen, J. Automatic continuity. In RogJD1980, Part 3.2, 337–398. 175, 177, 185, 187, 197

[HolN2012] Holá, L. and Novotný, B. Subcontinuity. *Math. Slov.* **62** (2012), 345–362. 242

[Holi2010] Holický, P. Preservation of completeness by some continuous maps. *Topol. Appl.* **157** (2010), 1926–1930. 220

[HoliP2010] Holický, P. and Pol, R. On a question by Alexey Ostrovsky concerning preservation of completeness. *Topol. Appl.* **157** (2010), 594–596. 196, 216, 219

[HruA2005] Hrušák, M. and Zamora Avilés, B. Countable dense homogeneity of definable spaces. *Proc. Am. Math. Soc.* **133** (2005), 3429–3435. 56

[HulLM2005] Hult, H., Lindskog, F., Mikosch, T. and Samorodnitsky, G. Functional large deviations for multivariate regularly varying random walks. *Ann. Appl. Probab.* **15**(4) (2005), 2651–2680. 278

[HurW1941] Hurewicz, W. and Wallman, H. *Dimension Theory.* Princeton University Press, 1941. 226

[HydLO2010] Hyde, J., Laschos, V., Olsen, L., Petrykiewicz, I. and Shaw, A. Iterated Cesàro averages, convergence, frequencies of digits and Baire category. *Acta Arith.* **144** (2010), 287–293. 225

[IhoS1989] Ihoda, J.I. (Judah, H.I.) and Shelah, S. Δ_2^1-sets of reals. *Ann. Pure Appl. Logic* **42** (1989), 207–223. 230

[IonT1961] Ionescu Tulcea, A. and Ionescu Tulcea, C. On the lifting property, I. *J. Math. Anal. Appl.* **3** (1961), 537–546. 104

[IonT1969] Ionescu Tulcea, A. and Ionescu Tulcea, C. *Topics in the Theory of Lifting.* Ergebnisse Math. **48**, Springer, 1969. 104

[Itz1972] Itzkowitz, G.L. A characterization of a class of uniform spaces that admit an invariant integral. *Pacific J. Math.* **41** (1972), 123–141. 196

[Jab2015] Jabłońska, E. Some analogies between Haar meager sets and Haar null sets in abelian Polish groups. *J. Math. Anal. Appl.* **421** (2015), 1479–1486. 249, 263, 264

[Jab2016] Jabłońska, E. A theorem of Piccard's type in abelian Polish groups. *Anal. Math.* **42** (2016), 159–164. 255

[Jak1959] Jakimovski, A. A generalization of the Lototsky method of summation. *Michigan Math. J.* **6**, 277–290. 16

[Jam1943] James, R.C. Linearly arc-wise connected topological Abelian groups. *Ann. Math.* **44**, (1943), 93–102. 86

[JamMW1947] James, R.C., Michal, A.D. and Wyman, M. Topological Abelian groups with ordered norms. *Bull. Am. Math. Soc.* **53** (1947), 770–774. 86

[Jan1997] Janson, S. *Gaussian Hilbert Spaces.* Cambridge Tracts in Mathematics **129**. Cambridge University Press, 1997. 260

[Jav1968] Javor, P. On the general solution of the functional equation $f(x + yf(x)) = f(x)f(y)$, *Aequat. Math.* **1** (1968), 235–238. 288

[Jec1973] Jech, Th.J. *The Axiom of Choice.* North-Holland, 1973. 20, 270

[Jec1997] Jech, Th.J. *Set Theory.* Springer, 1977. 277

[Jen1972] Jensen, R.B. The fine structure of the constructible hierarchy. *Ann. Math. Logic* **4** (1972), 229–308. 145

[JimSS2019] Jiménez-Garrido, J., Sanz, J. and Schindl, G. Indices of O-regular variation for weight functions and weight sequences. *Rev. R. Acad. Cienc. Ex. Fac. Nat. Ser. A (RACSAM)* **113** (2019), 3659–3697. 7

[Joh1982] Johnstone, P.J. *Stone Spaces.* Cambridge Studies in Advanced Mathematics **3**. Cambridge University Press, 1982. 149

[Jon1942a] Jones, F.B. Connected and disconnected plane sets and the functional equation $f(x + y) = f(x) + f(y)$. *Bull. Am. Math. Soc.* **48** (1942), 115–120. 145

[Jon1942b] Jones, F.B. Measure and other properties of a Hamel basis. *Bull. Am. Math. Soc.* **48** (1942), 472–481. 21

[Jon1975] Jones, F.B. Use of a new technique in homogeneous continua. *Houston J. Math.* **1** (1975), 57–61. 173

[Jor1974] Jordan, G.S. Regularly varying functions and convolutions with real kernels. *Trans. Am. Math. Soc.* **194** (1974), 177–194. 5

[JudR1993] Judah, H. and Rosłanowski, A. On Shelah's amalgamation. In *Set Theory of the Reals* (Ramat Gan, 1991), 385–414, Israel Math. Conf. Proc., **6**, Bar-Ilan Univ., Ramat Gan, 1993. 233

[JudS1993a] Judah, H.I. (Ihoda, J.I.) and Shelah, S. Δ^1_3-sets of reals. *J. Symbolic Logic* **58** (1993), 72–80. 231

[JudS1993b] Judah, H.I. and Shelah, S. Baire property and axiom of choice. *Israel J. Math.* **84** (1993), 435–450. 232, 233

[JudSp1997] Judah, H.I. (Ihoda, J.I.) and Spinas, O. Large cardinals and projective sets. *Arch. Math. Logic* **36** (1997), 137–155. 229, 233, 269

[Kah1985] Kahane, J.-P. *Some Random Series of Functions*, 2nd ed. Cambridge Studies in Advanced Mathematics **5**, Cambridge University Press. (1st ed. 1968, Heath, MA.) 224, 227, 229, 234

[Kah1997] Kahane, J.-P. A century of interplay between Taylor series, Fourier series and Brownian motion. *Bull. London Math. Soc.* **29** (1997), 257–279. 223, 224, 227

[Kah2000] Kahane, J.-P. Baire's category theorem and trigonometric series. *J. Anal. Math.* **80** (2000), 143–182. 138, 224, 227, 229

[Kah2001] Kahane, J.-P. Probabilities and Baire's theory in harmonic analysis. In J. S. Byrnes (ed), *Twentieth Century Harmonic Analysis, A Celebration*. Kluwer, 2001, 57–72. 138, 223, 224, 227, 229, 234

[KahQ1997] Kahane, J.-P. and Queffélec, H. Ordre, convergence et sommabilités des séries de Dirichlet. *Ann. Inst. Fourier* **47** (1997), 485–529. 227

[Kak1936] Kakutani, K. Über die Metrisation der topologischen Gruppen. *Proc. Imp. Acad. Tokyo* **12** (1936), 82–84. (Reprinted in *Selected Papers*, 1, R.R. Kallman (ed), 60–62. Birkhäuser, 1986.) 86, 91

[KalK2015] Kalemba, P. and Kucharski, A. Universally Kuratowski–Ulam spaces and open–open games. *Ann. Math. Siles.* **29** (2015), 421–427. 149

[Kall2002] Kallenberg, O. *Foundations of Modern Probability*, 2nd ed., Springer, 2002. (1st ed. 1997.) 240

[KaltPR1984] Kalton, N.J., Peck, N.T. and Roberts, J.R. *An F-space Sampler*, London Math. Soc. Lect. Notes Ser. **89**, Cambridge University Press, 1984. 174

[Kan2003] Kanamori, A. *The Higher Infinity. Large Cardinals in Set Theory from their Beginnings*, 2nd ed. Springer, 2003. (1st ed. 1994.) 62, 232, 277

[Kan2009] Kanamori, A. Set theory from Cantor to Cohen. In A. Irvine (ed), *Handbook of the Philosophy of Scienc. Philosophy of Mathematics*, Chapter 10. North-Holland, 2009. 36

[KanM1978] Kanamori, A. and Magidor, M. The evolution of large cardinal axioms in set theory. *Higher Set Theory* (Proc. Conf., Math. Forschungsinst., Oberwolfach, 1977), pp. 99–275, Lecture Notes in Math. **669**, Springer, 1978. 232

[KaniP1975] Kaniewski, J. and Pol, R. Borel-measurable selectors for compact-valued mappings in the non-separable case. *Bull. Acad. Polon. Sci. Sér. Sci. Math.* **23** (1975), 1043–1050. 209, 218

[KanoL2003] Kanovei, B.G. and Lyubetskii, V.A. On some classical problems of descriptive set theory. *Russian Math. Surveys* **58** (2003), 839–927. 230

[Kar2009] Karamata, J. *Selected Papers*, V. Marić (ed). Zavod za Udžbenike, Beograd, 2009. 8, 298

[KatS1974] Katznelson, Y. and Stromberg, K. Everywhere differentiable, nowhere monotone, functions. *Am. Math. Monthly* **81** (1974), 349–354. 113

[Kau1967] Kaufman, R. A functional method for linear sets. *Israel J. Math.* **5** (1967), 185–187. 223

[Kau1995] Kaufman, R. Thin sets, differentiable functions and the category method. In *J. Fourier Anal. Appl., Special Issue in honour of J.-P. Kahane*, 311–316. CRC Press, 1995. 223

[Kec1994] Kechris, A.S. Topology and descriptive set theory. *Topol. Appl.* **58** (1994), 195–222. 175

[Kec1995] Kechris, A.S. *Classical Descriptive Set Theory.* Graduate Texts in Mathematics **156**, Springer, 1995. xiii, 36, 38, 46, 47, 48, 61, 62, 79, 117, 121, 123, 154, 162, 167, 196, 203, 231, 232, 240, 241, 268, 274

[KecL1987] Kechris, A.S. and Louveau, A. *Descriptive Set Theory and the Structure of Sets of Uniqueness.* London Math. Soc. Lecture Note Series **128**, Cambridge University Press, 1987. 138, 228

[KecS1985] Kechris, A.S. and Solovay, R.M. On the relative consistency strength of determinacy hypotheses. *Trans. Am. Math. Soc.* **290**(1) (1985), 179–211. 232

[Kel1955] Kelley, J.L. *General Topology.* Van Nostrand, 1955. (2nd ed. Springer, 1975.) 36, 42, 59, 86, 136, 201, 219, 282

[Kell1953] Kellogg, O.D. *Foundations of Potential Theory.* Dover, 1953. (Originally published in Grundlehren Math. Wiss. **31** (1929), reprinted 1967, Springer.) 114, 115

[Kem1957] Kemperman, J.H.B. A general functional equation. *Trans. Am. Math. Soc.* **86** (1957), 28–56. 74, 241

[Ken1968] Kendall, D.G. Delphic semigroups, infinitely divisible regenerative phenomena and the arithmetic of *p*-functions. *Z. Wahrschein vew. Geb.* **9** (1968), 163–195. (Reprinted in *Stochastic Analysis*, E.F. Harding and D.G. Kendall (eds), 73–114. Wiley, 1973.) 18

[Kes1947a] Kestelman, H. The convergent sequences belonging to a set. *J. London Math. Soc.* **22** (1947), 130-136. 73, 162

[Kes1947b] Kestelman, H. On the functional equation $f(x + y) = f(x) + f(y)$. *Fund. Math.* **34** (1947), 144–147. 145

[Kha1998] Kharazishvili, A.B. *Transformation Groups and Invariant Measures. Set-Theoretic Aspects.* World Scientific, 1998. 240

[Kha2004] Kharazishvili, Alexander. *Nonmeasurable Sets and Functions.* Elsevier, 2004. 22, 172

[Kha2018] Kharazishvili, A.B. *Strange Functions in Real Analysis*, 3rd ed. Chapman and Hall, 2018. (2nd ed. 2006, 1st ed. 2000.) 113

[Kha2019] Kharazishvili, A. Some remarks on the Steinhaus property for invariant extensions of the Lebesgue measure. *Eur. J. Math.* **5** (2019), 81–90. 256

[KhiK1925] Khintchin, A.Y. and Kolmogorov, A.N. Über Konvergenz von Reihen, deren Glieder durch den Zufall bestimmt werden. *Mat. Sbornik* **32** (1925), 668–677. 224

[Khr2008] Khrushchev, S. *Orthogonal Polynomials and Continued Fractions, from Euler's Point of View.* Encycl. Math. Appl. **122**, Cambridge University Press, 2008. 225

[Kin1963] Kingman, J.F.C. Ergodic properties of continuous-time Markov processes and their discrete skeletons. *Proc. London Math. Soc.* **13** (1963), 593–604. 78, 82, 134

[Kin1964] Kingman, J.F.C. A note on limits of continuous functions. *Quart. J. Math.* **15** (1964), 279–282. 18, 78, 82

[Kle1952] Klee, V.L. Invariant metrics in groups (solution of a problem of Banach). *Proc. Am. Math. Soc.* **3** (1952), 484–487. 86, 89, 285

[Klei1977] Kleinberg, E.M. *Infinitary Combinatorics and the Axiom of Determinateness.* Lecture Notes in Math. **612**, Springer, 1977. 231

[Kne1963] Kneebone, G.T. *Mathematical Logic and the Foundations of Mathematics. An Introductory Survey.* Van Nostrand 1963. (Reprinted, Dover, 2001). 269

[Kod1941] Kodaira, K. Über die Beziehung zwischen den Massen und den Topologien in einer Gruppe. *Proc. Phys.-Math. Soc. Japan* **23** (1941), 67–119. 138

[KoeW2010] Koellner, P. and Woodin, W.H. *Large Cardinals from Determinacy.* In ForK2010. 232

[Kol1934] Kolmogorov, A.N. Zur Normierbarkeit eines allgemeinen topologischen linearen Raumes. *Studia Math.* **5** (1934), 29–33. 86

[Kom1971] Kominek, Z. On the sum and difference of two sets in topological vector spaces. *Fund. Math.* **71** (1971), 165–169. 239

[Kom1981] Kominek, Z. On the continuity of Q-convex and additive functions. *Aequationes Math.* **23** (1981), 146–150. 22

[Kom1988] Kominek, Z. On an equivalent form of a Steinhaus theorem, *Math. (Cluj)* **30** (53)(1988), 25–27. 255, 259

[Komj1988] Komjáth, P. Large small sets. *Colloq. Math.* **56** (1988), 231–233. 74

[KomT2008] Komjáth, P. and Totik,V. Ultrafilters. *Am. Math. Month.* **115** (2008), 33–44.

[Kor2004] Korevaar, J. *Tauberian Theory: A Century of Developments.* Grundl. Math. Wiss. **329**, 2004, Springer. 7, 13, 14, 15, 16

[Korn1995] Körner, T.W. Kahane's Helson curve. In *J. Fourier Anal. Appl., Special Issue in honour of J.-P. Kahane*, 325–346. CRC Press, 1995. 223, 227

[Kro1974] Krom, M.R. Cartesian products of metric Baire spaces. *Proc. Am. Math. Soc.* **42** (1974), 588–594. 160

[KucP2007] Kucharski, A. and Plewik, S. Game approach to universally Kuratowski–Ulam spaces. *Topol. Appl.* **154** (2007), 85–92. 149

[Kucz1985] Kuczma, M. *An Introduction to the Theory of Functional Equations and Inequalities. Cauchy's Functional Equation and Jensen's Inequality.* PWN, 1985. 20, 21, 123, 144, 146, 241, 260

[KuczS1976] Kuczma, M. and Smítal, J. On measures connected with the Cauchy equation. *Aequationes Math.* **14**(3) (1976), 421–428. 259

[Kun1970] Kunen, K. Some applications of iterated ultrapowers in set theory. *Ann. Math. Logic* **1** (1970), 179–227. 277

[Kun1983] Kunen, K. *Set Theory. An Introduction to Independence Proofs.* Studies in Logic and Foundations of Mathematics **102**, North-Holland, 1983. 268

[Kun2011] Kunen, K. *Set Theory.* Studies in Logic (London) **34**. College Publications, London, 2011. 268

[KunV1984] Kunen, K. and Vaughan, J.E. *Handbook of Set-Theoretic Topology.* North-Holland, 1984. 25, 300, 306, 322

[Kur1924] Kuratowski, C. Sur les fonctions représentables analytiquement et les ensembles de première catégorie. *Fund. Math.* **5** (1924), 75–86. 184

[Kur1966] Kuratowski, K. *Topology, I.* PWN, 1966. 38, 42, 122, 184, 195, 224

[Kur1968] Kuratowski, K. *Topology, II.* PWN, 1968. 186, 279

[KurM1968] Kuratowski, K. and Mostowski, A. *Set Theory.* North Holland, 1968. 199

[KurU1932] Kuratowski, C. and Ulam, S. Quelques propriétés topologiques du produit combinatoire. *Fund. Math.* **19** (1932), 247–251. 148

[LabR1995] Łabedzki, G. and Repický, M. Hechler reals. *J. Symbolic Logic* **60** (1995), 444–458. 122

[Lac1998] Laczkovich, M. Analytic subgroups of the reals. *Proc. Am. Math. Soc.* **126** (1998), 1783–1790. 140, 141

[Lar2010] Larson, P.B. A brief history of determinacy. In A.S. Kechris, B. Löwe, J.R. Steel (eds), *The Cabal Seminar Vol. 4.* Lecture Notes in Logic, **49**, Cambridge University Press, 2010, 3–60. 232

[LeP1973] LePage, R.D. Subgroups of paths and reproducing kernels. *Ann. Prob.* **1** (1973), 345–347. 260

[Lev1983] Levi, S. On Baire cosmic spaces. In *General Topology and Its Relations to Modern Analysis and Algebra, V*, (Prague, 1981), pp. 450–454. Sigma Ser. Pure Math. **3**, Heldermann, 1983. 56, 192, 193, 196, 198

[LevS1967] Lévy, A. and Solovay, R.M. Measurable cardinals and the Continuum Hypothesis. *Israel J. Math.* **5** (1967), 234–248. 274

[LiZ2017] Li, R. and Zsilinszky, L. More on products of Baire spaces. *Topol. Appl.* **230** (2017), 35–44. 160

[Lio1851] Liouville, J. Sur les classes très-étendues des quantités dont la valeur n'est ni algébrique, ni même réductible à des irrationelles algébriques. *J. Math. Pures et Appliquées* **16** (1851), 133–142. 223

[LiuR1968] Liu, T.S. and van Rooij, A. Transformation groups and absolutely continuous measures. *Indag. Math.* **71** (1968), 225–231. 243

[LiuRW1970] Liu, T.S., van Rooij, A. and Wang, J.-K. Transformation groups and absolutely continuous measures II. *Indag. Math.* **73** (1970), 57–61. 243

[Lit1944] Littlewood, J.E. *Lectures on the Theory of Functions.* Oxford University Press, 1944. 24

[LoeW2000] Loewen, P.D. and Wang, Xianfu. Typical properties of Lipschitz functions. *Real Anal. Exchange* **26** (2000), 717–726. 228

[Lou1976] Louveau, A. Une méthode topologique pour l'étude de la propriété de Ramsey. *Israel J. Math.* **23** (1976), 97–116. 121

[Lou1980] Louveau, A. A separation theorem for Σ_1^1 sets. *Trans. Am. Math. Soc.* **260** (1980), 363–378. 121

[Low1997] Lowen, R. *Approach Spaces. The Missing Link in the Topology–Uniformity–Metric Triad.* Oxford University Press, 1997. 87

[LukMZ1986] Lukeš, J. Malý, M., and Zajíček, L. *Fine Topology Methods in Real Analysis and Potential Theory*. Lecture Notes Math. **1189**, Springer, 1986. 72, 101, 105, 109, 110, 111, 117, 123, 124, 126, 127, 130, 256

[Lus1917] Lusin (Luzin), N.N. Sur la classification de M. Baire. *Comptes Rendus* **161** (1917), 91–94. 51

[Lus1930] Lusin (Luzin), N.N. *Leçons sur les Ensembles Analytiques*. Gauthier-Villars, 1930. 51

[LusS1923] Lusin (Luzin), N.N. and Sierpiński, W. Sur un ensemble non measurable B. *J. Math. (NS)* **2** (1923), 53–72. 51

[Mac1957] Mackey, G.W. Borel structure in groups and their duals. *Trans. Am. Math. Soc.* **85** (1957), 134–165. 253

[Mah1958] Maharam, D. On a theorem of von Neumann. *Proc. Am. Math. Soc.* **9** (1958), 987–994. 104

[Mar1935] Szpilrajn, E. (Marczewski, E.) Sur une classe de fonctions de M. Sierpiński et la classe correspondante d'ensembles. *Fund. Math.* **24** (1935), 17–34. 22

[MarS1949] Marczewski, E. and Sikorski, R. Remarks on measure and category. *Colloq. Math.* **2** (1949), 13–19. 33

[MartK1980] Martin, D.A. and Kechris, A.S. Infinite games and effective descriptive set theory. In RogJD1980, Part 4. 61, 121

[MartS1970] Martin, D.A. and Solovay, R.M. Internal Cohen extensions. *Ann. Math. Logic* **2** (1970), 143–178. 272

[Marti1961] Martin, N.F.G. Generalized condensation points. *Duke Math. J.* **28** (1961), 507–514. 127, 130, 256

[Marti1964] Martin, N.F.G. A topology for certain measure spaces. *Trans. Am. Math. Soc.* **112** (1964), 1–18. 105, 106, 107, 256

[Mat1979] Mathias, A.R.D. Surrealist landscape with figures (a survey of recent results in set theory). *Period. Math. Hung.* **10** (1979), 109–175. 234

[MatoZ2003] Matoušková, E. and Zelený, M. A note on intersections of non–Haar null sets. *Colloq. Math.* **96** (2003), 1–4. 255, 257

[Mau1981] Mauldin, D. *The Scottish Book*. Birkhäuser, 1981. 46

[Med2022] Medini, A. On the scope of the Effros theorem. *Fund. Math.* **258** (2022), 211–223. 173

[MeeS2001] Meerschaert, M.S. and Scheffler, H.-P. *Limit Distributions for Sums of Independent Random Vectors: Heavy Tails in Theory and Practice*. Wiley, 2001. 278

[Meh1964] Mehdi, M.R. On convex functions. *J. London Math. Soc.* **39** (1964), 321–326. 19, 21, 186, 283

[Mey1966] Meyer, P.-A. *Probability and Potentials*. Blaisdell, 1988. 117

[MeyK1949] Meyer-König, W. Untersuchungen über einige verwandte Limitierungsverfahren. *Math. Z.* **52** (1949), 257–304. 14

[Mic1940] Michal, A.D. Differentials of functions with arguments and values in topological abelian groups. *Proc. Nat. Acad. Sci. USA* **26** (1940), 356–359. 86

[Mic1947] Michal, A.D. Functional analysis in topological group spaces. *Math. Mag.* **21** (1947), 80–90. 86

[Mich1982] Michael, E. On maps related to σ-locally finite and σ-discrete collections of sets. *Pacific J. Math.* **98** (1982) 139–152. 208

[Mich1986] Michael, E. A note on completely metrizable spaces. *Proc. Am. Math. Soc.* **96** (1986), 513–522. 195, 219

[Mich1991] Michael, E. Almost complete spaces, hypercomplete spaces and related mapping theorems. *Topol. Appl.* **41** (1991), 113–130. 182, 195, 213

[Mil2004] van Mill, J. A note on the Effros Theorem. *Am. Math. Monthly* **111** (2004), 801–806. 59, 173, 175, 176, 177, 178

[Mil2008] van Mill, J. Homogeneous spaces and transitive actions by Polish groups. *Israel J. Math.* **165** (2008), 133–159. 173, 179

[Mil2009] van Mill, J. Analytic groups and pushing small sets apart. *Trans. Am. Math. Soc.* **361** (2009), 5417–5434. 175

[MilP1986] van Mill, J. and Pol, R. The Baire category theorem in products of linear spaces and topological groups. *Topol. Appl.* **22** (1986), 267–282. 139, 148, 175, 228

[Mill1989] Miller, A.W. Infinite combinatorics and definability. *Ann. Pure Appl Math. Logic* **41**(1989), 179-203. (See also updated web version at: `www.math.wisc.edu/\char126\relaxmiller/res/`.) 145

[Mill1995] Miller, A.W. *Descriptive Set Theory and Forcing*. Springer, 1995. 122, 145

[MillP2000] Miller, A.W. and Popvassilev, S.G. Vitali sets and Hamel bases that are Marczewski measurable. *Fund. Math.* **166** (2000), 269–279. 22

[Mille1989] Miller, H.I. Generalization of a result of Borwein and Ditor. *Proc. Am. Math. Soc.* **105** (1989), 889–893. 74, 162

[MilleO2012] Miller, H.I. and Ostaszewski, A.J. Group action and shift-compactness. *J. Math. Anal. Appl.* **392** (2012), 23–39. 177, 248, 279

[MilleMO2021] Miller, H.I., Miller-Van Wieren, L. and Ostaszewski, A.J. Beyond Erdős–Kunen–Mauldin: Singular sets with shift-compactness properties. *Topol. Appl.* **291** (2021), 107605. 73

[Mit2010] Mitchell, W.J. Beginning inner model theory. In ForK2010, pp. 1487–1594. 275, 276

[Mon1935] Montgomery, D. Non-separable metric spaces. *Fund. Math.* **25** (1935), 527–533. 96

[Mon1936] Montgomery, D. Continuity in topological groups. *Bull. Am. Math. Soc.* **42** (1936), 879–882. 197

[Mon1950] Montgomery, D. Locally homogeneous spaces. *Ann. Math.* **52** (1950), 261–271. 173

[Mos2009] Moschovakis, Y.N. *Descriptive Set Theory*, 2nd ed. Math. Surveys Monog. **155**, Am. Math. Soc., 2009. (1st ed. 1980). 61

[Mosp2005] Mospan, Y.V. A converse to a theorem of Steinhaus–Weil. *Real An. Exch.* **31** (2005), 291–294. 250, 251, 254

[Mue1965] Mueller, B.J. Three results for locally compact groups connected with the Haar measure density theorem. *Proc. Am. Math. Soc.* **16** (1965), 1414–1416. 256

[Mun1975] Munkres, J.R. *Topology, A First Course*. Prentice-Hall, 1975. 35

[Mut1999] Muthuvel, K. Application of covering sets. *Colloq. Math.* **80** (1999), 115–122. 117

[Myc1979] Mycielski, J. Finitely additive measures. *Coll. Math.* **42** (1979), 309–318. 240

[MycS1962] Mycielski, J. and Steinhaus, H. A mathematical axiom contradicting the Axiom of Choice. *Bull. Acad. Polon. Sci. Sér. Sci. Math. Astronom. Phys.* **10** (1962), 1–3. 231

[MycSw1964] Mycielski, J. and Świerczkowski, S. On the Lebesgue measurability and the axiom of determinateness. *Fund. Math.* **54** (1964), 67–71. 61, 231

[Nam1974] Namioka, I. Separate and joint continuity. *Pacific J. Math.* **51** (1974), 515–531. 197

[NamP1969] Namioka, I. and Pol, R. σ-fragmentability and analyticity. *Mathematika* **43** (1969), 172–181. 203

[Nee1997] Neeb, K.-H. On a theorem of S. Banach. *J. Lie Theory* **7** (1997), 293–300. 184, 189, 190, 191, 318

[Neem2010] Neeman, I. Determinacy in $L(\mathbb{R})$. In ForK2010, Chapter 21. 231

[Neu1954a] Neumann, B.H. Groups covered by finitely many cosets. *Publ. Math. Debrecen* **3** (1954/5), 227–242. 142

[Neu1954b] Neumann, B.H. Groups covered by permutable subsets. *J. London Math. Soc.* **29** (1954), 236–248. 142

[Nik1925] Nikodym, O. Sur une propriété de l'opération *A*. *Fund. Math.* **7** (1925), 149–154. 54

[Nol1990] Noll, D. A topological completeness concept with applications to the open mapping theorem and the separation of convex sets. *Topol. Appl.* **35** (1990), 53–69. 182

[Nol1992] Noll, D. Souslin measurable homomorphisms of topological groups. *Arch. Math. (Basel)* **59** (1992), 294–301. 217

[Ols2004] Olsen, L. Extremely non-normal numbers. *Math. Proc. Cambridge Phil. Soc.* **137** (2004), 43–53. 225

[OMa1977] O'Malley, R.J. Approximately differentiable functions: The *r* topology. *Pacific J. Math.* **72** (1977), 207–222. 122, 124

[Ost1976] Ostaszewski, A.J. On countably compact perfectly normal spaces. *J. London Math. Soc.* **14** (1976), 505–516. 145

[Ost1980] Ostaszewski, A.J. Monotone normality and G_δ-diagonals in the class of inductively generated spaces. In *Topology II*, p. 905–930. Colloq. Math. Soc. János Bolyai **23**, North-Holland, 1980. 44

[Ost1989] Ostaszewski, A.J. On how to trap a gap: "An Introduction to Independence for Analysts by H.G. Dales and W.H. Woodin". *Bull. London Math. Soc.* **21** (1989), 197–208. 232

[Ost2010] Ostaszewski, A.J. Regular variation, topological dynamics, and the uniform boundedness theorem. *Topol. Proc.* **36** (2010), 305–336. 278, 279

[Ost2011] Ostaszewski, A.J. Analytically heavy spaces: Analytic Cantor and analytic Baire theorems. *Topol. Appl.* **158** (2011), 253–275. 56, 57, 60, 174, 207

[Ost2012] Ostaszewski, A.J. Analytic Baire spaces. *Fund. Math.* **217** (2012), 189–210. 176, 183, 205

[Ost2013a] Ostaszewski, A.J. Almost completeness and the Effros open mapping principle in normed groups. *Topol. Proc.* **41** (2013), 99–110. 174, 178

[Ost2013b] Ostaszewski, A.J. Shift-compactness in almost analytic submetrizable Baire groups and spaces. *Topol. Proc.* **41** (2013), 123–151. 174, 175, 192, 195

[Ost2013c] Ostaszewski, A.J. Beyond Lebesgue and Baire III: Steinhaus' theorem and its descendants. *Topol. Appl.* **160** (2013), 1144–1154. 177, 197, 207, 213

[Ost2013d] Ostaszewski, A.J. The Semi-Polish Theorem: One-sided vs joint continuity in groups. *Topol. Appl.* **160** (2013), 1155–1163. 183, 213, 219

[Ost2015a] Ostaszewski, A.J. Beurling regular variation, Bloom dichotomy, and the Gołąb–Schinzel functional equation. *Aequationes Math.* **89** (2015), 725–744. 13, 21, 288

[Ost2015b] Ostaszewski, A.J. Effros, Baire, Steinhaus and non-separability. *Topol. Appl.* **195** (2015), 265–274. 173

[Ost2016a] Ostaszewski, A.J. Stable laws and Beurling kernels. *Advances in Applied Probability* **48(A)** (N. H. Bingham Festschrift, C.M. Goldie and A. Mijatović, eds), (2016), 239–248. 3

[Ost2016b] Ostaszewski, A.J. Homomorphisms from functional equations: The Goldie equation. *Aequationes Math.* **90** (2016), 427–448. 288

[Ost2017] Ostaszewski, A.J. Homomorphisms from functional equations in probability. In *Developments in Functional Equations and Related Topics*, pp. 171–213. Springer Optim. Appl. **124**, Springer, 2017. 283

[Ostr1929] Ostrowski, A. Mathematische Miszellen XIV: Über die Funktionalgleichung der Exponentialfunktion und verwandte Funktionalgleichungen, *Jahresb. Deutsch. Math. Ver.* **38** (1929), 54–62. (Reprinted in *Collected Papers of Alexander Ostrowski*, **4**, 49–57, Birkhäuser, 1984.) 19, 21

[Ostr1976] Ostrowski, A.M. On Cauchy–Frullani integrals. *Comment. Math. Helvet.* **51** (1976), 57–91. 11

[Oxt1946] Oxtoby, J.C. Invariant measures in groups which are not locally compact. *Trans. Am. Math. Soc.* **60** (1946), 215–237. 242

[Oxt1960] Oxtoby, J.C. Cartesian products of Baire spaces. *Fund. Math.* **49** (1960/1961), 157–166. 39, 40, 160

[Oxt1980] Oxtoby, J.C. *Measure and Category*, 2nd ed. Grad. Texts Math. **2**, Springer 1980. (1st ed. 1971.) 1, 33, 35, 37, 42, 43, 46, 104, 109, 110, 117, 137, 138, 139, 148, 156, 222, 223, 224, 228, 234

[PalW1934] Paley, R.E.A.C. and Wiener, N. *Fourier Transforms in the Complex Domain.* AMS Colloq. Publ. **XIX**, Am. Math. Soc., 1934. 5

[Par1967] Parthasarathy, K.R. *Probability Measures on Metric Spaces.* Academic Press, 1967. (Reprinted Am. Math. Soc., 2005.) 73, 161, 240

[Pat1988] Paterson, A.L.T. *Amenability*, Math. Surveys Monog. **29**, Am. Math. Soc., 1988. 246, 282

[Paw1985] Pawlikowski, J. Lebesgue measurability implies Baire property. *Bull. Sci. Math.* (2) **109** (1985), 321–324. 232

[Per1960] Perron, O. *Irrationalzahlen.* Walter de Gruyter, 1960. 223

[Pes1998] Pestov, V. Review of Nee1997, MathSciNet MR1473172 (98i:22003). 184

[Pet1950] Pettis, B.J. On continuity and openness of homomorphisms in topological groups. *Ann. Math.* **52** (1950), 293–308. 87, 99, 139, 173, 174, 236, 239, 240

[Pet1951] Pettis, B.J. Remarks on a theorem of E.J. McShane. *Proc. Am. Math. Soc.* **2** (1951), 166–171. 236, 239

[Pet1974] Pettis, B.J. Closed graph and open mapping theorems in certain topologically complete spaces. *Bull. London Math. Soc.* **6** (1974), 37–41. 188

[Pic1939] Piccard, S. *Sur les Ensembles de Distances des Ensembles de Points d'un Espace Euclidien.* Mém. Univ. Neuchâtel **13** (1939). 139, 173, 236, 240

[Pic1942] Piccard, S. *Sur des Ensembles Parfaites*, Mém. Univ. Neuchâtel **16** (1942). 139, 173, 236, 240

[PitP2016] Pitman, E.J.G. and Pitman, J.W. A direct approach to the stable distributions. In N. H. Bingham Festschrift, C.M. Goldie and A. Mijatović (eds), *Advances in Applied Probability* **48(A)** 2016, 261–282. 3

[Plo2003] Płotka, K. On functions whose graph is a Hamel basis. *Proc. Am. Math. Soc.* **131** (2003), 1031–1041. 22

[Pol1976] Pol, R. Remark on the restricted Baire property in compact spaces. *Bull. Acad. Polon. Sci. Sér. Sci. Math.* **24** (1976), 599–603. 39, 185

[Pol1979] Pol, R. Note on category in Cartesian products of metrizable spaces. *Fund. Math.* **102** (1979), 55–59. 139, 148, 159, 175, 228

[Poly1923] Pólya, G. Bemerkungen über unendliche Folgen und ganze Funktionen. *Math. Ann.* **88** (1923), 169–183. (Reprinted in *Collected Papers Vol. 1*, R.P. Boas (ed). MIT Press, 1974.) 6

[Pop1965] Popa, C. Gh. Sur l'equation fonctionnelle $f[x + yf(x)] = f(x)f(y)$, *Ann. Polon. Mathe.* **17** (1965), 193–198 288

[PorWW1985] W. Poreda, W., Wagner-Bojakowska E. and Wilczyński, W. A category analogue of the density topology. *Fund. Math.* **125** (1985), 167–173. 133

[PortS1978] Port, S.C. and Stone, C.J. *Brownian Motion and Classical Potential Theory.* Academic Press, 1978. 118

[Que1980] Queféllec, H. Propriétés presque sûres et quasi-sûres des séries de Dirichlet et des produits d'Euler. *Canad. J. Math.* **32** (1980), 531–558. 227

[Rad1919] Rademacher, H. Über partielle und totale Differenzierbarkeit I. *Math. Ann.* **79** (1919), 340–359. 228

[Rad1922] Rademacher, H. Einige Sätze über Reihen von allgemeinen Orthogonalfunktionen. *Math. Ann.* **87** (1922), 112–138. 224

[Rai1984] Raisonnier, J. A mathematical proof of S. Shelah's theorem on the measure problem and related results. *Israel J. Math.* **48** (1984), 49–56. 230, 233

[RaiS1983] Raisonnier, J. and Stern, J. Mesurabilité et propriété de Baire. *Comptes Rendus Acad. Sci. I. (Math.)* **296** (1983), 323–326. 232, 233

[RaiS1985] Raisonnier, J. and Stern, J. The strength of measurability hypotheses. *Israel J. Math.* **50** (1985), 337–349. 230

[Ram1951] Ramsey, A.S. *Dynamics, Part I*. Cambridge University Press, 1951. 114

[Rams1930] Ramsey, F.P. On a problem of formal logic. *Proc. London Math. Soc.* **30** (1930), 338–384. 274

[RaoR1974] Rao, K.P.S. Bhaskara and Rao, M. Bhaskara. A category analogue of the Hewitt–Savage zero–one law. *Proc. Am. Math. Soc.* **44** (1974), 497–499. 222

[RaoR1975] Rao, K.P.S. Bhaskara and Rao, M. Bhaskara. On the difference of two second category Baire sets in a topological group. *Proc. Am. Math. Soc.* **47** (1975), 257–258. 112

[Rea1996] Reardon, P. Ramsey, Lebesgue, and Marczewski sets and the Baire property. *Fund. Math.* **149** (1996), 191–203. 121

[Rog1970] Rogers, C.A. *Hausdorff Measures*. Cambridge University Press, 1970. 226

[RogJ1980] Rogers, C.A. and Jayne, J. K-Analytic sets. In RogJD1980, Part 1. xiii, 31, 42, 51, 52, 54, 56, 61, 65, 66, 123, 178, 181, 187, 188, 203, 204, 205, 212, 218, 236

[RogJD1980] Rogers, C.A., Jayne, J., Dellacherie, C., Topsøe, F., Hoffmann-Jørgensen, J., Martin, D.A., Kechris, A.S. and Stone, A.S. *Analytic Sets*. Academic Press, 1980. 62, 203, 241, 301, 308, 314, 319, 321, 322

[RogW1966] Rogers, C.A. and Willmott, R.C. On the projection of Souslin sets. *Mathematika* **13** (1966), 147–150. 52

[Roy1988] Royden, H.L. *Real Analysis*, 2nd ed. Prentice-Hall, 1988 (1st ed. 1963.) 24, 25

[Rud1966] Rudin, W. *Real and Complex Analysis*. McGraw-Hill, 1966. (3rd. ed. 1987, 2nd ed. 1974.) 269, 283

[Rud1973] Rudin, W. *Functional Analysis*. McGraw-Hill, 1973. (2nd ed. 1991.) 86, 90, 172, 182, 183

[Sai1983] Saint Raymond, J. Jeux topologiques et espaces de Namioka. *Proc. Am. Math. Soc.* **87** (1983), 499–504. 45, 47

[Sak1956] Sakovich, G.N. A single form for the conditions for attraction to stable laws. *Th. Prob. Appl.* **1** (1956), 322–325. 2

[Saks1964] Saks, S. *Theory of the Integral*, 2nd ed. Dover, 1964. (1st ed. 1937, Monografie Mat. **VII**, Instytut Matematyczny Polskiej Akademii Nauk.) 73, 104, 254

[Sal1966] Šalát, T. A remark on normal numbers. *Rev. Roum. Math. Pures Appl.* **11** (1966), 53–56. 225, 226

[Sch1971] Scheinberg, S. Topologies which generate a complete measure algebra. *Adv. Math.* **7** (1971), 231–239. 109, 130

[Schw1966] Schwartz, L. Sur le théorème du graphe fermé. *C. R. Acad. Sci. Paris Sér. A–B* **263** (1966), 602–605. 188

[Schw1973] Schwartz, L. *Radon Measures on Arbitrary Topological Spaces and Cylindrical Measures.* Tata Institute of Fundamental Research Studies in Mathematics **6**, Oxford University Press, 1973. 240

[Sco1961] Scott, D. Measurable cardinals and constructible sets. *Bull. Acad. Polon. Sci. Sér. Sci. Math. Astr. Phys.* **9** (1961), 521–524. 274

[Sem1971] Semadeni, Z. *Banach Spaces of Continuous Functions.* Monografie Matematyczne 55, PWN. Polish Scientific Publishers, 1971. 584pp.

[Sen1973] Seneta, E. An interpretation of some aspects of Karamata's theory of regular variation. *Publ. Inst. Math. Beograd* (NS) **15** (29) (1973), 111–119. 10

[She1969] Shea, D.F. On a complement to Valiron's Tauberian theorem for the Stieltjes transform. *Proc. Am. Math. Soc.* **21** (1969), 1–9. 4, 5

[Shel1984] Shelah, S. Can you take Solovay's inaccessible away? *Israel J. Math.* **48** (1984), 1–47. 230, 232, 233

[Shel1985] Shelah, S. On measure and category. *Israel J. Math.* **52** (1985), 110–114. 233

[ShelW1990] Shelah, S. and Woodin, H. Large cardinals imply that every reasonably definable set of reals is Lebesgue measurable. *Israel J. Math.* **70**(3) (1990), 381–394. 233

[ShiT1998] Shi, H. and Thomson, B.S. Haar null sets in the space of automorphisms on [0,1]. *Real Anal. Exchange* **24** (1998/99), 337–350. 241, 258

[Sho1984] Shortt, R.M. Universally measurable spaces: An invariance theorem and diverse characterizations. *Fund. Math.* **121** (1984), 169–176. 241

[Sie1920] Sierpiński, W. Sur l'equation fonctionelle $f(x + y) = f(x) + f(y)$. *Fund. Math.* **1** (1920), 116–122. (Reprinted in *Oeuvres Choisis II*, 331–336, PWN, 1975.) 21, 81

[Sie1935] Sierpiński, W. Sur deux ensembles linéaires singuliers. *Ann. Scuola Norm. Super. Pisa Cl. Sci* **4** (1935), 43–46. 21

[Sie1975] Sierpiński, W. *Oeuvres Choisies II: Théorie des Ensembles et ses Applications, Travaux des Années* 1908–1929, PWN, 1975. 137, 307

[Sie1976] Sierpiński, W. *Oeuvres Choisies III: Théorie des Ensembles et ses Applications, Travaux des Années* 1930–1966, PWN, 1976. 137

[Sil1971] Silver, J. Every analytic set is Ramsey. *J. Symbolic Logic* **35** (1971), 60–64. 121

[Silv1974] Silverstein, M.L. *Symmetric Markov Processes.* Lecture Notes in Math. **426**, Springer, 1974. 119

[Silv1976] Silverstein, M.L. *Boundary Theory for Symmetric Markov Processes.* Lecture Notes in Math. **516**, Springer, 1976. 119

[Sim1975] Simmons, S.M. A converse Steinhaus–Weil theorem for locally compact groups. *Proc. Am. Math. Soc.* **49** (1975), 383–386. 251, 254

[Simo2015] Simon, B. *Basic Complex Analysis.* A Comprehensive Course in Analysis, Part 2A, Am. Math. Soc., 2015. 229

[Sol2005] Solecki, S. Size of subsets of groups and Haar null sets. *Geom. Funct. Anal.* **15** (2005), 246–273. 241, 258

[Sol2006] Solecki, S. Amenability, free subgroups, and Haar null sets in non-locally compact groups. *Proc. London Math. Soc.* (3) **93** (2006), 693–722. 240, 241, 242, 245, 246, 248, 251, 259

[Sol2007] Solecki, S. A Fubini theorem. *Topol. Appl.* **154** (2007), 2462–2464. 241

[SolS1997] Solecki, S. and Srivastava, S.M. Automatic continuity of group operations. *Topol. Appl.* **77** (1997), 65–75. 197

[Solo1965] Solovay, R.M. $2^{\aleph_0}$ can be anything it ought to be. In J.W. Addison, L. Henkin and A. Tarski (eds), *The Theory of Models*, (Proc. 1963 Int. Symp. Berkeley). Studies in Logic and the Foundations of Mathematics, North-Holland, 1965, 435. 268

[Solo1969] Solovay, R.M. On the cardinality of Σ^1_2 sets of reals. In *Foundations of Mathematics* (Symposium Commemorating Kurt Gödel, Columbus, Ohio, 1966), pp. 58–73. Springer, 1969. 233

[Solo1970] Solovay, R.M. A model of set theory in which every set of reals is Lebesgue measurable. *Ann. Math.* **92** (1970), 1–56. 229, 230, 231, 232

[Solo1978] Solovay, R.M. The independence of DC from AD. In *The Cabal Seminar 76–77* (Proc. Caltech–UCLA Logic Sem., 1976–77), pp. 171–183. Lecture Notes in Math. **689**, Springer, 1978. 232

[Spe1994] Spencer, J. *Ten Lectures on the Probabilistic Method*, 2nd ed. CBMS-NSF Reg. Conf. Ser. Appl Math. **64**, SIAM, 1994 (1st ed. 1987.) 37

[Ste1970] Stein, E.M. *Singular Integrals and Differentiability Properties of Functions*. Princeton University Press, 1970. 104

[SteS2005] Stein, E.M. and Shakarchi, R. *Real Analysis: Measure Theory, Integration, and Hilbert Spaces*. Princeton University Press, 2005. 25, 269

[Stei1920] Steinhaus, H. Sur les distances des points de mesure positive. *Fund. Math.* **1** (1920), 93–104. 139, 236, 239

[Stei1930] Steinhaus, H. Über die Wahrscheinlichleit dafür, daß der Konvergenzkreis einer Potenzreihen ihre natürliche Grenze ist. *Math. Z.* **31** (1930), 408–416. 228

[Ster1985] Stern, J. Regularity properties of definable sets of reals. *Ann. Pure Appl. Logic* **29** (1985), 289–324. 232, 233

[Sto1963] Stone, A.H. Kernel constructions and Borel sets. *Trans. Am. Math. Soc.* **107** (1963), 58–70; errata, ibid. **107** (1963), 558. 37, 122, 128, 159

[Sto1980] Stone, A.H. Analytic sets in non-separable spaces. In RogJD1980, Part 5. 204

[Sze2011] Szenes, A. Exceptional points for Lebesgue's density theorem on the real line. *Adv. Math.* **226** (2011), 764–778. 167

[Tal1976] Tall, F.D. The density topology. *Pacific J. Math.* **62** (1976), 275–284. 110, 156

[Tal1978] Tall, F.D. Normal subspaces of the density topology. *Pacific J. Math.* **75** (1978), 579–588. 110

[TaoV2006] Tao, T. and Vu, V.N. *Additive Combinatorics*. Cambridge Studies in Adv. Math., **105**. Cambridge University Press, 2006. 141

[Tar1930] Tarski, A. Une contribution à la théorie de la mesure. *Fund. Math.* **15** (1930), 42–50. 232

[Tay2004] Taylor, S.J. John Hawkes (1944–2001). *Bull. Lond. Math. Soc.* **36** (2004), 695–710. 226

[Ten2015] Tenenbaum, G. *Introduction to Analytic and Probabilistic Number Theory*, 3rd ed. Grad. Studies Math. **193**, American Mathematical Society, 2015. (2nd ed. Cambridge Studies Adv. Math. **46**, Cambridge University Press, 1995.) 7, 8

[Tit1986] Titchmarsh, E.C. *The Theory of the Riemann Zeta-Function*, Second ed. (revised by Heath-Brown, D. R.), Oxford University Press, 1986. (1st ed. 1951.) 227

[TomW2016] Tomkowicz, G. and Wagon, S. *The Banach–Tarski Paradox*, 2nd ed. Encycl. Math. Appl. 163, Cambridge University Press, 2016. (1st ed., Encycl. Math. Appl. 24, Cambridge University Press, 1985.) 223, 257

[TopH1980] Topsøe, F. and Hoffmann-Jørgensen, J. Analytic spaces and their applications. In RogJD1980, Part 3. 95, 197

[Tor2013] Törnquist, A. Σ_2^1 and Π_1^1 mad families. *J. Symbolic Logic* **78** (2013), 1181–1182. 234

[Tra1987] Trautner, R. A covering principle in real analysis. *Quart. J. Math.* **38** (1987), 127–130. 162

[Ula1930] Ulam, S. Zur Maßtheorie in der allgemeinen Mengenlehre. *Fund. Math.* **16** (1930), 140–150. 200, 232, 273

[Ula1960] Ulam, S.M. *A Collection of Mathematical Poblems*. Wiley, 1960.

[Ung1975] Ungar, G.S. On all kinds of homogeneous spaces. *Trans. Am. Math. Soc.* **212** (1975), 393–400. 173

[WagW2000] Wagner-Bojakowska, E. and Wilczyński, W, Cauchy condition for the convergence in category. *Proc. Am. Math. Soc.* **128** (2000), 413–418. 30

[Wei1940] Weil, A. *L'intégration dans les Groupes Topologiques et ses Applications,* Actual. Sci. Ind. **869**, Hermann, Paris, 1940. (Republished, Princeton University Press, 1941.) 239

[Weis1984] Weiss, W. Versions of Martin's axiom. In KunV1984, Ch. 19, pp. 827–886. 272

[Whi1974] White, H.E. Topological spaces in which Blumberg's Theorem holds. *Proc. Am. Math. Soc.* **44** (1974), 454–462. 40, 111

[Whi1975] White, H.E. Topological spaces that are α-favorable for a player with perfect information. *Proc. Am. Math. Soc.* **50** (1975), 477–482. 56

[Wie1933] Wiener, N. *The Fourier Integral and Certain of Its Applications*. Cambridge University Press, 1933. (Reprinted, Cambridge Mathematical Library, 1988.) 17

[Wil2002] Wilczyński, W. *Density topologies*. In *Handbook of Measure Theory, Vol. I*, 675–702, North-Holland, 2002. 256

[WilW2007] Wilczyński, W. and Wojdowski, W. Complete density topology. *Indag. Math.* (NS) **18** (2007), 295–303. 133

[Will1972] Williamson, J. Meromorphic functions with negative zeros and positive poles and a theorem of Teichmüller. *Pacific J. Math.* **42** (1972), 795–810. 5

[Woo1988] Woodin, W.H. Supercompact cardinals, sets of reals, and weakly homogeneous trees. *Proc. Nat. Acad. Sci. USA* **85** (1988), 6587–6591. 271

[Woo1999] Woodin, W.H. *The Axiom of Determinacy, Forcing Axioms, and the Nonstationary Ideal*, de Gruyter, 1999. 275

[Xia1972] Xia, D.X. *Measure and Integration Theory on Infinite Dimensional Spaces. Abstract Harmonic Analysis.* Pure and App. Math. **48** Academic Press, 1972. 240, 253

[Zah1950] Zahorski, Z. Sur la première dérivée. *Trans. Am. Math. Soc.* **69** (1950), 1–54. 113

[Zap1999] Zapletal, J. Terminal notions. *Bull. Symbolic Logic* **5** (1999), 470–478. 64

[Zap2001] Zapletal, J. Terminal notions in set theory. *Ann. Pure Appl. Logic* **109** (2001), 89–116. 64

[Zel1960] Żelazko, B. A theorem on B_0 division algebras. *Bull. Acad. Plon. Sci.* **8** (1960), 373–375. 94

[ZelB1970] Zeller, K. and Beekmann, W. *Theorie der Limitierungsverfahren*, 2nd ed. Springer, 1970. 16

[Zem2002] Zeman, M. *Inner Models and Large Cardinals.* De Gruyter Series in Logic and its Applications, **5** de Gruyter, 2002. 271

[Zsi2004] Zsilinszky, L. Products of Baire spaces revisited. *Fund. Math.* **183** (2004) 115–12. 160

[Zyg1988] Zygmund, A. *Trigonometric Series, Vols. I, II*, 2nd ed. Cambridge University Press, 1988. (3rd ed., 2002, with foreword by R. Fefferman, 2002.) 138

Index

Printed in the United States
by Baker & Taylor Publisher Services